Water Vapor Measurement

Water Vapor Measurement

Pieter R. Wiederhold
Wiederhold Associates
Boston, Massachusetts

CRC Press
Taylor & Francis Group
Boca Raton London New York

CRC Press is an imprint of the
Taylor & Francis Group, an **informa** business

Materials formerly available for this book
on CD-ROM are now available for
download from our website
www.crcpress.com

Go to the *Downloads & Updates Tab* on the book's page

CRC Press
Taylor & Francis Group
6000 Broken Sound Parkway NW, Suite 300
Boca Raton, FL 33487-2742

First issued in paperback 2019

ISBN-13: 978-0-8247-9319-7 (hbk)
ISBN-13: 978-0-367-40101-6 (pbk)

Library of Congress Cataloging-in-Publication Data

Wiederhold, Pieter, R.
   Water vapor measurement: methods and instrumentation / Pieter R. Wiederhold.
    p. cm.
   Includes bibliographical references and index.
   ISBN 0-8247-9319-6 (alk. paper)
   1. Hygrometers. 2. Humidity—Measurement I. Title.
QC916.W54   1997
681'.2—dc21

                                         97-1910
                                         CIP

Visit the Taylor & Francis Web site at
http://www.taylorandfrancis.com

and the CRC Press Web site at
http://www.crcpress.com

Preface

During my more than 20 years in the field of humidity measurements and instrumentation, I have given numerous lectures, seminars, and presentations covering the many types of instruments and sensors that are in use today, their advantages and limitations, and applications where certain instruments and sensors should or should not be used.

Not long after I started this work, it became evident to me that measuring humidity is far more difficult than, for example, measuring pressure or temperature. A humidity range of 0–100% RH corresponds to a water vapor pressure span of about 0.8×10^{-6} mm Hg at 1 part-per-billion by volume, to 760 mm Hg for saturated steam, i.e., a dynamic range of about 10^9. No single sensor is capable of covering this entire range. Furthermore, humidity instrumentation requires far more care in terms of installation, maintenance, and calibration than most other types of measurement devices such as for flow or temperature detection. A lack of knowledge on the part of an operator often leads to unsatisfactory results.

Over the years a great many sensing methods have been developed for measuring a variety of humidity related parameters. Each of these has unique advantages and limitations and is suitable only in certain applications. The range of applications where humidity measurements are needed is almost endless and is increasing with time as new information is acquired on the effects of humidity on quality, cost, safety, comfort, and human health. More than 60 such applications are discussed in Chapter 12 and this is only a small cross-section. Hence, it is important that sensors be properly matched to the application. This involves rather extensive knowledge of the subject.

Few engineers and technicians who are assigned to humidity measurement projects have adequate knowledge of this kind and this has resulted in many disappointments with instruments that have been misapplied. During the last two decades there has been a great demand for more information on humidity measurements. This high level of interest motivated me to undertake the task of writing this book.

The book is intended to assist the many engineers and technicians who are involved in making humidity measurements, and who have to select the proper instrumentation to do so. It should also be useful for test, calibration, and quality control engineers, as well as students in these disciplines.

Humidity can be expressed in many different parameters and most of such measurements are also dependent on temperature and pressure. Conversions involve complex mathematical equations or the use of a multitude of charts, tables, and graphs which are presented in Chapter 13. For those who have access to a computer, these conversions can be greatly simplified using the accompanying computer disk.

I would like to acknowledge the support and encouragement received from many of my friends and associates. I would especially like to mention:

General Eastern Instruments Company, where I spent so many rewarding years of my working career and where I obtained the knowledge presented in this book;

Sumner Weisman of General Eastern, who provided me with valuable information based on his experience as marketing director and lecturer on humidity measurements;

Other former associates at General Eastern with whom I spent so many exciting years developing new and better methods for measurement and solving problems in industrial, laboratory, and environmental areas;

Arden Buck of Buck Research, who provided me with much technical information and photos covering his extensive work on cryogenic and Lyman–Alpha hygrometers. Arden also, together with Sumner Weisman, generated the software for the "Humidity Parameter Conversion Program for Windows" (HCON), a copy of which is included on the accompanying disk;

Michael Scelzo of Panametrics, who offered me much useful information and applications information and who reviewed drafts of the section on aluminum oxide hygrometers;

Colin Blakemore of AMETEK, for the technical information he provided on piezoelectric hygrometers, and on semi-conductor and natural gas applications;

Beverly White of MEECO, for providing much information on electrolytic hygrometers, some of which was excerpted in Chapter 5, and for reviewing my drafts on this subject;

Frank Cooper of Protimeter, for the many stimulating discussions we had on the CCM hygrometer and for providing materials and photos for the book;

Dr. Franz-Josef Hoffman of Endress & Hauser, for contributing materials and technical information on aluminum oxide systems, review, and comments;

Dr. Peter Huang of NIST, for reviewing and commenting on the section covering NIST calibration;

Dr. Stephanie Bell and *Mark Stevens* of NPL, for many contributions, photos, and comments. Their permission to reproduce certain NPL information and portions of relevant publications is greatly appreciated;

Robert Pragnell and *Steve Lower* of Sira, for providing information on Sira and SIREP, and for reviewing my drafts describing these organizations in the UK;

Dr. Bernard Cretinon of CETIAT in France, for contributing information and review of the section on calibration standards in France;

Dr. Teruko Inamatsu of NRLM in Japan, for generously contributing information on the humidity calibration facilities at NRLM in Japan.

I also want to recognize many of my former associates and friends who have encouraged me to complete this book and whose many valuable comments are greatly appreciated.

Pieter R. Wiederhold

CONTENTS

Preface   iii

List of Figures   xiii

List of Tables   xvii

1. Introduction   1
    I. What Is Humidity?   1
    II. Importance of Humidity   2
    III. Units Used   3
    IV. Historical Background   3
    V. Water Vapor and Moisture Measurement   5

2. Definitions and Fundamentals   7
    I. Temperature, Pressure, Humidity and Gases   7
        A. Temperature   7
        B. Pressure   8
        C. Gases   10
        D. Humidity   13
    II. Definitions and Fundamental Relationships   14
        A. Symbols   14
        B. Generally Used Humidity Terms   14
        C. Vapor Pressure   15
        D. Mixing Ratio   17
        E. Latent Heat   17
        F. Percent Saturation   17
        G. Humidity Parameters   18
        H. Pressure Effects and Dew Point Measurements   21
        I. Pressure Units   23
    III. Psychrometric Chart   24

3. Chilled Mirror Hygrometers   27
    I. Introduction   27
    II. Conventional Chilled Mirror Hygrometer   28
        A. Sensitivity to Contaminants   29
        B. Self-Standardization of Chilled Mirror Hygrometers   29
    III. Cycling Chilled Mirror Dew Point Hygrometer (CCM)   36
        A. Mirror Cycling   36
        B. CCM Sensor   38
        C. Dew Point/Frost Point Conversion   38
        D. Maintenance Requirements   39
        E. Benefits of the CCM Hygrometer   40
        F. Limitations of the CCM Hygrometer   41
        G. High Temperature Fiber Optic Hygrometers   42
    IV. Dew Point Measurement Range   43
    V. Chilled Mirror Dew Point Transmitters   46

VI. Summary of Balancing or Self-Calibration Methods   48
     A. Manual Balance   48
     B. Automatic Balance Control (ABC)   48
     C. PACER Cycle   49
     D. Continuous Balance   49
     E. Cycling Chilled Mirror (CCM) Technique   50
     F. CCM With Sapphire Mirror and Wiper   50
VII. Dew Cup   50
VIII. Sampling Systems   51
     A. Design of Sampling System   52
     B. Problems to Avoid   55
     C. Conclusions   57
IX. Error Analysis   57
     A. General Principles   57
     B. Mirror Surface Errors   57
X. Cryogenic Dew/Frost Point Hygrometer   61
     A. Purpose   62
     B. Description of Cryogenic Hygrometer   62

4. Relative Humidity   69
I. General   69
II. Bulk Polymer Humidity Sensor   69
     A. Resistive Polymer Sensor   70
III. Dunmore Cell   77
IV. Pope Cell   77
V. Capacitive Polymer Sensor   78
     A. Operation   79
     B. Temperature Dependence   80
     C. Performance   80
     D. Advantages   81
     E. Limitations   81
VI. Displacement (Mechanical) Hygrometers   82
VII. Percent RH Transmitters and Instrumentation   82
VIII. Summary   84

5. Trace Moisture Instrumentation   85
I. Aluminum Oxide Hygrometers   85
     A. General   85
     B. Aluminum Oxide Instrumentation   87
     C. Advantages   88
     D. Limitations   89
     E. Calibration   89
     F. Applications   91

Contents

II. Silicon Oxide Hygrometers   91
   A. Advantages   91
   B. Limitations   92
III. Piezoelectric Sensor   92
   A. Operation of a Typical Piezoelectric Moisture Analyzer   93
   B. Response Time   95
   C. Sensitivity   96
   D. Advantages   96
   E. Limitations   97
   F. Applications   97
   G. Summary   97
IV. Electrolytic Hygrometer   98
   A. Introduction   99
   B. Electrolytic Cell   99
   C. Theory of Operation   101
   D. Conventional Implementations   103
   E. Application Problems   104
   F. Applications   105
   G. Advantages   106
   H. Limitations   106

6. Optical Absorption Hygrometers   **109**
I. Infrared Hygrometer   109
   A. Operation   109
   B. Advantages   111
   C. Limitations   111
   D. Applications   112
II. Lyman–Alpha Hygrometer   113
   A. Moisture Measurement   113
   B. Windows   113
   C. Detectors   114
   D. Sources   115
   E. Calibration   116
   F. Performance   116
   G. Summary   117
   H. Advantages   119
   I. Disadvantages   119
   J. Applications   119

7. Dry/Wet-Bulb Psychrometer   **121**
I. General Description   121
II. Operation   123
   A. Theory   123
   B. Advantages   124
   C. Limitations   124

III. Error Analysis   125
   A. Temperature Errors   125
   B. Pressure Errors   125
   C. Radiation Errors   125
   D. Errors Arising From Other Sources   125
   E. Accuracy   125
IV. Applications   126
V. Summary   126

8. Other Humidity Instruments   **127**
  I. Saturated Salt (Lithium Chloride) Sensor   127
   A. General Discussion   127
   B. Theory of Operation   127
   C. Advantages   129
   D. Limitations   129
   E. Applications   129
  II. Fog Chamber   130
  III. Impedance-Based Ceramic Sensors   131
  IV. Fiber Optic Humidity Analyzer   132
  V. Other Types   134

9. Meteorological Systems   **137**
  I. General   137
  II. Weather Stations   137
   A. Psychrometers   137
   B. Saturated Salt Sensor (Dew Cell)   138
   C. Electrical RH Sensors   138
   D. Polymer RH Sensors   139
   E. Aluminum Oxide   140
   F. Chilled Mirror Hygrometer   141
   G. Infrared Hygrometer   142
  III. Noise Pollution Measurements   143
  IV. Communications and Cloud Studies   143
  V. Upper Atmosphere Studies   144

10. Calibration   **145**
  I. Importance of Calibration   145
   A. Traceability to National Standards   145
   B. Calibration Standards   146
   C. Uncertainty Versus Accuracy   146
   D. Types of Standards Used   147
   E. Field Applications   149
   F. Methods of Calibration   151
   G. Accuracy   153

II. National Standards Laboratories   154
   A. NIST (USA)   156
   B. National Physical Laboratory (NPL) United Kingdom   166
   C. Calibration Facilities at CETIAT, France   173
   D. Ecole Polytechnique, Two-Temperature
      Calibration System, France   175
   E. National Research Laboratory for Metrology (NRLM), Japan   178
   F. Physicalish-Technische Bundesanstalt (PTB)   181
   G. Other National Standards Laboratories   184
III. Commercially Available Primary Standards   185
   A. Dew Point Calibration Chambers   185
   B. Relative Humidity Calibration Chambers   185
IV. Calibration Using Saturated Salt Solutions   187
V. Calibration in the Low PPM Range   190

11. Water Vapor Pressure Tables   **193**
I. General   193
II. Smithsonian Tables   193
III. Vapor Pressure of Water Above 100°C   193

12. Applications   **217**
I. Heat Treating   217
   A. Humidity Instruments Used for Heat Treating   218
   B. Heat Treatment of Steel   219
   C. Other Metal Operations   224
   D. Summary   225
II. Semiconductors   225
   A. General   225
   B. Moisture Penetration   225
   C. Moisture Contamination   226
   D. Damage Caused By Moisture   226
   E. Instrumentation   228
III. Water Activity Measurements   229
   A. Definition   229
   B. Measuring Water Activity   229
   C. Instrumentation for Water Activity Measurement   233
IV. Natural Gas   234
   A. General   234
   B. Measurement Technology   235
V. Medical Applications   236
   A. Perspiration Measurements   236
   B. Incubators   237
   C. Artificial Hearts   237
   D. Medical Gases   237
   E. ETO Sterilizers   240

VI. Museums   241
VII. Dryers   242
   A. General   242
   B. Nitrogen Polyester Chip Dryer   244
   C. Nylon Chip Dryer   244
   D. Drying Plastic Resin Pellets   245
   E. Dual Tower Regenerative Desiccant Dryer   246
   F. Paper Dryers   247
   G. Dry Snack Food Processing   247
VIII. Gases   248
   A. Carbon Dioxide Gas   248
   B. Gases Containing Hydrogen Sulfide   248
   C. Steam Leaks in Process Gas Lines   249
   D. Argon in Glove Boxes   249
   E. Controlled Atmosphere Glove Boxes   249
   F. Blanketing Gas in Radionuclide Glove Boxes   250
   G. Hydrogen-Rich Hydrocarbon Streams   250
   H. Ethylene Gas for Polyethylene Production   251
   I. Recycle Gas in Catalytic Reforming Processes   251
   J. Cylinder Gases   251
IX. Meteorological Applications   252
   A. Weather Forecasting   252
   B. Airports   252
   C. Plant Site Locations   253
   D. Aircraft Noise Pollution   253
   E. Upper Atmosphere Measurements   253
X. Other Applications   254
   A. Industrial   254
   B. Automotive   264
   C. Laboratory Standards   266
   D. Nuclear Reactors   266
   E. Computers   267
   F. Data Communications Through Telephone Cables   268
   G. Buildings and Construction   269
   H. Relative Humidity Measurements   276
   I. Plant Growth Chambers   278
   J. Waste Products   279

13. Charts, Graphs and Tables   **281**
   I. General Discussion   281
   II. Psychrometric Charts   282
   III. Relative Humidity Conversions   282
   IV. Dew Point Conversions   283
   V. Moisture Content Tables and Charts   283
   VI. Pressure Conversions   283

Contents

VII. Flow Conversions   283
VIII. Unit Conversions   283

14. Laboratories and Test Facilities   **327**
    I. National Calibration Laboratories   327
    II. Certification and Accreditation Organizations   329
       A. Europe   329
       B. United States   329

15. References and Sources for Further Information   **331**
    I. Introduction   331
    II. Definitions   331
    III. Chilled Mirror   332
    IV. Relative Humidity   334
    V. Trace Moisture   335
    VI. Optical Absorption Hygrometers   338
    VII. Psychrometers   340
    VIII. Other Types   341
    IX. Meteorological Instruments   342
    X. Calibration   343
    XI. Water Vapor Pressure Tables   348
    XII. Applications   348

Appendix   **351**

Index   **353**

LIST OF FIGURES

Figure 2.1	Saturation vapor pressure for water.	16
Figure 2.2	Saturation vapor pressure for ice and supercooled water.	16
Figure 2.3	Dew point–pressure conversions.	22
Figure 2.4	The psychrometric chart.	25
Figure 3.1	Schematic of conventional chilled mirror sensor.	28
Figure 3.2	Error caused by mirror contamination	30
Figure 3.3	Typical PACER cycle.	32
Figure 3.4	Mirror condition before and after PACER cycle.	32
Figure 3.5	Error elimination with the PACER cycle.	32
Figure 3.6	Twin beam dual mirror sensor.	33
Figure 3.7	Microprocessor based multi-parameter hygrometer.	34
Figure 3.8	Microprocessor based hygrometer with continuous balance.	34
Figure 3.9	Precision chilled mirror hygrometer.	35
Figure 3.10	Precision low frost point hygrometer.	35
Figure 3.11	Dew point determining cycle.	37
Figure 3.12	Typical CCM sensor.	38
Figure 3.13	Schematic of typical cycling hygrometer sensor.	40
Figure 3.14	CCM hygrometer with dew point, temperature, and pressure sensors.	42
Figure 3.15	Depression of single and double stage CCM sensors.	44
Figure 3.16	Single-stage, two-stage, and four-stage remote sensors.	44
Figure 3.17	Two-stage sensors for remote or integral mounting.	46
Figure 3.18	Low cost chilled mirror transmitter (Dew-10).	47
Figure 3.19	Dew track transmitter.	47
Figure 3.20	Dew Tector transmitter.	48
Figure 3.21	Schematic of typical sampling system.	52
Figure 3.22	Heated sampling system components.	53
Figure 3.23	Multi-point sampling system.	53
Figure 3.24	Effects of sample line material on response time.	54
Figure 3.25	Curves for pressure correction of flow meter readings.	54
Figure 3.26	Curves for correcting flow meter readings for gases other than air.	56
Figure 3.27	Block diagram of cryogenic dew/frost point hygrometer.	62
Figure 3.28	Cryogenic sensing chamber and mirror assembly.	63
Figure 3.29	Response time of cryogenic hygrometer.	65
Figure 3.30	Response to step change.	66
Figure 3.31	Cryogenic hygrometer with dewar.	66
Figure 3.32	Cryogenic hygrometer with cryo-pump.	67

Figure 4.1	Resistive bulk polymer RH sensor.	70
Figure 4.2	%RH versus output, sensor type A.	75
Figure 4.3	%RH versus output, sensor type B.	76
Figure 4.4	Schematic of capacitive polymer sensor.	78
Figure 4.5	Typical capacitance change as a function of RH.	81
Figure 4.6	Typical %RH transmitters.	82
Figure 4.7	RH transmitter with remote probe.	83
Figure 4.8	Typical polymer RH instrumentation.	83
Figure 5.1	Construction of aluminum oxide sensor.	85
Figure 5.2	Aluminum oxide sensor.	86
Figure 5.3	Microprocessor based aluminum oxide hygrometer.	86
Figure 5.4	Aluminum oxide hygrometer and sensor.	87
Figure 5.5	Aluminum oxide sensor calibration laboratory.	90
Figure 5.6	Sensing crystal oscillator unit.	93
Figure 5.7	Piezoelectric process measurement cycle.	94
Figure 5.8	Piezoelectric moisture analyzer.	98
Figure 5.9	Electrolytic cell.	98
Figure 5.10	Electrolytic sensing cell.	100
Figure 5.11	Inner construction of electrolytic cell.	100
Figure 5.12	Schematic of a typical electrolytic hygrometer.	104
Figure 5.13	Electrolytic hygrometer.	107
Figure 5.14	Electrolytic low ppm/ppb hygrometer.	107
Figure 6.1	Water vapor absorption spectrum.	110
Figure 6.2	Absorption spectrum for uniformly mixed gases.	111
Figure 6.3	Environmental infrared hygrometer.	112
Figure 6.4	Block diagram of Lyman–Alpha absorption hygrometer.	113
Figure 6.5	Schematic of atmospheric Lyman–Alpha hygrometer.	114
Figure 6.6	Magnesium fluoride transmission curve.	115
Figure 6.7	Cut-away view of ionization chamber.	116
Figure 6.8	Photograph of an ionization chamber.	116
Figure 6.9	UV hygrometer.	117
Figure 6.10	Lyman–Alpha instrument for aircraft use.	118
Figure 6.11	Lyman–Alpha hygrometer for use on a tower.	118
Figure 6.12	Commercial Lyman–Alpha hygrometer.	119
Figure 7.1	Sling psychrometer.	121
Figure 7.2	Assman psychrometer.	122
Figure 8.1	Saturated salt sensor.	128
Figure 8.2	Response of lithium chloride sensor.	130
Figure 8.3	Fog chamber.	131
Figure 8.4	Fiber optic sensor.	132
Figure 8.5	Relative humidity versus wavelength.	133
Figure 8.6	Dew point versus wavelength.	134

Figures

Figure 9.1	Meteorological RH/temperature sensor probe.	140
Figure 9.2	Meteorological solar shield for sensors.	140
Figure 9.3	Meteorological chilled mirror hygrometer.	142
Figure 10.1	Dew point sensor accuracies.	155
Figure 10.2	Typical accuracies for various measurement methods.	155
Figure 10.3	Hierarchy of humidity standards.	156
Figure 10.4	Schematic diagram of NIST gravimetric hygrometer.	157
Figure 10.5	Schematic of two-temperature method.	160
Figure 10.6	Block diagram of NIST two-pressure humidity generator.	161
Figure 10.7	Schematic diagram of two-pressure humidity generator.	162
Figure 10.8	PGH hygrometer.	166
Figure 10.9	NPL humidity calibration facility.	166
Figure 10.10	Schematic of NPL gravimetric hygrometer.	168
Figure 10.11	Schematic of NPL humidity generator.	170
Figure 10.12	Recirculation humid air generator for dew point calibration.	173
Figure 10.13	Recirculation humid air generator for %RH calibration.	174
Figure 10.14	Schematic diagram of calibration system.	176
Figure 10.15	Low humidity calibration system.	177
Figure 10.16	Flow chart of NRLM humidity standard.	178
Figure 10.17	Schematic diagram of NRLM gravimetric hygrometer.	179
Figure 10.18	Schematic of NRLM humidity generator.	181
Figure 10.19	PTB gravimetric hygrometer.	182
Figure 10.20	Schematic of PTB two-pressure generator.	182
Figure 10.21	Flow sheet of the PTB coulometric trace humidity generator.	183
Figure 10.22	Self-contained commercial two-pressure generator.	185
Figure 10.23	Two-pressure, two-temperature humidity generator.	186
Figure 10.24	Relative humidity generator.	186
Figure 10.25	Block diagram of divided flow humidity generator.	187
Figure 10.26	Wet/Dry air selection for RH generator.	187
Figure 10.27	Temperature dependence of saturated salts.	190
Figure 12.1	Chilled mirror hygrometer for water activity measurements.	231
Figure 12.2	%RH instrument for measuring water activity.	231
Figure 12.3	Operating room with gas delivery system.	238
Figure 12.4	CCM medical gas hygrometer.	239
Figure 12.5	Conventional medical gas hygrometer.	239
Figure 12.6	Moisture measurement in ozone purification of water.	259
Figure 12.7	Measuring moisture in pellet dryer.	261
Figure 12.8	Hygrometer for automobile emissions testing.	265
Figure 12.9	Schematic of CPU cooling system.	269
Figure 12.10	RH measurement in concrete slab.	270
Figure 12.11	Typical IAQ chart for US government building.	274
Figure 12.12	Typical IAQ chart for US suburban office building.	275

Figure 12.13   Impact of relative humidity on air quality.   276

Figure 12.14   Dew point measurement in plant growth chamber.   279

Figure 13.1   ASHRAE psychrometric chart—Metric units.   284

Figure 13.2   Psychrometric table.   285

Figure 13.3   % Relative humidity versus dew point and temperature.   286

Figure 13.4   % Relative humidity versus absolute humidity
                from 50°C to 350°C at 1 bar pressure.   288

Figure 13.5   % Relative humidity versus absolute humidity
                from 100°F to 700°F at a pressure of 29.921 in. Hg.   289

Figure 13.6   % Relative humidity versus wet- and dry-bulb temperature
                difference at 20°F to 198°F and pressure of 29.921 in. Hg.   290

Figure 13.7   % Relative humidity versus wet- and dry-bulb temperature
                difference at 0°F to 80°F and pressure of 29.921 in. Hg.   291

Figure 13.8   % Relative humidity versus wet- and dry-bulb temperature
                difference at –50°F to 0°F and pressure of 29.921 in. Hg.   292

Figure 13.9   Absolute humidity (PPM_w) versus dew point
                at different pressures.   302

Figure 13.10   Moisture content of air and perfect gases
                at atmospheric pressure in English units.   303

Figure 13.11   Dew point–pressure–PPM_v conversion chart.   308

Figure 13.12   Dew point–pressure conversions at 0-400 psi_g
                and –160°F to +100°F.   309

Figure 13.13   Dew point–pressure conversions, at 0–500 psi_g
                and –60°F to 160°F.   310

Figure 13.14   Saturated water content of high pressure air
                at low temperatures in English units.   311

Figure 13.15   Saturated water content of high pressure air
                at moderate temperatures in English units.   312

Figure 13.16   Gas flow readings, corrected for pressure.   320

Figure 13.17   Equivalent gas flow corrected for specific gravity.   320

LIST OF TABLES

Table 2.1	Pressure conversion constants.	9
Table 2.2	Units of pressure referred to 1 psi.	9
Table 2.3	Values for R for some typical units.	12
Table 3.1	Dew point–frost point conversion.	39
Table 3.2	Typical depression and range of chilled mirror sensors.	45
Table 3.3	Typical chilled mirror hygrometer errors.	60
Table 4.1	Environmental tests.	72
Table 4.2	Environmental/Field tests.	74
Table 10.1	Comparison of non-gravimetric national standards.	146
Table 10.2	Instrument classifications.	149
Table 10.3	Typical application accuracies.	153
Table 10.4	Equilibrium relative humidity of saturated salts.	189
Table 11.1	Definitions and specifications.	194
Table 11.2	Saturation vapor pressure formulas.	198
Table 11.3	Saturation vapor pressure over water—Metric units.	199
Table 11.4	Saturation vapor pressure over water—English units.	202
Table 11.5	Saturation vapor pressure over ice—Metric units.	208
Table 11.6	Saturation vapor pressure over ice—English units.	210
Table 11.7	Vapor pressure of water above 100°C.	213
Table 12.1	Proposed SEMI standards for moisture content of gases.	227
Table 12.2	Contaminant levels.	238
Table 13.1	% Relative humidity at certain temperatures and dew points.	287
Table 13.2	% Relative humidity at certain temperatures versus dry/wet bulb depression, in °C.	293
Table 13.3	% Relative humidity at certain temperatures versus dry/wet bulb depression, in °F.	294
Table 13.4	Dew point–frost point conversions in °C.	295
Table 13.5	Dew point–frost point conversions in °F.	296
Table 13.6	Dew point relationships with common humidity parameters.	297
Table 13.7	Dew point versus vapor pressure, PPM_v, % relative humidity, and PPM_w from −150°C(−238°F) to 60°C(140°F).	298
Table 13.8	Moisture content, vapor pressure, and dew point in Metric units.	300
Table 13.9	Moisture content, vapor pressure, and dew point in English units.	301
Table 13.10	Moisture content of dry air—Metric units.	305

Table 13.11	Moisture content of dry air—English units.	306
Table 13.12	Mass of water vapor (g/m3) of saturated air @ 101.3 Pa.	307
Table 13.13	Vacuum conversion table.	313
Table 13.14	Altitude pressure table.	314
Table 13.15	Pressure conversion factors.	315
Table 13.16	Detailed pressure conversion chart.	316
Table 13.17	Flow equivalents.	321
Table 13.18	°C–°F conversions.	322
Table 13.19	Metric–English unit conversions.	324
Table 13.20	Unit conversion factors.	325

Water Vapor Measurement

1

INTRODUCTION

I. What Is Humidity?

The term "humidity" refers to water vapor, i.e., a gas. It is water in gaseous form. This book does not cover "moisture," which relates to water in liquid form that may be present in solid materials or liquids. However, the term "moisture" is frequently used in this book and in practice relating to measurements that are in fact water vapor measurements. For example, the term "trace moisture" is commonly used rather than "trace humidity" or "trace water vapor." There are also many industrial applications where moisture in a solid material or its surface is measured, by calculating the water content from humidity measurements in the immediate vicinity of the object, such as moisture in a container filled with pellets and in grain silos.

Humidity is present everywhere in the earth's atmosphere. Even in extremely dry areas and in boil-off from liquefied gases, there are traces of water vapor which in some applications could cause problems. Measurement of humidity is more difficult than the measurement of most other properties such as flow, temperature, level, and pressure. One reason for this is the extremely broad dynamic range, which could start from 1 part-per-billion or less (−112°C frost point) representing a partial vapor pressure of about 0.8×10^{-6} mm Hg, to saturated steam at 100°C (212°F) representing a partial vapor pressure of 760 mm Hg. This amounts to a dynamic range of about 10^9. Another reason is that measurements may have to be made in widely varying atmospheres, for example, from temperatures of −80°C (−112°F) to 1000°C (1832°F), in the presence of a wide range of gases which could be corrosive or non-corrosive, and in the presence of a variety of contaminants, of particulate and/or chemical nature.

Humidity measurements play an ever-increasing role in industrial, laboratory and process control applications by allowing improvements in quality of product, reduction of cost, or increasing human comfort. In the tobacco industry increasing proper humidity control greatly improves the quality of tobacco products. In warehouses humidity control protects corrosive or humidity sensitive materials, such as coils of steel, food, and dried milk. Cost savings applications include industrial, textile, and paper dryers. If humidity in the dryer is monitored, the dryer can be turned off as soon as the humidity is below a specified level. This could save large amounts of money in energy costs compared to the traditional way of running the dryer for a sufficient length of time to assure, with some safety margin, that the product is dry. Examples of human comfort and health are found in humidity-controlled hospital operating rooms, incubators, air-conditioning, meteorological applications, auto emissions, air pollution, ozone depletion studies, and many other areas.

These measurements are so important that the older and simpler humidity or dew point detectors such as dry/wet bulb psychrometers, hair hygrometers, and dew cups are no longer considered sufficiently accurate or suitable for most industrial applications. This has led to the use of more sophisticated, often micro-processor based, instrumentation to meet rigid requirements and government regulations of the Environmental Protection Agency (EPA), Food and Drug Administration (FDA), Federal Aviation Administration (FAA), and nuclear agencies.

The average person with a humidity measurement problem is confronted with a multitude of available measurement techniques and instrument types. The problem is to decide which sensor types are best suited for a particular situation, and of these, which is most useful for the application at hand, considering cost, maintenance, calibration requirements, ease of installation, and service.

Measurement of humidity is not a trivial task, nor is it intuitively understood. It is the intent of this book to give the potential user of humidity instrumentation the fundamental tools to make an informed decision as to which measurement technique is the best choice.

II. Importance of Humidity

Humidity measurements have become increasingly important, especially in the industrialized world since it has been recognized that humidity has a significant effect on quality of life, quality of products, safety, cost, and health. This has brought about a significant increase in humidity measurement applications and concurrent with this, an increase in research and development activities to improve measurement techniques, accuracy and reliability of instrumentation.

Despite the increase in research and development activities during the last two decades to improve humidity sensors, the present state of the art is still such that humidity measurements require more care, more maintenance, and more calibration than other analytical measurements. Furthermore, there is no sensor available that can even come close to covering the full dynamic range of water vapor levels. For this reason, many different measurement methods and sensors have been developed through the years, each having certain advantages and limitations and suitable for some, but not all, applications. For someone not skilled in the field of humidity, it is in many cases very difficult to make an intelligent choice of sensor, and when this is not done, disappointing results will often occur.

There are no manufacturers who manufacture all available types of sensors and instruments. Some offer only one type to meet requirements of a certain market segment or application. Others offer a broad range to cover a wider variety of applications. Manufacturers with the broader product ranges are generally in a better position to assist a user in selecting the right instrument and sensor for an application. However, for those involved in humidity measurements, or in new applications requiring such measurements, it is important to be able to obtain adequate knowledge of this subject. Though many articles and papers have been written on humidity measurements, this book is the first to cover a full range of sensors and measurement methods as needed to meet most water vapor measurement applications.

The book also covers some theory and definitions of humidity, calibration techniques, national and international standards, and many other topics. It is written for a broad audience, including educators, teachers, researchers, and academics, as well as the practicing engineer in industry or in the field.

Also included is a wealth of practical topics on various sensors and instruments, together with typical applications, conversion tables, charts, and other data. A computer disk (in Windows 95) is included in the back cover of the book for the convenience of readers who need to make conversions and have the availability of a DOS type computer. For the practicing engineer it is not necessary to dig through the theoretical portions of the book, while the more academically inclined can find what is needed without paying much attention to the practical parts of the book. Since humidity measurements are difficult to make with great accuracy and often have to be certified, calibration is an important part of humidity technology. A considerable part of the book has therefore been devoted to calibration, calibration standards, methods and standards employed by the National Institute of Standards and Testing (NIST), formerly the National Bureau of Standards (NBS), as well as similar institutions overseas such as the National Physical Laboratory (NPL) in the United Kingdom, CETIAT in France, and the National Research Laboratory of Metrology (NRLM) in Japan.

III. Units Used

Many different units are generally used to express humidity, pressure, temperature, and physical measurements. Although it could lead to some confusion, the most customary units are used to make the book practical and easy to work with for the reader. In some cases a conversion into another generally used parameter will be shown. The metric system is used throughout the book with the equivalent English units in parenthesis next to it. In the more scientific parts, such as calibration standards, in some cases only metric units are used. Tables and charts in Chapter 13 can be used to convert from one unit to another.

IV. Historical Background

The first known reference to humidity measurements was found in an ancient oriental document describing the use of a "balance-type" hygrometer for meteorological measurements in China. This was in the years before Christ.

In 1450, Nicholas Cryfts described the first hygroscope: "If someone should hang a good deal of wool, tied together on one end of a large pair of scales, and should balance it with stones at the other end in a place where the air is temperate, it would be found that the weight of the wool would increase when the air becomes more humid, and decrease when the air tends to dryness". The device was improved in 1550 by Leonardo da Vinci in his Codex Atlanticus, substituting a sponge for the wool. This formed the basis of our current national standard hygrometer, a gravimetric type. In the latter case, a known mass of air is passed through a desiccant (P_2O_5) that is weighed before and after the measurement. When great care is used, this method can be very accurate and it is fundamental.

The next basic type of hygrometer, attributed to Santorio Santorre in about 1614, involved the contraction or elongation of cords. A cord or lyre string was stretched between two fixed points and a weight was attached to the center. As the humidity increased, the cord contracted and the weight rose; as it dried out, the weight sank. This type of hygrometer quickly proliferated, with the substitution of paper, and other materials such as hair, followed later by strips of nylon and acetate. The theory of humidity did not keep pace with the instrumentation. It was in 1637 that René Descartes proposed that water was a distinct substance composed of long atoms. He compared the process of evaporation to raising dust from the feet of a passer-by. Even though the water becomes invisible, its particles, he maintained, keep their distinctive form, which is quite different from that of air particles. In 1660 the dew point hygrometer was discovered. It was reported that water formed from the air on rainy days on the outside of an ice-packed container.

During most of the eighteenth century, the fashionable opinion was that water dissolves in air in exactly the same way that a solid dissolves in a liquid. The main competing theory was that water vapor was the chemical combination of the two elements, fire and water. The real answers, provided by Bernoulli in 1738 in his kinetic theories of gases, were generally ignored. By 1790 this began to change. Jean Andrea Deluc had established an important principle stating that aqueous vapors have the properties of gases, "and exercise them in complete independence of these fluids. The product of evaporation is always of the same nature, namely an expandable fluid, which either alone or mixed with air, affects the manometer by pressure, and the hygrometer by moisture, without any difference arising from the presence or absence of air". Horace Benedict de Saussure also demonstrated that water vapor exercises a pressure of its own, its maximum value depending on the temperature. As data accumulated, scientists began to get a feel for the relationship between humidity and temperature. De Saussure adopted the method of stating humidity in grains of water per cubic foot and produced an extensive table of readings from his hair hygrometer for various humidities at various temperatures. He found that the relative weights of water vapor at various temperatures were about the same for all readings of his hygrometer, but he missed the simplifying concept of relative humidity. Deluc also came close when he wrote that "the quantities of steam correspondent to its different maxima according to temperature will afford a coefficient for the ratio observed." L.W. Gilbert finally put the issue quite clearly in 1803: "The degree of humidity depends on the ratio of the vapor actually present to that which is possible." Another flurry of activity occurred in humidity science during the twentieth century, due in large part to the explosive growth in the field of electronics. The advances were centered on various sensing techniques since the theories had solidified by the beginning of the nineteenth century. In rapid succession, the Dunmore sensor, the Pope cell, the "dew cell" (saturated salt), and the capacitive types were developed in the 1930s, the carbon type in the 1940s, the electrolytic, infrared, and Lyman–Alpha instruments in the 1950s, and the automatic chilled mirror hygrometer, the aluminum oxide hygrometer, and some other humidity measurement techniques in the 1960s. Each of these methods is still in use today, and many companies are actively pushing the state of the art. Every type of sensor has found a niche market where its performance characteristics apply. But new requirements for humidity measurement are coming up all the time where humidity was not of any concern in the past. In these applications, the user must make the right choices, so it is important to learn what type of sensor to use, and where and how to use it.

With regard to humidity standards, the NBS (now NIST) US Standard Gravimetric Hygrometer was built in the 1950s. This was followed by the NRLM Gravimetric Hygrometer and Constant Humidity Generator in Japan around 1962 and the NPL Gravimetric Hygrometer in England during the early 1980s. Other standards laboratories in the world today include the Standards Laboratory in Australia, The National Research Center in China, KRISS in Korea, IMGC in Italy, which also has a Two-Temperature generator, and the Centre for Metrology and Accreditation (CMA) in Finland.

V. Water Vapor and Moisture Measurement

Water is essential for all known forms of life, and is also always present in the earth's environment. Water molecules are highly polar—there is a nonuniform distribution of electric charge between the hydrogen and oxygen atoms. The results of this are many, but one of primary importance in moisture measurement is that water molecules attach themselves tenaciously to surfaces. As a result, the surfaces in any system are normally contaminated with water from the atmosphere and it is a prerequisite for accurate measurement to ensure that all surfaces are cleaned or purged beforehand. In addition, moisture must not be allowed to leak into the measurement system or otherwise interfere with the measurement. Saturated air contains about 17g (0.6 oz.) water in a cubic meter (about 10^{-3} lbs. per cubic foot) at 20°C (68°F). Fortunately, many sensors for use with liquids or solids are only marginally affected by gaseous moisture, which greatly simplifies on-line determinations. Another result of the highly polar nature of water molecules is the formation of so-called "hydrogen bonding". This means that water in liquids and solids possesses ill-defined structures of great importance in life and related processes. It exists in many forms within a solid or a solid–liquid mixture. The determination of the concentration of these various forms on-line is becoming increasingly important in industrial measurement, but will not be further discussed in this book which is only concerned with *water vapor* measurements.

2

DEFINITIONS AND FUNDAMENTALS

Water vapor measurements are closely related to—and dependent on—temperature and pressure. Water vapor is a form of gas and therefore follows the traditional gas laws. This chapter discusses definitions, equations and formulas showing the interrelationships. Much of the information presented in this chapter was derived from General Eastern's Handbook[4].

I. Temperature, Pressure, Humidity and Gases

Every substance can exist as a solid, liquid or gas, depending on the temperature and pressure that it experiences. At low temperatures all substances become solids. At some intermediate temperature these substances become liquids, and at sufficiently high temperatures they become gaseous.

Temperatures needed to achieve a given physical state in one substance do not necessarily yield the same state for another. Melting point and boiling point tables have been tabulated for many substances and can be found in chemistry handbooks.

A substance in the solid state has a definite shape, which is rigid and tends to resist change. Solids form a crystalline lattice where each molecule is fixed in space relative to each other.

A liquid substance will conform to any confining volume, flows easily, and is relatively incompressible. It is the state of matter intermediate between solid and gas. Liquids can also possess a free surface.

Gaseous substances can expand readily and without limit to fill any confining volume. They have the lowest density of all the physical states. The following discussion will concentrate specifically on the gaseous state of matter.

A. Temperature

Temperature is defined as:

- The degree of hotness or coldness of a body or environment

- A specific degree of hotness or coldness as indicated on, or referred to, a standard scale. A scale must be independent of the size of the system and must determine the direction of heat flow between any two systems that are in thermal contact.

Several temperature scales are commonly used in industry today. The best known are the Fahrenheit and Celsius scales, labeled °F and °C, respectively. Both scales are linear and can be interchanged by using the following expressions:

$$°F = 32 + 1.8°C \tag{2.1}$$

$$°C = \frac{(°F - 32)}{1.8} \tag{2.2}$$

The other two temperature scales, used mostly in scientific applications, are the Kelvin and Rankine scales, labeled K and °R, respectively. Note that there is no degree sign on the Kelvin scale; i.e., 100 K would be read as 100 Kelvins. These two scales are known as absolute temperature scales. At 0 K or 0°R, all molecular motion stops and there is no heat energy. It is the lowest possible temperature attainable in the universe. Absolute zero can be equated to the Fahrenheit and Celsius scales by the following equations:

$$K = 273.16 + °C \tag{2.3}$$

$$°R = 459.69 + °F \tag{2.4}$$

The sizes of the Kelvin and the °R temperatures are the same as the sizes of the °C and the °F respectively. Note that the absolute temperatures are never negative, while the °C and °F scales can be negative.

B. Pressure

Pressure is defined as the stress or force that is applied uniformly in all directions. It is the force per unit of area, e.g., kg per cm^2, Pascal, or pound per square inch (psi). The most common type of pressure experienced is atmospheric or barometric pressure. Like all matter on earth, air molecules are subject to the earth's gravitational pull. This causes the environmental air to exert a pressure of approximately 1.03 kg/cm^2 = 101.35 kPa = 14.7 pounds per square inch absolute (psi_a). Atmospheric pressure is not constant. It varies with height above sea level. For example, New Orleans, LA at sea level has an environmental pressure of 1.03 kg/cm^2 or 101.4 kPa (14.7 psi_a). In Denver, CO, which is about 1.5 km (1 mile) above sea level, the atmospheric pressure is approximately 0.95 kg/cm^2 or 93 kPa (13.5 psi_a).

Pressure can also be caused by enclosing a gas inside a volume. The gas exerts pressure on the walls of the container because gas molecules are randomly bombarding the walls. This can be observed using a pressure gage on the compressor tank. As the compressor pumps more air into the tank, pressure rises. This occurs because the number of molecules of air is increased. With more molecules there are more collisions against the tank wall, hence higher pressure.

Pressure can be measured in many different units, which leads to some confusion. In the United States, the most commonly used unit is pounds per square inch (psi). In Europe the Pascal or kPa unit is most commonly used as the standard. Regardless of which unit is used, pressure is a force per unit area and units should be chosen to be consistent with all other parameters that are to be measured. Pressure is normally measured with respect to a perfect vacuum. That is, when there are no gas molecules within a given volume, there is no pressure exerted on the walls of that volume. Measurements with respect to a perfect vacuum are referred to as absolute pressure (P_{abs}). Very often in industry it is more convenient to measure pressures referenced to the barometric (atmospheric) pressure. Measurements of this type are referred to as gage pressure (P_{gage}).

The following relationship equates P_{abs} and P_{gage}:

$$P_{abs} = P_{gage} + P_{barometric} \qquad (2.5)$$

To convert from one unit of pressure to another, reference is made to Table 2.1. More extensive conversion charts are presented in Chapter 13.

Table 2.1   Pressure conversion constants.

	psi[1]	In. H_2O[2]	In. Hg[3]	kPa	millibar	cm H_2O[4]	mm Hg[5]
psi[1]	1.000	27.680	2.036	6.8947	68.947	70.308	51.715
In. H_2O[2]	3.6127×10^{-2}	1.000	7.3554×10^{-2}	0.2491	2.491	2.5400	1.8683
In. Hg[3]	0.4912	13.596	1.000	3.3864	33.864	34.532	25.400
kPa	0.14504	4.0147	0.2953	1.000	10.000	10.1973	7.5006
millibar	0.01450	0.40147	0.02953	0.100	1.000	1.01973	0.75006
cm H_2O[4]	1.4223×10^{-2}	0.53525	3.9370×10^{-2}	0.09806	0.9806	1.000	0.7355
mm Hg[5]	1.9337×10^{-2}	0.53525	3.9370×10^{-2}	0.13332	1.3332	1.3595	1.000

Notes: (1) psi–pounds per square inch   (2) at 39°F   (3) at 32°F   (5) at 0°C.

Units of pressure related to psi (pounds per square inch) are shown in Table 2.2. In the following discussions all pressures and temperatures will be in absolute units.

Table 2.2   Units of pressure referred to 1 psi.

1 psi = 27.68 in. w.c. (inches of water column)
1 psi = 2.036 in. Hg (inches of mercury)
1 psi = 51.715 mm Hg or torr
1 psi = 0.06804 atmospheres
1 psi = 6.8947 kilo Pascals (kPa) = 0.068947 bar
(Note that 1 bar is not exactly 1 atmosphere)
1 psi = 0.0703 kg/cm^2
1 psi = 2.307 ft. H_2O

C. Gases

1. General Properties

If a solid is heated it will melt and become liquid. If a liquid is heated to a sufficiently high temperature, it will begin to boil and evaporate into the gaseous or vapor phase. This process causes gas molecules to exist in a much greater volume than when in the liquid state. It follows that if gas molecules are the same as liquid molecules, they must be of equal size. The only difference between the liquid and gaseous states is that distances between molecules are different.

With such large spatial distances, it is evident that gases can be readily compressed. If a gas is compressed at a constant temperature, there is an increase in the pressure of the container because molecules don't have to travel as far before they collide with the container wall or other molecules. Gases can be expanded, virtually, without limit.

Gases can also mix readily to form homogeneous solutions. Gases don't settle out or stratify, like some liquids do.

2. Gas Laws

a. Boyle's Law

In 1660 Robert Boyle performed some experiments with air as his primary gas and developed the following relationship between pressure and volume:

The volume of a given mass of a gas at constant temperature varies inversely with the pressure.

Therefore:

The product of the volume and the pressure of a given mass of a gas are a constant at constant temperature.

Mathematically this can be expressed as:

$$P = \frac{K_1}{V} \qquad \text{or} \qquad PV = K_1 \qquad (2.6)$$

where K_1 is a constant depending on the mass of the gas and the ambient temperature.

b. Charles and Gay-Lussac Law

It was about 100 years after Boyle's law was published that the next significant gas law was developed. Boyle observed that there are large changes in the volume of a gas if a candle is held close to the gas. However, he could not quantify the change in volume caused by the added heat because the concept of temperature, as we understand it today, was not fully formulated. Temperature effects were considered when good laboratory-grade thermometers became available. This occurred in about 1750. At this time two French scientists independently arrived at the law that bears their names. This law equates the volume of a gas and its temperature, and it states:

At constant pressure, the volume of a given amount of gas is directly proportional to its absolute temperature

or mathematically:

$$V = K_2T \qquad \text{or} \qquad \frac{V}{T} = K_2 \qquad\qquad (2.7)$$

The value of K_2 depends on the gas, its pressure, and units used. The temperature T is in degrees Rankine or Kelvins depending on the units used for T.

c. Avogadro's Law

An Italian physicist, Amadeo Avogadro, found that he could explain Gay-Lussac's findings if he assumed that gases of equal volume, at constant conditions of temperature and pressure, have equal numbers of molecules. This is expressed mathematically as:

$$V = \frac{n}{K_3} \qquad\qquad (2.8)$$

where K_3 is a constant and n is the number of moles of gas.

Avogadro defined a mole to be equal to 6.023×10^{23} molecules and that this quantity of molecules could be found in one gram equivalent weight (g.e.w.) of a substance. One g.e.w. is equal to the atomic weight of a substance measured in grams and 6.023×10^{23} is known as Avogadro's number. He also found that with certain gases he needed twice the g.e.w. to have the same reactive properties of a mole, and these he named diatomic gases. Some of these are nitrogen, oxygen, hydrogen and chlorine, and in nature they are found as N_2, O_2, H_2, and Cl_2, respectively. The subscripts denote the number of atoms.

d. Ideal Gas Law

Boyle's, Charles', and Avogadro's laws express gas volume as a function of pressure, temperature, and quantity and are respectively:

Boyle's law $\qquad\qquad\qquad V = \dfrac{K_1}{P} \qquad\qquad (2.9)$

Charles' law $\qquad\qquad\qquad V = K_2T \qquad\qquad (2.10)$

Avogadro's law $\qquad\qquad\qquad V = \dfrac{n}{K_3} \qquad\qquad (2.11)$

If pressure, temperature and quantity of a gas are all proportional to constants as described above, volume should be able to be combined with pressure, temperature and quantity of a gas into one expression. This is indeed the case and Equations 2.9, 2.10, and 2.11 can be combined into the following equation:

$$PV = nRT \qquad\qquad (2.12)$$

This is known as the ideal gas law, where:

P = Absolute pressure
V = volume
n = Number of moles of gas
T = Absolute temperature
R = Universal gas constant

Values for R, in some selected units, can be found in Table 2.3.

Table 2.3   Values for R for some typical units.

VOLUME	TEMP.	ABSOLUTE PRESSURE					
		Atm	Psi$_a$	mm Hg	in. Hg	in. H$_2$O	ft. H$_2$O
ft^3	K	0.00290	0.0426	2.20	0.0867	1.18	0.0982
	°R	0.00161	0.02366	1.22	0.0482	0.655	0.0546
cm^3	K	85.05	1206	62,400	2450	33,400	2780
	°R	45.6	670	34,600	1360	18,500	1550
liters	K	0.08205	1.206	62.4	2.45	33.4	2.78
	°R	0.0456	0.670	34.6	1.36	18.5	1.55

The Ideal Gas law is often not convenient to use as stated by Equation 2.12. Frequently the number of moles of a gas or the value of the gas constant is not readily known, but the pressure, volume, and temperature are easily measurable.

Equation 2.12 can be rewritten as follows:

$$\frac{PV}{T} = nR \quad \text{or} \quad \frac{PV}{T} = K \tag{2.13}$$

where K is a constant equal to the product of the gas constant and the number of moles of gas.

It can be readily observed that if up to two of the measurable parameters are varied, the third can be easily calculated using Equation 2.13, which is often written as:

$$\frac{P_1 V_1}{T_1} = \frac{P_2 V_2}{T_2} \tag{2.14}$$

where the subscripts 1 and 2 denote different pressure, volume, and temperature cases of the same gas.

e. Dalton's Law

In 1807 John Dalton, a British chemist, considered what would happen if a volume contained not just one type of gas, but a mixture of gases. How would this affect the ideal gas equation? It is known that:

$$PV = nRT \qquad \text{or} \qquad P = \frac{nRT}{V} \qquad (2.15)$$

Dalton supposed that if he had a mixture of gases a, b, and c to be combined into a known volume at a known temperature, then their individual contributions to the pressure in that volume could be written as:

$$P_a = \frac{n_a RT}{V}, P_b = \frac{n_b RT}{V}, P_c = \frac{n_c RT}{V} \qquad (2.16)$$

These he called "partial pressures" due to the fact that he considered each separately or as a partial portion of the volume. He proposed that the total pressure should be the sum of all the partial pressure constituents. Hence:

$$P_{tot} = P_a + P_b + P_c \qquad (2.17)$$

This is Dalton's law and can be expressed as follows:

The total pressure of a mixture of gases is equal to the sum of the pressures of the constituent gases, if each were individually to occupy that same volume, at that same temperature.

Or stated otherwise:

The total pressure of a gas mixture is equal to the sum of the partial pressures of the component gases.

D. Humidity

Water is present everywhere. One of the unique properties of our earth and its environment is that it contains water, which may be in solid (ice), liquid (water), or gaseous (water vapor) state. It is to be found in the air we breathe and the food we eat. It composes a significant portion of our body weight and is one of the most abundant resources on this planet. It affects nearly everything in life, i.e., comfort, quality of life, safety, health, etc. Humidity is the name given to water in the gaseous state (water vapor).

Since water vapor is found in gas mixtures, it behaves in accordance with the gas laws and exerts a partial pressure in the gas mixture per Dalton's law. This law is perhaps the most important gas law applicable to humidity measurements. Based on this law the measurement of humidity basically amounts to the measurement of the partial pressure of the water vapor component of the mixture of gases.

One of the most common and fundamental ways of measuring a partial pressure is by measuring the condensation temperature of that gaseous component of the gas. Hence the most fundamental way of measuring water vapor partial pressure is by cooling a sample of the gas until water condenses and then determine the exact temperature at which this occurs.

This is known as the dew or frost point temperature. There is a direct correlation between dew point temperature and water vapor partial pressure, also called vapor pressure. This is the reason why the chilled mirror hygrometer, discussed in Chapter 3, is generally considered the most fundamental measurement technique.

II. Definitions and Fundamental Relationships

A. Symbols

The following symbols are commonly used and are used in this chapter.

$$e = \text{Vapor pressure, millibars}$$
$$e_i = \text{Vapor pressure with respect to Ice, millibars at } T_f$$
$$e_w = \text{Vapor pressure with respect to Water, millibars at } T_d$$
$$e_{is}\, e_{ws} = \text{Saturation vapor pressure, millibars}$$
$$P = \text{Total pressure, millibars}$$
$$T = \text{Temperature, °C}$$
$$T_a = \text{Ambient or dry bulb temperature, °C}$$
$$T_d = \text{Dew point temperature, °C}$$
$$T_f = \text{Frost point temperature, °C}$$
$$T_w = \text{Wet bulb temperature, °C}$$
$$W = \text{Humidity ratio}$$
$$\%RH = \text{Percent relative humidity}$$
$$\%Sat = \text{Percent saturation}$$

B. Generally Used Humidity Terms

Many terms specific to humidity are derived from the ancient Greek words *hydor* (water), *hygros* (wet), and *psychros* (cold). Some vocabulary often used with regard to humidity are:

- Absorption   Retention of water vapor by penetration into the bulk material

- Adsorption   Retention of water vapor in a surface layer on a material

- Condensate   Condensed material, i.e., liquid water or ice

- Desiccant   A substance which exerts a drying action by chemically absorbing water vapor

- Desorption   Release of adsorbed or absorbed substance

- Humidity   The presence of water vapor in a gas. The word "humidity" is sometimes used to express relative humidity only. However, strictly speaking, "humidity" refers to all expressions related to water vapor.

- Hygrometer   An instrument for measuring humidity

- Hygrometry   The subject of humidity measurement

- Hygroscopic   Material prone to absorb water vapor

- Inert Gas   A chemically non-reactive gas, such as nitrogen, helium, argon, etc.

- Moisture   This term is commonly used to refer to liquid water or water vapor in any form. The term "moisture" is also often used to indicate water that is absorbed or bound into a material or substance.

- Probe   The part of an instrument that houses the main body of the sensor, i.e., at the end of a connecting electrical cable. In some cases the word "probe" may be used to refer to an entire hygrometer. The term "probe" is also loosely used inter changeably with "sensor" and "transmitter." Finally, "probe" could refer to a tube used to extract gas for a measurement.

- Sensor   The active or sensing part of a measuring instrument. There are some cases where a whole hygrometer is referred to as a "sensor." It is also often loosely used interchangeably with "probe" and "transmitter."

- Transmitter   An instrument which normally gives an electrical output, analog or digital, rather than a display on the instrument panel. The sensing head may be an integral part of the trans- mitter or may be connected via an external cable.

C. Vapor Pressure

Vapor pressure is that part of the total pressure that is contributed by the water vapor. It is expressed in units of pressure, i.e., in Pascals (Pa), or in units such as millibars (mbar), millimeters of mercury (mm Hg), inches of mercury (in. Hg), or psi (gage or absolute).

The saturation vapor pressure with respect to water is a function of temperature only and can be represented as:

$$e_{ws} = (1.0007 + 3.46 \times 10^{-6} P)\, 6.1121\, \text{EXP}\, [17.502\, T/(240.97 + T)] \qquad (2.18)$$

Saturation vapor pressure with respect to ice requires a minor adjustment of the constants in Equation 2.18 as given by the following:

$$e_{is} = (1.0003 + 4.18 \times 10^{-6}\, P)\, 6.1115\, \text{EXP}\, [22.452\, T/(272.55 + T)] \qquad (2.19)$$

In addition to yielding saturation vapor pressure as a function of ambient temperature, Equations 2.18 and 2.19 also yield actual vapor pressure as a function of dew or frost point temperature.

Total pressure of a gas mixture is equal to the sum of the partial pressures each constituent gas would exert, were it to occupy the same total volume, according to Dalton's law. The first term (in parentheses) in Equations 2.18 and 2.19 is the enhancement factor, and corrects for the slight difference between the ideal behavior of pure water and the behavior of water vapor as a constituent of air. An illustrative curve for water vapor pressure versus temperature for water is shown in Figure 2.1 and for supercooled water and ice in Figure 2.2.

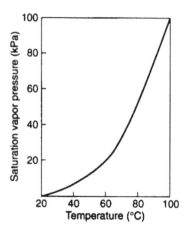

Figure 2.1   Saturation vapor pressure for water.

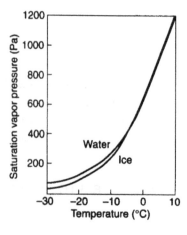

Figure 2.2   Saturation vapor pressure for ice and supercooled water.

D. Mixing Ratio

Mixing Ratio, also called humidity ratio or moisture content, is defined as the ratio of the mass of water to the mass of a dry carrier gas:

$$W = \frac{M_w}{M_g} \tag{2.20}$$

In terms of vapor pressures this can be written as follows:

$$W = 0.62198 \times \frac{e}{P - e}$$

$$W = 0.62198 \times \frac{e_w}{P - e_w} \tag{2.21}$$

$$W = 0.62198 \times \frac{e_i}{P - e_i}$$

The constant 0.62198 is the ratio of molecular weight of water to that of dry air and should be modified if carrier gas is other than air as follows:

$$mol.\ wt.\ ratio = \frac{mol.\ wt.\ H_2O}{mol.\ wt.\ dry\ gas} \tag{2.22}$$

Note: The humidity ratio is a dimensionless quantity as written above; it is usually followed by units such as:

$$\frac{lbs.\ H_2O}{lbs.\ air} \quad \text{or} \quad \frac{g\ H_2O}{g\ air}$$

E. Latent Heat

Latent heat is the heat stored in a substance, but not directly related to its temperature. For example, heat is stored in a gas because this heat was originally supplied to evaporate it. "Latent" means "hidden." Latent heat is expressed in energy per unit mass of substance, i.e., joules per kilogram (J kg^{-1}).

F. Percent Saturation

Percent saturation is the ratio of the actual mixing ratio to the mixing ratio at the same temperature, expressed as a percentage. It can be expressed as follows:

$$\% \ Sat = 100 \times \frac{e \ (P - e_s)}{e_s \ (P - e)}$$

$$\% \ Sat = 100 \times \frac{e_w \ (P - e_{ws})}{e_{ws} \ (P - e_w)} = 100 \ \frac{W_d}{W_s} \qquad (2.23)$$

Note:

$$\% \ Sat = \% \ RH \times \frac{(P - e_s)}{(P - e)}$$

However, at ordinary temperatures the partial pressure of atmospheric moisture, even at saturation, is so small compared to the total pressure that the term $(P - e_s)/(P - e)$ is substantially equal to unity, and for practical purposes, the percentage saturation and the percentage relative humidity are approximately equal. Obviously, for a condition of saturation, relative humidity is identical to the percentage saturation, in this case $e = e_s$.

It must be realized that the two terms "percentage saturation" and "percentage relative humidity" are not synonymous, but fundamentally different, and that for a given humidity, the numerical difference between percent saturation and percent relative humidity increase as temperature is increased. Often in psychrometric charts the label of %RH is used whereas the actual parameter is % saturation as defined above.

G. Humidity Parameters

1. Dew Point

The temperature to which a volume of gas must be cooled such that it becomes saturated with respect to liquid water.

It is the solution to Equation 2.18 where $T = T_d$ at a particular vapor pressure, and

$$e_{ws} = e_w, \text{ at pressure } P.$$

2. Frost Point

The temperature to which a volume of gas must be cooled, such that it becomes saturated with respect to ice.

It is the solution to Equation 2.19 where $T = T_f$ at a particular vapor pressure, and

$$e_{is} = e_i \text{ at pressure } P.$$

3. Relative Humidity

Relative humidity is the ratio of the actual partial vapor pressure to the saturation vapor pressure of the gas, multiplied with 100% at the prevailing ambient temperature.

For an actual vapor pressure e, and saturation vapor pressure, e_s,

$$\% \text{ Relative Humidity} = \frac{e}{e_s} \times 100\% \qquad (2.24)$$

4. Absolute Humidity

Absolute humidity is defined as water vapor density and is expressed as water vapor mass per unit volume of dry air.

Absolute humidity can be expressed as follows:

$$g/m^3 = 216.7 \, e_w/(T_a + 273.16) = 216.7 \, e_i/T_a + 273.16 \qquad (2.25)$$

Water vapor content can be expressed as parts per million by volume (ppm_v) as follows:

$$PPM_v = 10^6 \, e/(P - e) \qquad (2.26)$$

Part per million by weight (PPM_w), or mixing ratio, is derived by multiplying the above by the ratio of the molecular weight of water to that of air as given as follows:

$$PPM_w = 0.62198 \times 10^6 \times e/(P - e) \qquad (2.27)$$

$$PPM_w = 0.62198 \times PPM_v \qquad (2.28)$$

$$PPM_w = W \times 10^6 \qquad (2.29)$$

In short, PPM_v and PPM_w are defined as:

a. Parts per million by volume (PPM_v)

Volume of water vapor per total volume of gas, for an ideal gas. Sometimes expressed relative to the total volume of moist gas (mole fraction times one million) or sometimes relative to the total dry gas. For small numbers of parts per million, the two are almost identical; at higher humidities they become significantly different.

Or in terms of vapor pressures:

$$PPM_v = \frac{\text{Partial Pressure of } H_2O}{\text{Total Pressure} - \text{Partial Pressure of } H_2O} \times 10^6$$

b. Parts per million by weight (PPM_w)

Sometimes used to express the amount (mass) of water vapor relative to the total dry gas (mixing ratio times one million), but sometimes to express the amount relative to the total moist gas (specific humidity times one million). For small numbers of parts per million, the two are almost the same; at higher humidities they become significantly different.

$$PPM_w = \frac{\text{Mass of Water Vapor}}{\text{Mass of Carrier Gas}} \times 10^6$$

$$PPM_w = PPM_v \times \frac{\text{Molecular Weight of Water Vapor}}{\text{Molecular Weight of Carrier Gas}}$$

For example:

$$\text{Molecular weight of water } (H_2O) = 18$$

$$\text{Molecular weight of air} = 29$$

Hence for air:

$$PPM_w = PPM_v \times 18/29 = 0.62068699 \, PPM_v$$

Molecular weights for some other gases are:

Gas	Molecular Wt.	Gas	Molecular Wt.
Air	29	Hydrogen	2
Water	18	Ethylene	28
Acetylene	26	Helium	4
Ammonia	17	Methane	17
Argon	40	Nitrogen	28
CO_2	44	Oxygen	32
CO	28	Sulfur Dioxide	64

5. Wet Bulb Temperature

When unsaturated air is passed over a wet thermometer bulb, water evaporates from the wet surface and latent heat absorbed by the vaporizing water causes the temperature of the wetted surface and enclosed thermometer bulb to fall. As soon as the temperature of the wet surface drops below that of the surrounding atmosphere, heat begins to flow from warmer air to the cooler surface, and the quantity of heat transferred in this manner increases with an increasing drop in temperature. On the other hand, as temperature drops, the vapor pressure of water becomes lower and, hence, the rate of evaporation decreases. Eventually a temperature is reached where the rate at which heat is transferred from the air to the wet surface by convection and conduction is equal to the rate at which the wet surface loses heat in the form of latent heat of vaporization. Thus no further drop in temperature can occur. This equilibrium temperature is known as the wet bulb temperature.

A relationship between wet bulb temperature T_w, dry bulb temperature T_d, and the humidity ratio W is as follows:

$$W = \frac{(2501 - 2.381\ T_w)\ W_W - (T_a - T_w)}{2501 + 1.805\ T_a - 4.186\ T_w}$$

(2.30)

where W_w is the humidity ratio at T_w.

H. Pressure Effects and Dew Point Measurements

Dew and frost point are measures of the partial pressure of water vapor in a gas mixture. This means that there is a one-to-one correspondence between vapor pressure and condensation temperature.

Since water vapor is a gas, it must obey the gas laws as outlined in the previous section. Water vapor is a constituent of a gas mixture so it must obey Dalton's law of partial pressures as follows:

$$P_t = P_a + P_b + P_c + \ldots + P_w$$

(2.31)

where:

$$P_t = \text{Total pressure}$$
$$P_{a,\ b,\ c} = \text{Partial pressure of various constituent gases}$$
$$P_w = \text{Water vapor pressure}$$

If the total pressure is raised, each partial vapor pressure is raised in the same amount. This can be written as:

$$K\,P_t = K(P_a + P_b + P_s + \ldots + P_w)$$

(2.32)

Because water vapor pressure is a function of dew/frost point temperature, when the total pressure is increased, the dew/frost point increases. Likewise, if the pressure goes down, the dew/frost point goes down. Therefore, the ratio of the dew point partial pressure to the total pressure must be constant, hence:

$$\frac{P_w}{P_t} = C$$

(2.33)

This can be written as:

$$\frac{P_{w1}}{P_{t1}} = \frac{P_{w2}}{P_{t2}}$$

(2.34)

where the subscripts 1 and 2 denote different pressure cases of the same gas volume. The chart in Figure 2.3 shows how dew points change with pressure and how a dew point at one pressure can be converted to dew point at another pressure.

EFFECTS OF PRESSURE ON DEW POINT

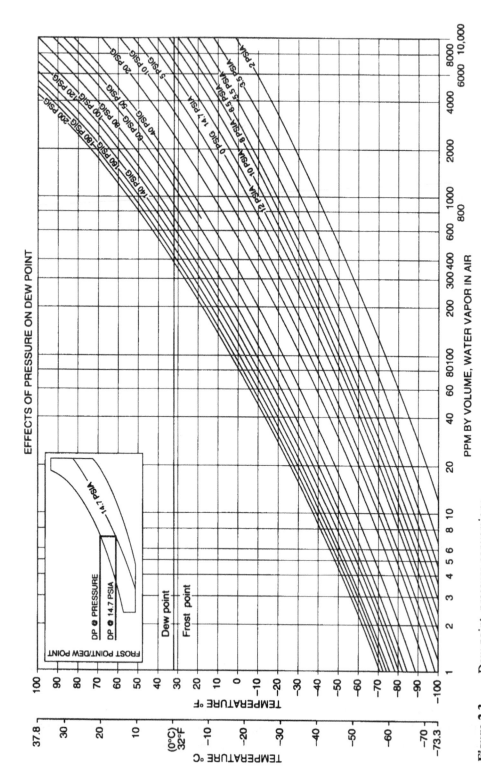

Figure 2.3   Dew point–pressure conversions.

I. Pressure Units

Most of the western world is now using, or at least converting to, the Metric system and the Pascal (Pa) or kilo Pascal (kPa) are the standard units, except in meteorological applications where the bar, millibar, and mm Hg are still widely used. However, much of the existing literature, especially in the English speaking world (US and UK), is still using the English units. Therefore all units in this book are shown first in Metric units and then between parenthesis in English units.

Pressure can be expressed in many different Metric and English units. This often leads to confusion. A brief discussion is therefore in order.

$$1 \text{ Newton } (N) = 10^{-5} \text{ dynes} \qquad\qquad 1 m^2 = 10^4 \text{ cm}^2$$

Pascal (Pa) is the universal metric unit for pressure.

$$1 \text{ Pa} = 1 N/m^2 = 10 \text{ dynes/cm}^2 = 10^5 \text{ dynes/m}^2$$

$$1 \text{ kg/cm}^2 = 9.807 \times 10^4 \text{ Pa} = 98.07 \text{ kPa}$$

In meteorology, the units commonly used are "bar," "atm," "millibar," "inches of Hg," "mm Hg," and "inches of water."

$$1 \text{ bar} = 10^6 \text{ dynes/cm}^2$$
$$1 \text{ mbar} = 10^3 \text{ dynes/cm}^2$$

Therefore:

$$1 \text{ bar} = 10^5 \text{ Pa} = 100 \text{ kPa} = 10^6 \text{ dynes/cm}^2 = 10^{10} \text{ dynes/m}^2$$

$$1 \text{ mbar} = 10^7 \text{ dynes/m}^2 = 10^3 \text{ dynes/cm}^2 = 100 \text{ Pa}$$

$$1 \text{ mbar} = 133.3 \text{ Pa} = 133.3 \ N/m^2 = 1.333 \times 10^3 \text{ dynes/cm}^2$$

$$1 \text{ inch Hg (@ 32°F)} = 25.4 \text{ mm Hg} = 3.386 \text{ kPa}$$

$$1 \text{ bar} = 14.504 \text{ psi} = 0.98692 \text{ atm}$$

$$1 \text{ atm} = 14.7 \text{ psi}$$

Hence:

$$1 \text{ atm} = \text{close to, but not exactly, 1 bar}$$

Where no exact numbers are required, it is convenient to use the following approximations:

$$1 \text{ bar} = 100 \text{ kPa}$$

$$1 \text{ psi} = 6.895 \text{ kPa} = \text{approx. } 7 \text{ kPa}$$

$$1 \text{ inch Hg} = 3.386 \text{ kPa} = \text{approx. } 3.5 \text{ kPa}$$

$$1 \text{ inch water} = 0.249 \text{ kPa} = \text{approx. } 0.25 \text{ kPa}$$

$$1 \text{ mm Hg} = 0.1333 \text{ kPa} = \text{approx. } 0.125 \text{ kPa}$$

$$1 \text{ atm} = \text{approx. } 1 \text{ bar}$$

Please note that the approximate orders of magnitude between these units are 100: 7: 3.5: 1/4: 1/8. The confusion between relative magnitudes disappears when Metric units are used exclusively. Because of the importance of these relationships, it may be useful to list the units in reverse as follows:

$$1 \text{ kPa} = 7.501 \text{ mm Hg} = \text{approx. } 8 \text{ mm Hg}$$

$$1 \text{ kPa} = 4.015 \text{ inch H}_2\text{O} = \text{approx. } 4 \text{ inch H}_2\text{O}$$

$$1 \text{ kPa} = 0.295 \text{ inch Hg} = \text{approx. } 0.3 \text{ inch Hg}$$

$$1 \text{ kPa} = 0.145 \text{ psi} = \text{approx. } 0.15 \text{ psi}$$

$$1 \text{ kPa} = 0.01 \text{ bar}$$

The approximate numbers are given as a convenient way to remember the orders of magnitude. The Pascal (Pa) unit is often too small for general use. In most cases where Metric units are used, pressures are expressed in kPa or MPa (Million Pa).

III. Psychrometric Chart

The chart shown in Figure 2.4 represents many of the parameters defined in this chapter. Easy conversions can be made using such a psychrometric chart, but one drawback is that the chart is always presented for a certain pressure, usually atmospheric pressure, at sea level. Pressure corrections must therefore be made if the chart is used for barometric pressures other than at sea level.

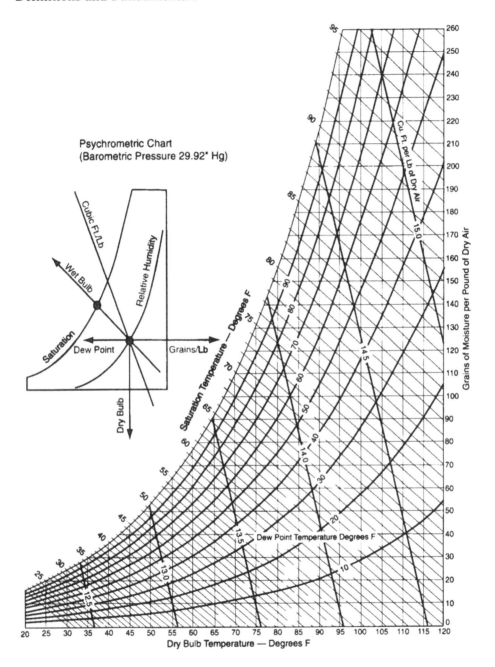

Figure 2.4   The psychrometric chart relating all of the basic humidity parameters. Directions for reading the chart are illustrated by the insert. For more detailed psychrometric charts, see Chapter 13.

3

CHILLED MIRROR HYGROMETERS

I. Introduction

The chilled mirror hygrometer, sometimes called an optical condensation hygrometer, is the most accurate, reliable, and fundamental hygrometer commercially available and is therefore widely used as a calibration standard. The first commercial applications of the hygrometer started in the early 1960s after it became practical to use thermoelectric coolers and electro-optical detection methods. Prior to 1960, the use of chilled mirror hygrometers was primarily confined to laboratories because of the difficulties in cooling the mirror with cryogenic liquids, controlling its temperature, and detecting the onset of dew, usually through visual means.

Precise and fundamental measurements are often so important that the older and simpler humidity or dew point detectors, such as dry/wet bulb psychrometers and dew cups, are no longer considered to provide the accuracy required for most laboratory and industrial applications.

A drawback of the chilled mirror instrument is that it is more expensive than most other types, and requires maintenance by skilled personnel, monitoring, and care of installation (such as providing proper sample flow). A new chilled mirror hygrometer, using a unique cycling mirror control technique, was developed during the 1980s. The method of mirror cycling reduces the sensor's sensitivity to contaminants, but at the expense of some of its fundamental measurement properties and accuracy. Several instruments incorporating this technology have been successfully applied in some selected applications in Europe and the US. They are especially useful in cases where low-maintenance, trouble-free, and user-friendly operation are more important than maximum precision.

The largest source of error in a condensation hygrometer is the difficulty of accurately measuring the condensation surface temperature. Special industrial and laboratory versions of the instrument can be made with accuracies as high as $\pm 0.1°C$ ($\pm 0.2°F$) over a wide temperature span. Most instruments offer an accuracy of $\pm 0.2°C$ ($\pm 0.4°F$). Chilled mirror hygrometers can be made very compact by using solid state optics and thermoelectric cooling, and offer many capabilities through the use of microprocessors.

II. Conventional Chilled Mirror Hygrometer

In its most fundamental form, dew point is detected by cooling a reflective condensation surface (mirror) until water begins to condense, and by detecting condensed fine water droplets optically with an electro-optic detection system. The signal is fed into an electronic feed back control system to control the mirror temperature, which maintains a certain thickness of dew at all times.

Since the introduction of the chilled mirror hygrometer, this type of instrumentation has become widely accepted for precision laboratory calibrations and standards despite its higher cost. Chilled mirror hygrometers have also been used in many industrial applications where high accuracy and traceability are required. This has resulted in many improvements, new developments, and several patents.

Operation of the basic optical dew point hygrometer is shown in Figure 3.1. The surface temperature of a small gold or rhodium-plated copper mirror is controlled by a thermoelectric cooler (heat pump). A high intensity light-emitting diode (LED) illuminates the mirror. The quantity of reflected light from the mirror surface is detected by a photo-transistor or optical detector.

A separate LED and photo-transistor pair are used to compensate for thermally-caused errors in the optical components. The photo-transistors are arranged in an electrical bridge circuit with adjustable balance which controls the current to the thermoelectric mirror cooler and, therefore, the mirror temperature.

Reflectance is high when the mirror surface temperature is above the dew point (dry mirror) and maximum light is received by the optical detector. However, when the thermoelectric cooler reduces the mirror temperature below the dew or frost point (if below 0°C or 32°F), moisture condenses on the surface, causing light scattering, thereby reducing the amount of light received by the detector.

The system is designed so that the bridge is balanced and stable only when a predetermined layer of dew or frost is maintained on the mirror surface. Under these equilibrium conditions, the surface temperature is precisely at the dew point of the gas passing over the mirror. A precision NIST-traceable platinum or equivalent thermometer

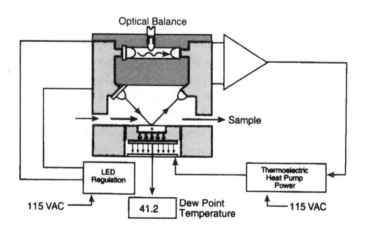

Figure 3.1   Schematic of conventional chilled mirror sensor.

is imbedded within the mirror surface to measure its surface temperature. The dew point temperature is displayed on the front panel of the instrument. When the mirror is clean, a perfect layer of condensation can be maintained and high accuracy and repeatability results.

A. Sensitivity to Contaminants

Like any other humidity instrumentation, the main difficulty with chilled mirror hygrometers is that the sensor is sensitive to contaminants. Even in clean applications, ultimately some contaminants will appear on the mirror surface, thereby influencing the optical detector and servo balancing functions of the instrument. A common practice to minimize this problem is to periodically open the servo feedback loop, causing the mirror surface to heat to a dry state, and then re-adjusting the "balance" or reference of the optical circuit to compensate for the reduced reflectance of the mirror.

In earlier models this procedure was performed manually. Subsequent improvements have been the so-called "automatic balance" and "continuous balance" by EG&G (now Edgetech), "automatic balance" by MBW and Michelle in Europe, and "PACER" by General Eastern. In all of these methods, the mirror is continuously operated at the dew point temperature and has at all times, except during the short balancing cycle, a dew or frost layer on it.

A new type of hygrometer was developed and patented by Protimeter plc in the UK, which cycles the mirror temperature and only briefly keeps the mirror at the dew point. As a result, the mirror is kept in the "dry" state about 95% of the time. The advantage of a dry mirror is that it does not hold on to particulate contaminants as readily as a wet mirror and does not cause chemical reactions between chemical constituents and the water droplets on the mirror. Even though very high accuracies can be obtained, some fundamental measurement features are sacrificed in the CCM system, which makes this type of instrument less suitable for use as a fundamental calibration standard.

B. Self-Standardization of Chilled Mirror Hygrometers

1. Automatic Balance Control (ABC)

The automatic balance control (ABC) method of standardizing a chilled mirror hygrometer was first developed and patented by EG&G (formerly Cambridge Systems). This method, with some variations, was subsequently adopted by most other manufacturers and became a widely accepted method of checking and standardizing the mirror of a condensation hygrometer.

The ABC method involves the use of circuitry to automatically interrupt the control circuit at a prescribed interval selected by the user. These intervals can be, for example, once every 12, 24 or 48 hours, depending on the expected contamination level. The mirror is then heated to a temperature above ambient, typically 60°C to 80°C (140°F to 176°F), causing any dew to be removed. Reflectance from the dry, but contaminated mirror, is then compared with a standard, based on a dry but clean mirror, and a correction is made to "standardize" the mirror without cleaning it. The mirror is then cooled down again to the dew point and goes into control like it was before the automatic balance cycle. This operation typically takes three to eight minutes, depending on how low a dew or frost point is being monitored. A longer time is required for low frost points because it takes longer for a mirror to go into the control mode when the mirror has to be cooled far below ambient temperature and when there is little water vapor in the sample flow.

The automatic balance cycle does not remove or disturb any contaminants on the mirror. It only allows the instrument to be re-standardized based on a contaminated mirror, thereby extending the time period between required mirror cleanings. Soluble contaminants can also contribute to errors in measurement as will be discussed later. The automatic balance reduces, but does not eliminate, such errors.

2. PACER Circuit

The automatic balance circuit extends the mirror cleaning cycle and reduces contaminant-induced errors. The PACER technique, acronym for "Programmable Automatic Contaminant Error Reduction," permits accurate long-term operation of a thermo-electrically-cooled, optically-detected dew point hygrometer without need for shutdown to clean the mirror. After operating the sensor for some time, water-soluble salts and other contaminants, if present in the sample gas, can accumulate on the mirror surface. They create two error-causing effects, illustrated graphically in Figure 3.2.

The operating dew density decreases as contamination in the dew layer increases. The light received by the detector becomes more and more dependent on the layer of contaminants rather than the dew layer and eventually the hygrometer becomes inoperative. At this point the system must be shut down and the mirror cleaned to attain proper operation.

Soluble contaminants, after dissolving in the operating dew layer, modify the vapor pressure. The control loop compensates for this by elevating the mirror temperature. Since the thermometer measures the actual mirror surface temperature, a measurement error results. The temperature gradually increases, sometimes by as much as 5°C (9° F), which

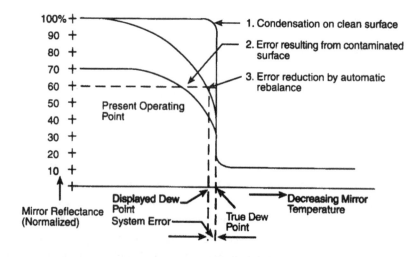

Figure 3.2   Error caused by mirror contamination.

is typically undetected. Then it finally takes a dramatic jump at which point it is observed and the mirror is to be cleaned.

Systems outfitted with automatic balance correct the first type of error-causing effect only, which is reduced mirror reflectance due to contaminants. This has been accomplished by periodic automatic or manual interruption of the temperature control loop, and heating the mirror to evaporate dew. The reflectance of the dry mirror, although reduced by contaminant deposits, is then compensated by re-balancing. This is effective in maintaining control because it re-establishes the operating dew layer. However, it does not address the measurement error problem (Raoult effect) that occurs before the system goes out of control.

3. PACER Technique*

Using the PACER technique, the time interval between mirror cleanings can usually be significantly extended. The PACER circuit, unlike conventional automatic balance circuits, first causes the mirror to cool below the prevailing dew point, which results in an excess of water deposited on the mirror surface. In essence, it allows a "puddle" to form on the mirror. This excess of solvent (water) encourages soluble materials to go into solution.

The PACER circuit then causes the mirror to quickly heat, evaporating the water. The large puddle breaks up into small puddles which shrink in size, holding higher and higher concentrations of solute in solution, and leaves bare, uncontaminated surfaces behind.

Eventually each puddle becomes saturated and the solute begins to precipitate out in polycrystalline clusters. The amount of salt has been redistributed or clustered in such a fashion that most of the mirror surface is clean. The resulting reduction in reflectance is only 15% or 20%, even in severely contaminated situations.

The PACER circuit automatically adjusts the control loop offset to operate at a level of condensate which corresponds to a preset reduction of reflectance below that of a clean mirror. The temperature control loop, calling for this reflectance level, causes the mirror to be maintained at true dew point because it is at that temperature only that such a reduction can be realized. A typical PACER cycle is shown in Figure 3.3.

Figure 3.4 shows a contaminated mirror surface before and after the PACER cycle. In this figure, a mirror that has been allowed to become contaminated for a period of one week in a heavily laden lithium-chloride air stream is shown after the drying process of a typical "balancing" procedure (Figure 3.4a). The salt is uniformly distributed over the entire mirror surface. The reflectance of this surface is only 40% of that of a clean mirror in the dry state.

In Figure 3.4b, the mirror has first been cooled for thirty seconds, allowing the salt to dissolve into the puddle, and then heated to the dry state. The salt is now grouped into isolated colonies and the reflectance of the surface is 85% of that of a clean mirror. The result is a hygrometer with improved long-term accuracy in the presence of high levels of soluble contaminants.

Patented by General Eastern Instruments.

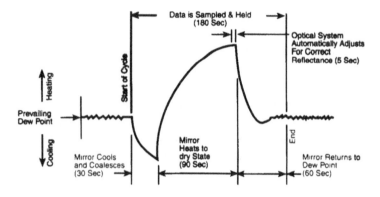

Figure 3.3   Typical PACER cycle.

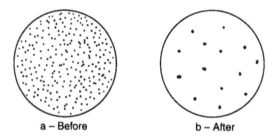

a – Before   b – After

Figure 3.4   Mirror condition before and after PACER cycle.

Figure 3.5 shows graphically the elimination of dew point error. Note that the operating point after the PACER cycle and re-balancing has returned to the true dew point.

Typically, a PACER cycle can be programmed automatically every 2, 6, 12, or 24 hours; cooling and heating times are also programmable. During the short interval of the PACER cycle, the output data and display are held at the last prevailing dew point.

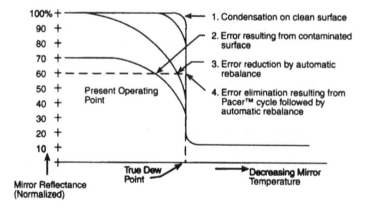

Figure 3.5   Error elimination with the PACER cycle.

4. Continuous Balance/Dual Mirror Twin Beam Sensor

As indicated above, optical condensation-type hygrometers must be periodically re-adjusted or "balanced" versus a reference to compensate for contaminants which always deposit on the reflectance mirror of the sensor. In conventional systems this is done manually or automatically with time intervals of a few times a day to once every several days depending on contamination levels. Using the "PACER" circuit described above, a "cool" or "coalescence" cycle is added to redistribute soluble contaminants on the mirror and thereby prevent measurement errors.

In a dual mirror sensor, one mirror is always kept at a temperature above the dew point (dry mirror), while the other is controlled at the dew point (wet mirror). The sample gas is passed over both mirrors. Two light sources are used one for each mirror (see Figure 3.6).

Reflectance from the dry mirror is used as a reference for controlling the wet mirror and it is therefore no longer necessary to periodically interrupt the control circuit for a "balance" cycle (typically 2 to 8 minutes) and the method offers the advantage that there is no loss of data during this 2 to 8 minute cycle every 12 or 24 hours. At very low frost points a conventional balance cycle could take 30 minutes or more and in such low frost point applications the dual mirror method offers more significant advantages. Limitations of this method are the following:

- It is evident that when a dry and a wet mirror are exposed to the same contaminated gas stream, the wet mirror will collect contaminants at a faster rate than the dry mirror so that effective balance control is only possible for a limited period of time. Some improvements may be possible by keeping the temperatures of both mirrors in close proximity, i.e., keeping the dry mirror temperature at just above the dew point. However, the problem of contaminant imbalance cannot be eliminated, which is a disadvantage of this method.

- The continuous balance method does not address the soluble contaminant error problem described above. Contaminants remain on the mirror and are undisturbed by the continuous balance operation. Long-range measurement errors are therefore to be expected and time intervals between mirror cleanings are usually shorter than for instruments outfitted with the "PACER" circuitry.

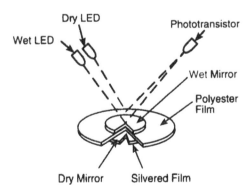

Figure 3.6   Twin beam dual mirror sensor.

Typical conventional chilled mirror hygrometers are shown in Figures 3.7, 3.8, 3.9, and 3.10.

Figure 3.7   Microprocessor based multi-parameter hygrometer.
(Courtesy General Eastern Instruments)

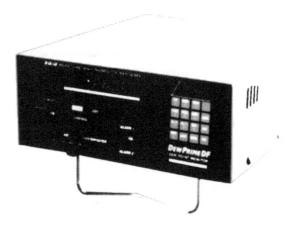

Figure 3.8   Microprocessor based hygrometer with continuous balance.
(Courtesy EG&G, presently Edgetech)

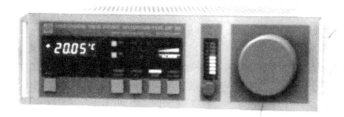

integral sensor

Figure 3.9   Precision chilled mirror hygrometer.
            (Courtesy MBW Elektronik AG, Switzerland)

integral sensor

Figure 3.10   Precision low frost point hygrometer for measurements to –95°C at pressures to 10 Bars (1 MPa).
            (Courtesy MBW Electronik AG, Switzerland)

The low frost point hygrometer shown in Figure 3.10 is specially designed for measuring very low frost points, reportedly down to –95°C. This is much lower than the bottom range of conventional chilled mirror hygrometers which are rarely used below –75°C, even when outfitted with sensors using three- to five-stage coolers and when cooled with chilled glycol. The instrument shown in Figure 3.10 is provided with an integral cooling system and is quite expensive. Its response time at this low temperature is slow.

In the early 1990s a cryogenic hygrometer was developed for the purpose of measuring frost points to –100°C and –110°C using the fundamental condensation technique, but offering a much faster response time. This instrument type is discussed in Section X of this chapter. Since very few cryogenic hygrometers have been built as of this date, their cost is presently very high.

III. Cycling Chilled Mirror Dew Point Hygrometer (CCM)*

A significant departure from the above chilled mirror systems is the Cycling Chilled Mirror (CCM) control system, often used with an integral filter, offering the user a sensor requiring considerably less mirror or filter maintenance. In addition, the new sensing technique makes it possible to always measure dew point below 0°C (32°F), solving the well known problem of dew versus frost point uncertainties which could cause interpretation errors of 1°C to 3°C (1.8°F to 5.4°F) in conventional hygrometers. Signal processing is often done inside the sensor allowing longer and lower cost cables to be used when the sensor is operated remotely. For high temperature applications a sensor has been developed utilizing fiber optic bundles which keep the temperature sensitive electro-optical components away from the high temperature environment.

Almost all chilled mirror hygrometers that have been on the market prior to 1980 are of the standard, or continuous condensation type. The condensation surface (reflective mirror) is maintained electronically in vapor pressure equilibrium with the surrounding gas, while surface condensation is detected by optical techniques. The surface temperature is then the dew point temperature by definition. Typical industrial versions of the instrument are accurate to ± 0.2°C (0.36°F) over wide temperature and dew point spans. Such instruments provide fundamental measurements which can be made traceable to NIST (National Institute for Standards Testing) or other national standards.

The CCM hygrometer uses a cycling chilled mirror method. The mirror temperature is lowered at a precisely controlled rate until the formation of dew is detected. Before the dew sample is able to form a continuous layer on the mirror, the mirror is heated and the dew on the mirror surface is evaporated. Hence the mirror is almost always (95% of the time) in the dry state and contains a dew layer for only a short time (5% of the time), when a dew point measurement is made. Typically the measurement cycle is once every 20 seconds. Faster rates are possible. Another important benefit is that the dew layer is maintained for such a short time that it will never convert into frost, even when well below 0°C (32°F). This eliminates the frequently encountered problem with conventional optical hygrometers, that one cannot be certain whether the instrument reads dew point or frost point when below 0°C. The reason is that due to super cooling, a dew layer below 0°C (32°F) often remains for a long time, or even continuously, depending on flow rate, mirror condition, contaminants and other factors.

A. Mirror Cycling

Figure 3.11 shows how the mirror is cycled in four specific steps. Step one consists of rapid cooling of the mirror from a level above ambient temperature, to approximately 1.5°C (2.7°F) above the last known dew point. This reduces total cycle time, especially when the difference between the dew point temperature and ambient temperature is large. During step two, starting at about 1.5°C (2.7°F) above the dew point, the cooling rate is slowed to approach and cross the dew point as slowly as possible, allowing dew to form in a uniform and repeatable manner, so that the correct formation of dew is detected.

Patented by Protimeter, plc.

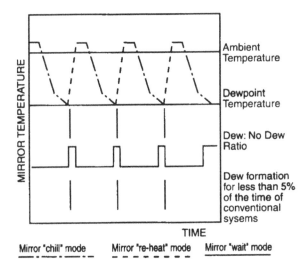

Figure 3.11   Dew point determining cycle.

Step three starts when the dew detection is completed. The current to the cooling device is reversed, which causes the mirror to rapidly rise in temperature to a few degrees above ambient temperature.

The fourth and final stage of the cycle represents a short period of time between successive cooling stages which allows the mirror to stabilize. This period is usually a few seconds and could be varied from seconds to minutes if desired.

It is evident from Figure 3.11 that dew is present on the mirror surface for only a very short time, causing contaminant buildup on the mirror to be kept at an absolute minimum. Also much less power is used and therefore the dissipated energy is very low. This means that any heating of the sensor or surrounding air is kept to a minimum. This is especially important when the sensor is operated in a small confined area, such as when used for water activity measurements in small containers.

The optical system is arranged in such a way that the absolute light level is unimportant. Only changes in light level, as to whether the mirror is dry or contains dew, are important. Although the mirror is allowed to cool below the dew point, the correct dew point is stored in memory and will be used if the downward optical trend persists long enough to be real dew formation.

All measured signals from the sensor are converted into current levels within the sensor, which is connected to the instrument. This arrangement of current drive makes it possible to operate the sensor over a long distance from the readout unit, up to 300 m (1000 ft.) while using low-cost, non-shielded cables. A disadvantage is that the sensor electronics are exposed to the same temperature range, typically –20°C to +80°C (–4°F to 176°F), which causes some temperature dependence and possible errors related to ambient temperature.

B. CCM Sensor

Top and bottom views of an industrial CCM sensor are shown in Figure 3.12. A pressure tight, O-ring sealed, pressure cap covers and protects the mirror and filter compartment. A cylindrical filter surrounds the mirror and can easily be replaced if needed. The optical bridge assembly is mounted in a separate, removable unit which plugs into the sensor mirror compartment.

Chilled mirror sensors are generally available in anodized aluminum or stainless steel construction for corrosive or low frost point applications. Most chilled mirror sensors, conventional as well as CCM, are designed for pressures up to about 20 bar (300 psi). The sensor shown is rated for 12 bar (180 psi).

In the sensor shown, three electronic circuit boards are mounted in the bottom part of the sensor. One is the optics control board, the second a temperature board, and the third a pressure compensation board. The sensor can be installed for on-line measurements or for use with a sampling system. Because of temperature limitations of the optical components and the thermoelectric cooler, this CCM sensor is not suitable for operation at ambients above 80°C (176°F). Most conventional (continuous) sensors are rated for operation to 95°C or 100°C (203°F or 212°F).

C. Dew Point/Frost Point Conversion

Misinterpretation of dew versus frost point could result in errors of up to 3.7°C (6.7°F) at −40°C (−40°F), as is shown in Table 3.1.

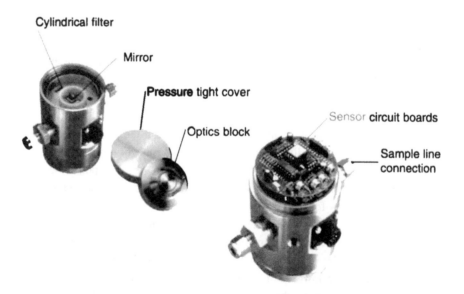

Cylindrical filter

Mirror

Pressure tight cover

Sensor circuit boards

Optics block

Sample line connection

Figure 3.12   Typical CCM sensor showing top view with mirror and integral filter (left) and bottom view showing integral signal processing electronics. (*Courtesy Protimeter, Inc.*)

Table 3.1   Dew point–frost point conversion.

Frost Point	Dew Point	Deviation
0°C	0°C	0°C
–5°C	–5.6°C	0.6°C
–10°C	–11.2°C	1.2°C
–15°C	–16.7°C	1.7°C
–20°C	–22.2°C	2.2°C
–25°C	–27.7°C	2.7°C
–30°C	–33.1°C	3.1°C
–35°C	–38.4°C	3.4°C
–40°C	–43.7°C	3.7°C

In conventional continuous chilled mirror hygrometers, when measuring below 0°C (32°F), initially a dew layer is established on the mirror and the instrument reads dew point. It is generally assumed that the dew layer converts to frost within a short time. However, this could in fact take several minutes, or even several hours and is unpredictable. In some cases, the dew layer continues to exist indefinitely, even when the equilibrium temperature is well below 0°C (32°F). This is called "super-cooled dew," a condition that could exist because of certain types of mirror contamination, a high flow rate or a number of other reasons. In "continuous" chilled mirror hygrometers, it is usually "assumed" that when the mirror temperature is below 0°C, the measurement represents frost point, but this is not at all certain. Even if this were the case, it could take a long time before the real frost point is measured. In the interim, the measurements could be in error without the user realizing it. When such errors cannot be tolerated, it is customary to use a sensor microscope mounted on top of the sensor, making it possible to observe the mirror and determine whether dew or frost is measured. This is impractical and not an attractive industrial procedure. The CCM hygrometer therefore offers a significant advantage when measuring below 0°C because one can be "sure" that the measurement is "dew point." If the user wants to read frost point, it is a simple matter to program the on-board microprocessor to display frost point.

D. Maintenance Requirements

The combined use of CCM technology and the circular filter (shown in Figure 3.13) in the industrial sensor, which is not consumed by the sample flow, results in significantly less maintenance, i.e., less mirror cleaning. In cases where typical conventional hygrometers require mirror cleaning at intervals of a few days or weeks, the CCM hygrometer has sometimes been used without mirror cleaning for periods of up to, or more than one year, while maintaining good accuracy. A typical CCM hygrometer sensor is shown schematically in Figure 3.13.

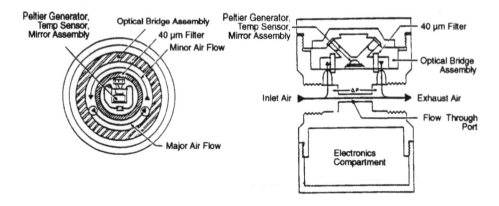

Figure 3.13   Schematic of typical cycling hygrometer sensor with integral cylindrical filter.

Surrounding the mirror is a cylindrical polyethylene, 40 micron filter. This filter, unlike the in-line filters used in conventional hygrometer systems, does not require 100% of the total sample gas to pass through its element. Instead, sample gas circulates around the outside of the element, and it is measured by means of convection across the filter element. Because most particulates circulate freely around the filter and exit the measurement chamber, the filter is not "consumed" by the sample flow, and therefore does not contaminate as rapidly. Furthermore, the usual slow-down of the sensor response time as a result of filter contamination is reduced, causing this kind of sensor to have a faster response than many conventional systems using an in-line filter.

E. Benefits of the CCM Hygrometer

The advantages of CCM technology are:

- Incorporation of an optical slope detection system which virtually eliminates the need for regular re-calibration or re-standardization of optics. The mirror contains dew for less than 5% of the time and is therefore almost always dry. Particulate contaminants do not readily attach to the mirror and less mirror cleaning is required.

- A unique integral cylindrical 40 micron filter which is not "consumed" by the sample air as in conventional systems, and which causes most of the sample air to bypass the mirror and filter and thereby further reduce mirror contaminants. Mirror cleaning cycles as long as one year in some cases.

- A chilled mirror sensor design (shown in Figure 3.12) which includes signal processing electronics. This eliminates the need for heavy, costly shielded cables and allows the use of low-cost, standard non-shielded cables at lengths of 300 m (1000 ft.) or more as compared to the 90 m (300 ft.) cable length limit in most conventional systems. However, this is accomplished at the expense of introducing a temperature coefficient, resulting in somewhat lower accuracy at extreme temperatures.

- Dew point measurements guaranteed at all times, even below 0°C (32°F). A common drawback of the chilled mirror hygrometer is that at dew/frost points below 0°C (32°F) it is unclear whether super-cooled dew or frost is measured. The CCM slope detection system solves this problem.

In general, these advantages eliminate some of the traditional operational problems and make the CCM type of hygrometer a potentially more user-friendly instrument. The CCM hygrometer is believed to be suitable for many industrial applications, but has had few applications to date. Potential applications include heat treating, automobile emissions testing, dryers, humidity chambers, environmental areas, plant air systems, process control, and semiconductor manufacturing, where low maintenance and ease of operation are more important than attaining the highest possible accuracy. Current successful applications include monitoring of medical gases and dew point measurements in gas turbines.

F. Limitations of the CCM Hygrometer

The following limitations should be considered when selecting a CCM hygrometer:

- The dew point determining cycle must be accurately adjusted and locked in for each instrument, using an accurate calibration system. Although the cycle rarely changes in a measurable way, a yearly calibration against a transfer standard is recommended. Hence measurements made with a CCM hygrometer are not fully "fundamental," and for this reason the CCM hygrometer is less desirable for use as a laboratory transfer calibration standard.

- Although high accuracies are attainable and can be maintained for long periods of time, the instrument is not as accurate as the conventional continuous chilled mirror hygrometer. As a result, both types of instruments can be considered complementary rather than competitive for most applications.

- Incorporating signal processing electronics inside the sensor offers the advantage of being able to use much longer cables, but a disadvantage is that the maximum operating temperature is somewhat lower, i.e., typically 80°C (176°F) versus 95°C (203°F) for a conventional sensor. This is because the electronic components used in the sensor are only rated for operation below 80°C (176°F).

A multipurpose CCM Hygrometer with dew point sensor, temperature probe, and pressure detector is shown in Figure 3.14.

Figure 3.14   CCM hygrometer with dew point, temperature, and pressure sensors
for humidity determination in terms of any selected parameter.
(Courtesy Protimeter, Inc.)

G. High Temperature Fiber Optic Hygrometers

The electro-optical components and the thermoelectric cooler used in a chilled mirror sensor are temperature sensitive and generally cannot be operated at temperatures in excess of 100°C (212°F). For example, the sensor cannot be used in applications inside a hot oven where the surrounding temperature is in excess of 100°C. In these situations, the sensor must be used outside the chamber at an ambient temperature of less than 100°C and sample air must be brought to the sensor. In most cases, this is not a major problem, but merely an inconvenience. However, if the chamber temperature is close to 100°C and the humidity close to 100% RH, such a measurement becomes very difficult, because outside the chamber the sensor must be operated at a temperature higher than the dew point to be measured. If the dew point is above ambient, the sensor and sample lines must be heated. The sample lines must also be heated because if any part of the system that comes in contact with the gas stream is at a temperature below the dew point, condensation (flooding) occurs, rendering any measurement meaningless.

A CCM hygrometer has been developed using fiber optic light bundles, allowing the sensor optics and electronics to be placed outside the chamber while the mirror is inside the chamber. However, the Peltier cooler is part of the mirror and must also be inside the hot chamber. This cooler is sensitive to excessive temperatures (temperatures above 100°C or 212°F), thus the advantages of a fiber optic sensor are limited and the fiber optic chilled mirror hygrometer has found few applications.

IV. Dew Point Measurement Range

The dew point measurement range of a chilled mirror sensor is dependent on the type of Peltier cooler used, the mode of operation, gases to be monitored, and the ambient or sensor body temperature.

Most sensors are built with so-called single or dual stage coolers. A single stage sensor uses a cooler with one cooling stage consisting of an array of thermoelectric junctions. Typically such a sensor has a measurement span or depression of 40°C to 45°C (72°F to 81°F). Two stage sensors use coolers with two thermoelectric units in tandem. Though the number of junctions is doubled, the increase in range is only about 50%, i.e., the span increases to 60°C to 65°C (108°F to 117°F). Likewise there are three, four, and five stage sensors on the market. With increasing number of stages, the span, also called "depression," increases. However, due to the low efficiencies of Peltier elements, the amount of power needed to operate multi-stage coolers increases rapidly with the number of stages, causing self heating of the sensor and thus defeating the purpose. To solve this problem, multi-stage sensors are always air or water cooled, which adds cost and is an inconvenience especially in industrial processes.

The depression numbers given above are for monitoring air and most other inert gases with a similar heat capacity, such as nitrogen, argon, CO_2, etc. For gases with a higher heat capacity, such as helium and especially hydrogen, the depression is considerable lower because of the heat load caused by the surrounding gas. Typically, the depression of a chilled mirror sensor when used to measure dew point in hydrogen is at least 10°C (18°F) less.

The temperature at which the sensor is operated also affects the depression. The depressions given above are based on a sensor operating temperature of 25°C or 75°F. If the sensor is operated at a higher temperature, the depression is usually higher. When operated at a lower temperature, some depression is lost.

The depression numbers vary from sensor to sensor and depend on the sensor design (heat transfer) and the mirror material used. The best mirror material for attaining the highest depression is copper, coated with gold or rhodium. For sensors using graphite mirrors, the depression is about 10°C (18°F) less. A good rule-of-thumb is that for every 3°C above 25°C ambient, the depression increases by 1°C and for every 3°C below 25°C ambient, 1°C of depression is lost. A typical depression curve for a CCM sensor is shown in Figure 3.15.

It must be emphasized that the depression data shown is based on the sensor operating temperature, which is not necessarily the same as the sensor ambient temperature. In the above examples it was assumed that the sensor temperature is the same as the ambient temperature. However, operation of the cooler requires power which could result in significant self heating of the sensor. A well designed and properly installed sensor will dissipate this energy efficiently to the environment or base plate, and self heating can be kept at a minimum. If this is not the case, the sensor could heat up to several degrees above ambient and the depression must then be measured against the higher sensor body temperature, which reduces the range that the sensor can cover.

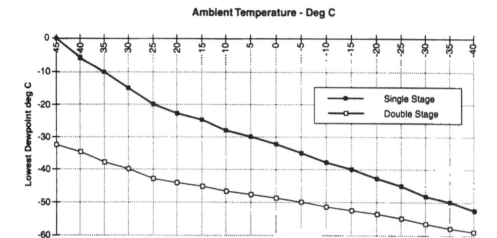

Figure 3.15   Depression of single and double stage CCM sensors.

Table 3.2 shows typical depressions and ranges for conventional and CCM chilled mirror sensors. However, this could vary between different manufacturers and the table is only intended as a general guide. Typical remote single-, two-stage, and four-stage sensors are shown in Figure 3.16. It is more common to use remote sensors, connected to the instrument housing by means of a suitable cable. Some manufacturers offer chilled mirror instruments with integral sensors. This does not affect the depression capabilities of the sensor unless internal heating of the instrument causes the sensor body temperature to increase. Typical two-stage sensors which can be operated as integral sensors or mounted remotely, are shown in Figure 3.17.

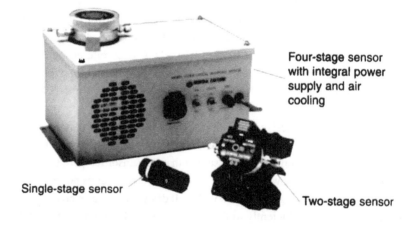

Figure 3.16   Single-stage, two-stage, and four-stage remote sensors.
                     (Courtesy General Eastern)

Table 3.2   Typical depression and range of chilled mirror sensors.

Dew Point Depression

	1-Stage	2-Stage	2-Stage (1) Fortex	3-Stage Air cooled	3-Stage (2) Water cooled	4-Stage Air cooled	4-Stage (3) Water cooled	5-Stage (3) Water cooled
Depression @ 25°C	45°C	65°C	75°C	85°C	90°C	90°C	95°C	100–105°C
Depression @ 75°F	80°F	120°F	135°F	155°F	160°F	160°F	170°F	180°–190°F
Lowest Dew Point								
Sensor @ 50°C	0°C	-22°C	-35°C	-45°C				
@ 40°C	-8°C	-29°C	-41°C	-51°C				
@ 30°C	-16°C	-36°C	-47°C	-57°C				
@ 25°C	-20°C	-40°C	-50°C	-60°C	-65°C	-65°C	-70°C	-75 to -80°C
@ 20°C	-24°C	-44°C	-43°C	-63°C				
@ 10°C	-32°C	-51°C	-59°C	-69°C				
@ 0°C	-40°C	-58°C	NA	-75°C				
@ 120°F	30°F	-15°F	-30°F	-52°F				
@ 100°F	14°F	-29°F	-43°F	-64°F				
@ 80°F	-2°F	-43°F	-57°F	-77°F				
@ 75°F	-5°F	-45°F	-60°F	-80°F	-85°F	-85°F	-95°F	-105 to -115°F
@ 70°F	-8°F	-47°F	-63°F	-83°F				
@ 60°F	-16°F	-54°F	-68°F	-88°F				
@ 40°F	-32°F	-58°F	-71°F	-100°F				

Note: Using water cooled sensors, the lowest frost point that can be measured depends on the efficiency of cooling and the cooling liquid temperature. In practice the lowest measurable dew/frost points are usually about 5°C (8°F) higher.
(1) Fortex Cooled. Currently available in CCM Sensor only. 7 bar (100 psi) pressure.   (2) Depression depends on temperature of cooling liquid.
(3) Extensive chilled water cooling needed. Sensor failure if not cooled.

Remote or integral sensors

Sensor with water cooled base

Figure 3.17   Two-stage sensors for remote or integral mounting.
                    (Courtesy EG&G/Edgetech)

V. Chilled Mirror Dew Point Transmitters

The cost of chilled mirror hygrometers is prohibitive for many applications such as air lines, air conditioning, and room dew point monitoring. In these applications, RH transmitters using polymer resistive or capacitive sensors are very popular. Although these sensors have attained a high degree of performance and accuracy, they are secondary measurement devices requiring periodic calibration against a certified calibration standard. This is time consuming and costly. Furthermore, these sensors often cannot be relied upon if measurements have to be "traceable" and "certified." In such instances, the fundamental chilled mirror sensor is the preferred choice.

To accommodate applications where no readout and only a current or voltage signal is needed, several manufacturers now offer chilled mirror transmitters. Although they are still more expensive than polymer RH transmitters, chilled mirror transmitters are considerably less expensive than the conventional systems and offer fundamental and traceable measurements.

General Eastern Instruments was the first company to offer this kind of product. Its Dew-10, shown in Figure 3.18, is still the lowest cost chilled mirror transmitter on the market, but lacks some features of the later product entries. The sensor is sensitive to contaminants, flow rates, and other environmental and operating conditions.

The Edgetech/EG&G Dew Track, shown in Figure 3.19, offers more features and offers better performance in a number of applications, but is substantially more expensive and is therefore not widely used.

The most recent introduction has been by Protimeter, which offers the "Dew Tector" transmitter with CCM control and with the additional new improvement of a scratch-resistant sapphire mirror and an integral wiper system to automatically clean the mirror at pre-determined intervals. This system is shown in Figure 3.20. The combination of CCM and the sapphire mirror and wiper system has resulted in a product offering the potential of virtually maintenance-free chilled mirror measurements, at a cost similar to, Edgetech/EG&G Dew-Track. Since the Dew Tector is quite new, field applications have been limited as of 1996.

Figure 3.18   Low cost chilled mirror transmitter (Dew-10).
(Courtesy General Eastern Instruments)

Figure 3.19   Dew track transmitter.
(Courtesy Edgete)

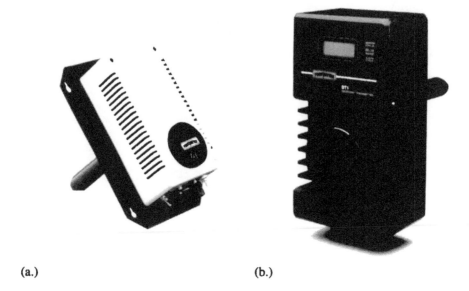

(a.)   (b.)

Figure 3.20   Dew Tector transmitter (a.) indoor unit, (b.) outdoor unit.
                    (Courtesy Protimeter)

VI. Summary of Balancing or Self-Calibration Methods

A. Manual Balance

The advantages of manual balance are:

- It is simple and reliable.

- It is the least expensive method.

The limitations are:

- It is not suitable for automatic industrial applications since an operator is required at frequent intervals for service.

- The hygrometer does not measure dew point during the balance cycle.

This method of balancing has been almost completely abandoned in favor of one of the automatic systems listed below.

B. Automatic Balance Control (ABC)

The advantages include:

- No operator is required until the system goes out of control or indicates that it is about to.

- In modern systems, a warning signal is given in advance to indicate that mirror cleaning is required.

The limitations of ABC are:

- Contaminants on the mirror are not removed even though the time interval between mirror cleanings is extended by using the ABC method. Measurement errors resulting from contaminants (Raoult effect) are not eliminated.

- The hygrometer does not measure dew point during the ABC cycle which is performed at prescribed intervals regardless of whether important measurements are being made when the cycle is initiated.

C. PACER Cycle

The advantages of the PACER are:

- The "cool" cycle added to the automatic balance cycle redistributes contaminants on the mirror and thereby reduces measurement errors.

- The time interval between mirror cleanings is generally longer than using the ABC circuit.

- Contaminant errors including the Raoult effect are for the most part eliminated.

Limitations of the PACER are:

- The system must be operated to control on a relatively thick dew layer, which in some situations could cause "flooding" of the mirror and a "crash" of the control system.

- The system has proven to be less reliable in outdoor, meteorological applications, mainly due to "flooding" problems, especially during fog conditions.

- Like with ABC, no measurements can be made during the short PACER cycle.

D. Continuous Balance

The advantages of the continuous balance method are:

- The hygrometer makes measurements continuously. The 2 to 8-minute interruption during balancing is eliminated.

- For low frost point measurements, it offers a significant advantage since in a conventional system it takes a long time to heat the mirror, balance it, and then cool it down again and stabilize the mirror at the very low frost point. When using continuous balance, the mirror is always kept at low temperature.

The limitations are:

- Like with ABC, contaminants on the mirror are not disturbed and hence contaminant (Raoult effect) errors are not reduced or eliminated.

- The time interval between mirror cleanings is generally not reduced significantly since a dry and a wet mirror contaminate at different levels and at different speeds.

This technique, patented by Edgetech/EG&G, has had limited use, and is used primarily for low frost point measurements where it has the advantage of faster response.

E. Cycling Chilled Mirror (CCM) Technique

The advantages of CCM are:

- The mirror is "dry" for 95% of the time and therefore contaminates at a much slower rate.

- The time interval between mirror cleanings is much longer than for any of the other balancing methods.

The limitations are:

- The cycling parameters must be carefully set.

- The instrument must be calibrated against a NIST-traceable hygrometer to insure accurate measurements.

The method uses a basically fundamental principle. However, it has lost some of the fundamental properties of a conventional, (continuous dew layer) hygrometer and is thus less suitable for use as a laboratory calibration standard. By its nature, the CCM hygrometer is of greater importance in industrial applications than for use as a laboratory calibration standard, where the continuous chilled mirror method is usually the preferred choice.

F. CCM With Sapphire Mirror and Wiper

The advantages of this method are the same as in (5) above, plus the additional feature of automatic mirror cleaning and a virtually indestructible sapphire mirror.

VII. Dew Cup

The dew cup is another, older type of chilled mirror instrument. It offers perhaps the simplest technique for the measurement of dew point. The gas sample is drawn from the source across the outside of a polished cup made of chromium-plated copper. The cup is enclosed in a glass container so that the moisture can be observed condensing on the cup surface when the dew point is reached. The cup surface is cooled progressively by dropping small pieces of dry ice in acetone (or methanol) until the dew point is reached, as indicated by condensation on the cup surface at a temperature indicated by a thermometer in the acetone. The dew cup is most accurate for dew points above the freezing point of water. At dew points lower than 0°C (32°F), there is the possibility of super cooling, with a resulting low dew point reading.

Making accurate measurements with a dew cup requires a considerable amount of skill and consistency on the part of the operator, and is not recommended where accuracy is paramount. Incorrect dew point readings may result if the atmosphere is sooty, if there are leaks in the dew cup or sampling lines, if the flow of atmosphere is too fast, if the temperature is lowered too fast, or if the lighting conditions are poor in the area where the dew point is being observed. A dew cup measurement is a onetime measurement and is not suitable for continuous monitoring. Measurements may be repeated one or more times a day. The accuracy of a dew cup measurement depends strongly on the skill of the operator.

VIII. Sampling Systems

When designing systems to make dew point measurements in a duct, room, process, or chamber, engineers often are not sure where to locate the sensor. The question is, should it be installed directly in the measurement area (in situ) or should a sampling system be used to carry a gas sample to a remote sensor? A sampling system is in many cases a must, for example, when the temperature of the gas being measured exceeds the maximum temperature rating of the dew point sensor. Sampling is then needed to lower the gas temperature enough to prevent sensor damage. Because the dew point temperature is independent of the gas temperature (if not saturated), it can be measured at the other end of the sample line, after the gas has cooled. In many cases, either type of installation, if carefully designed, will provide reliable, continuous dew point measurements. However, there are several advantages to setting up a sampling system, even when an in-situ measurement is possible.

A sampling system often provides better, more reliable, long-term dew point measurements than other systems.

The advantages are:

- Incoming gas can be filtered and undesirable pollutants removed before contaminating the sensor.

- The flow rate can be measured and controlled at the optimum rate for the sensor in use.

- A remote external sensor located in a relatively constant ambient will provide more stable measurements when measuring dew point in an environmental chamber which is programmed for wide and rapid temperature excursions.

- A multiplexing sampling system can be made using manually operated valves or solenoid valves to sequentially select the points. Thus, one hygrometer could take the place of several. This is shown in Figure 3.23.

- Measurement errors will be minimal because all sampling points are measured with the same sensor.

A. Design of Sampling System [13, 14]

A frequently used sampling system is shown schematically in Figure 3.21.

　　　The selection of components is an important part of the sampling system design. For example, a pump is not always needed. If the system is under pressure, a quantity of gas can be bled, via a needle valve, through the dew point sensor. If a high flow rate exists in a pipe or duct, a section of tubing, pointing upstream, often can be inserted to act as a "pitot tube" and capture an adequate gas sample. A pump is normally only required when the source of gas to be measured is either at atmospheric pressure or under vacuum. If under vacuum, it is necessary that the pump is large enough to overcome the vacuum and draw out the sample. A flow gage and needle valve, either built-in or separate, are required to measure and properly control the flow. The flow gage does not have to be very accurate because the sensor can be operated satisfactorily over a wide range of flow settings. Figure 3.21 shows a typical sampling system. Commercially available heated sampling components are shown in Figure 3.22. In some cases, like in heat treating, it is necessary to monitor dew point in several locations. To limit the cost of using one sensor for each point, a sequencing scanner can be used to select the location from which the sample air is monitored at one particular time. The scanner can be set to scan at any desired speed. Such a system is shown schematically in Figure 3.23.

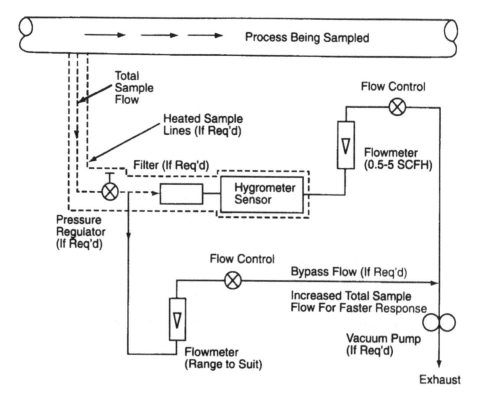

Figure 3.21　　　Schematic of typical sampling system.

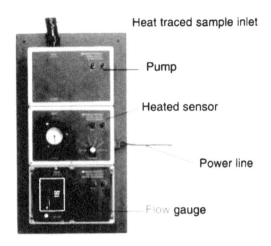

Figure 3.22   Heated sampling system components.
                   (Courtesy General Eastern)

The type of sample tubing should be carefully considered. Outside diameter tubing of 6 mm (1/4 in.) is commonly used. Soft plastic tubing, such as tygon or rubber should be avoided because these materials are highly hygroscopic and will therefore absorb water and outgas for long periods of time. But hard plastic tubing such as nylon, Teflon, and polypropylene, along with copper tubing, can be used for most measurements. For very low frost point measurements below $-30°C$ ($-22°F$), only stainless steel tubing should be considered. This is because the other materials mentioned are too hygroscopic, which would lead to water vapor outgassing, resulting in a slow response and measurement errors. Typical response times as a function of the sample line material used are shown in Figure 3.24. Pressure corrections for flow rates are shown in Figure 3.25.

Perhaps the most important question to ask is, "How long a sample line is needed?" If the source temperature is higher than the maximum allowable sensor temperature, the minimum sample line length is that which allows the gas temperature to decrease to a safe level for the sensor. The maximum line length is equally important. The sample gas temperature at the outlet must be higher than the dew point; otherwise saturation will

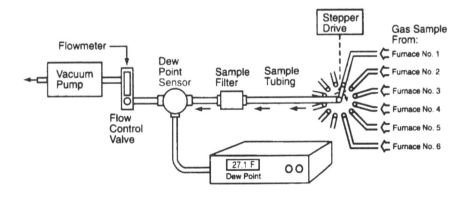

Figure 3.23   Multi-point sampling system.

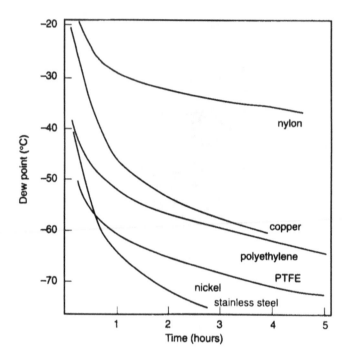

Figure 3.24   Effects of sample line material on response time.

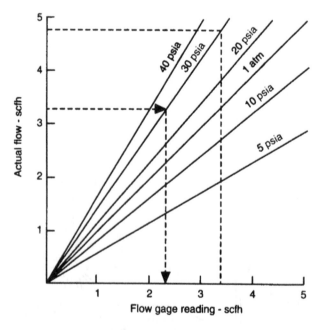

Figure 3.25   Curves for pressure correction of flow meter readings.

occur and the sample lines will fill with water. A sampling system is a heat exchanger and the temperature will decrease to ambient if initially it is higher than ambient. If the dew point is higher than the ambient temperature, the tubing must be heat traced. In addition, the dew point sensor must be heated to eliminate the possibility of condensation inside the detector. Insulation is also important. If the dew point is lower than the ambient temperature, the sample line length should be kept at a minimum because it has a significant effect on the system response due to the hygroscopic properties of the sample line material.

There are other important considerations. If possible, the sample line should be sloped downward from the sensor to the source so that any condensation, if present, will run back to the source. It is also good practice to insert a tee fitting and shutoff valve before the sensor. This makes it possible to isolate the sensor should the line need to be purged. If the sample gas is merely bled to atmosphere after passing through the sensor, the outlet port may be left open.

Measuring dew points which are substantially below that of the surrounding atmosphere requires a three or four-foot coiled length of tubing on the downstream side of the sensor. This prevents water vapor in the room air from backing up into the sensor and causing a false (high) reading. If the flow rate to the sensor varies widely, a pressure regulator may be installed on the upstream side. When filterable contaminants are present, filters with disposable elements should be located at convenient locations between the source and the sensor. Checking these for cleanliness should become part of a preventative maintenance program.

B. Problems to Avoid

Careful consideration should be given to a number of error-causing problems. Some of these are listed below:

- **Creating an artificial pressure or vacuum**
  Because the dew point increases with pressure, accidentally increasing the line pressure will cause false high readings. For example, sample gas should not be pumped into a sensor that is restricted at the outlet side with a needle valve. Instead, a valve should be placed between the pump and the sensor.

- **Laminar flow errors**
  Most ducts and pipes have laminar flow characteristics. A sampling tube should be inserted into the center, not merely near the wall, or a measurement error could result.

- **Sampling system leaks**
  The lower the dew point, the more important it is to eliminate all leaks. All fittings should be leak checked. Pressurizing a dry system above atmosphere will not eliminate leaks into the system from a higher water vapor pressure in the room. This is true because the water vapor pressure component of the total pressure can be lower inside the sensor chamber and higher outside the chamber.

- **Improper flow rates**
  Specifications of the dew point sensor in use should be checked and the flow rate controlled at an optimum range for the sensor. For some types of sensors, there are limitations on both the low and high extremes. Other types can operate properly even in stagnant air, although response time suffers. It should also be remembered that the contamination rate is proportional to flow rate.

- **Percent relative humidity (RH) errors**
  If the dew point data and gas temperature (dry bulb) data are used to convert to percent RH readings, it is important to measure the gas temperature at the source. If it is measured at the dew point sensor location where the gas temperature has been modified by flowing through the sampling system, resulting RH calculations will in all likelihood be incorrect.

- **Pneumatic scanning errors**
  Sufficient equilibration time should be allowed when switching the sampling system to a new source. To speed up the system, larger diameter piping and higher flow rates should be used.

- **Flow gage errors**
  Most flow gages are calibrated for air. If other gases are measured, a flow rate correction should be made. Figure 3.26 shows a number of curves for making such corrections. Most flow gages are calibrated at one atmosphere. A correction must also be made for other pressure or vacuum conditions as shown in Figure 3.25.

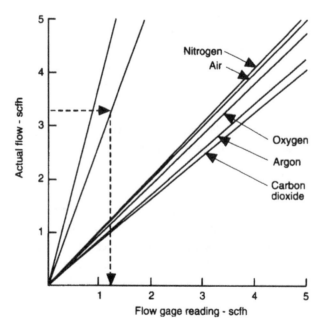

Figure 3.26   Curves for correcting flow meter readings for gases other than air.

- **Condensation problems**

  Although the average temperature in an area may be well above the dew point, a cold region could exist close to a location were the sample line is installed. If the temperature at any point of the sample line is below the dew point, problems will result. Condensation could occur in front of an air conditioning outlet, or near a cold window in northern areas. A small amount of condensation reaching the sensor will cause serious measurement problems. When in doubt, insulate the entire sample line.

C. Conclusions

The use of a dew point sampling system, even when not mandatory, can be advantageous. If properly installed, the system can provide multiple measurements from a single sensor, as well as more stable and accurate readings.

IX. Error Analysis

A. General Principles

Conventional chilled mirror dew point hygrometers are fundamental instruments which utilize a thermo-electrically cooled optically-sensed mirror to directly measure the temperature at which the vapor adjacent to the surface is in equilibrium with the vapor in the sample gas. This is by definition the dew point temperature of the gas. From the dew point measurement, the partial pressure of the water vapor can easily be determined from the vapor pressure tables shown in Chapter 11.

Errors associated with the measurement of dew point in these systems can be separated into two categories: those errors which occur at the surface itself, and those errors which involve the measurement and display of the surface temperature.

B. Mirror Surface Errors

1. The Kelvin Effect

Operation of the instrument does not provide the precise dew point temperature when the water surface on the mirror is uneven. The effect of a small droplet is to increase mirror vapor pressure due to surface tension forces of the droplet and to reduce the mirror temperature required for a given gas sample vapor content. This is called the Kelvin effect. The magnitude of this effect actually present in a hygrometer depends to a large extent on the condition or treatment of the mirror surface and on the fraction of dry mirror reflectance chosen for the operating point. A small fractional reflectance is obtained with a larger mass of water, resulting in reduced error due to the Kelvin effect. This is referred to as operating the instrument with a thick dew layer. The estimated value of the error due to this and all other effects is listed in Table 3.3.

2. The Raoult Effect

As discussed in the section on "PACER," the presence of soluble contaminants, especially salts, on the mirror surface decreases the vapor pressure at the mirror surface for a given temperature. This causes the sensor to control at a temperature above the dew point temperature, and is known as the Raoult effect. The magnitude of the effect is proportional to the concentration of solute and is dependent on its chemical composition. As previously discussed, errors due to the Raoult effect can be as high as 5°C (9°F) under extreme conditions. For a given amount of contaminant present on the mirror surface, a greater mass of water, i.e., operation on a heavy dew layer, will help to reduce this effect and the resulting errors.

3. Mirror Temperature Measurement Errors

Common sources of errors in temperature measurement and display are:

- Temperature gradient from water surface to mirror surface

- Temperature gradient from mirror surface to thermometer

- Temperature error from self-heating of the thermometer due to the measurement current

- Temperature error from thermal conductance of thermometer leads

- Temperature error due to thermometer calibration limitations

- Temperature error due to thermometer non-linearity

- Temperature error due to readout calibration limitations

- Temperature error due to readout calibration linearity

To evaluate the above-listed errors, specific operating parameters must be evaluated. The mirror temperature, sample temperature, sample flow rate, and hygrometer base temperature influence many of the above errors. To avoid consideration of all combinations of these factors, typical worst case values are used. The values of the errors listed above are shown in Table 3.3.

4. Sampling System Errors

The categories of error discussed above result in limitations to the accuracy of measurement with respect to the dew point of the sample actually passing over the mirror surface of the hygrometer. In some hygrometer installations a large error is possible as a result of the sampling. Some factors contributing to such sampling errors are:

- Sample line leaks

- Absorption and adsorption at sample line surfaces

- Desorption at sample line surfaces

- Diffusion through sampling materials

- Condensation in the sampling system (moisture traps)

5. Dew/Frost Point Interpretation Errors

As was discussed in the section on CCM hygrometers, it is usually assumed that for measurements below 0°C (32°F), the condensate deposit is ice, and the instrument is assumed to be measuring the frost point temperature which corresponds to the vapor pressure over ice, rather than water. In the region between 0°C (32°F) and about –30°C (–22°F), it is possible for a deposit to exist in the water phase for an undetermined period of time or permanently, thereby introducing a significant error if it is assumed that the instrument is measuring frost point while in fact determining dew point. This is particularly true if the mirror is exceedingly clean, or contaminated with salty deposits. There is no way to determine for certain whether dew or frost point is measured other than inspecting the mirror surface with a microscope, supplied as an option on most instruments used for calibration work, and visually determining the phase of the deposit. As has been discussed, the CCM hygrometer offers the advantage in that this system appears to always measure dew point because of the very short time that the dew layer exists on the mirror surface. For conventional hygrometers there is another operational procedures which can be used to eliminate this error. This involves forced cooling of the mirror to well below the frost point by overriding the automatic control system. The mirror must be kept at this very low temperature for a significant period of time and the sensor must have a cooler of sufficient depression, to make certain that the deposit is positively "frozen," and that measurements below 0°C (32°F) are, in fact, frost point determinations. A further complication is that when the mirror is kept for a long time at well below frost point, an excessive water or ice deposit accumulates on the mirror and when the procedure is ended, this excess must be burned off and could cause "flooding" on the mirror surface. This solution is far from ideal and rarely used.

6. Errors From Other Vapor Constituents

Though the condensation hygrometer is a fundamental instrument, it cannot distinguish between water vapor and other gaseous vapors. When the mirror is cooled, in most cases it is water vapor which is the first to saturate and condense on the mirror surface, and it is usually assumed that the instrument will sense and control on water or ice. In cases where the gas sample contains components which condense at a higher temperature than the water vapor, the instrument can control on vapors other than water. Some dew point hygrometers are actually used in this manner to measure and control vapors such as carbon tetrachloride, freon and ethers. In most cases the instrument will control on the highest condensable vapor, which is not necessarily water vapor. When this is a problem, an electrolytic hygrometer which is indeed specific to water vapor, may be a better solution for the measurement.

7. Summary of Hygrometry Errors

Errors occurring at the mirror surface interface include:

- The Kelvin Effect   Typical Dew Point Error: −0.005°C to −0.01°C
(−0.01°F to −0.02°F)

- The Raoult Effect   Typical Dew Point Error: +0.05°C to 0.5°C*
(+0.1°F to 0.9°F)

Errors in temperature measurement are:

- Temperature gradient from water surface to mirror for air flow of 300m/min. (1000 ft./min.) at 50% RH   +0.002°C (+0.004°F)

- Temperature gradient from mirror surface to thermometer based on air flow of 300m/min. (1000 ft./min.) at 50% RH   +0.001°C (+0.002°F)

The above errors are general. Other errors must be added depending on the particular type of temperature sensor and readout equipment used.

Typical instrument errors are shown in Table 3.3 below:

Table 3.3   Typical chilled mirror hygrometer errors.

	Platinum Resistance (100 ohm)	Thermometer
Temperature error from self heating		+0.04°C (+0.09°F)
Temperature error from thermal conduction of thermometer leads		+0.005°C (+0.01°F)
Temperature error from thermometer calibration limitation and interchangeability		+0.05°C (+0.09°F)
Temperature error from temperature readout calibration limitation		+0.03°C (+0.054°F)
Temperature error due to readout calibration linearity		+0.015°C (+0.027°F)
Error Extremes:		+0.14°C (+0.25°F)

---

* In extreme cases this type of error can be as high as 5°C (9°F).

At dew points below –40°C (–40°F), errors increase. The above data does not include errors introduced by various readout instruments such as recorders and digital panel meters. It is pointed out, that these are worst case conditions. In actuality, instruments are sometimes claimed to be more accurate which may be borne out by calibration reports furnished by NIST or other calibration facilities.

X. Cryogenic Dew/Frost Point Hygrometer [30, 31]

The chilled mirror hygrometer is widely used as a laboratory transfer standard and for industrial applications requiring high accuracy and traceability. The instrument has a good response time in the range of –20°C to 100°C (–4°F to 212°F). In the range of –20°C to –40°C (–4°F to –40°F), the response is slower but good enough for most applications. Below –40°C the response becomes significantly longer as the frost point is lowered. The response time is also influenced by the temperature of the sample gas, flow rate, size of the mirror, and some other considerations. Typically, the response time at –75°C (–103°F) could be a few hours and at –80°C (–112°F), many hours. This represents a significant drawback, especially for field applications. For this reason chilled mirror hygrometers are not often used in industry for measurements much below –60°C (–76°F). In laboratories this type of instrument is often used for frost point measurements down to –70°C (–94°F) and sometimes to –75°C (–103°F) when the long response time can be tolerated. Frost point measurements below –75°C (–103°F) are very time consuming and cumbersome to make.

Cryogenic dew point hygrometers for aircraft use were built in the 1960s by Cambridge Systems (later acquired by EG&G, Inc. and now operating under the name Edgetech). In the late 1960s Robert Mastenbrook[3*, 4*] of the Naval Research Laboratories (1968) developed a freon-cooled hygrometer for use in high altitude sounding balloons. These instruments were manufactured by General Eastern Instruments, and are now used by NOAA for high altitude balloon measurements (Mastenbrook and Oltman, 1983)[8*]. The Mastenbrook design formed the basis for an aircraft Instrument developed at NCAR (Spyers-Duran, 1991), and a balloon-borne hygrometer with microprocessor control developed at Sandia Laboratory (Brown, 1988).[11*]

The cryogenic hygrometer discussed in this section is based on instrumentation developed by Buck Research. Its design is an outgrowth of the original Mastenbrook design. It has been extensively modified to obtain greater stability over a wider range than the earlier balloon-borne hygrometers, and incorporates a faster output, improved reliability, and safety to accomplish the required major thermal and electronic redesign. The instrument is most useful in laboratory and field applications involving very low frost points, where a fast response is desired and where long-term (many weeks or months) of unattended operation is needed.

The cryogenic hygrometer, although also a condensation mirror type, operates in a different mode from the standard chilled mirror hygrometer in that the mirror is at all times cooled to a temperature well below its measurement range, usually at liquid nitrogen

---

References from Chapter 15, Section IX.

temperature. The mirror will therefore always be covered with a frost layer when not in operation. To make a measurement, the mirror is heated by means of a heater attached to the mirror, until the frost layer has disappeared. The exact temperature at which the frost layer disappears is the frost point.

A schematic representation of the cryogenic hygrometer is shown in Figure 3.27.

A. Purpose

The ability to rapidly measure dew/frost points down to very low values is important for a number of research applications. In the laboratory, the capability to accurately measure low humidities at faster speed than can be accomplished with a conventional chilled mirror hygrometer, is useful for calibrating secondary humidity instruments, verifying the water vapor content of supposedly dry gases, and a number of other uses. Aircraft observations up to the tropopause can involve frost points down to –80°C (112°F) and below, and a conventional hygrometer would not be useful for this purpose.

B. Description of Cryogenic Hygrometer

1. General

In a cryogenic hygrometer, rather than driving a Peltier junction, the mirror is cooled by freely boiling a cryogenic liquid such as freon or nitrogen through an attached rod having a carefully tailored degree of thermal conductivity. The frosted mirror is raised to the dew point by a heater element wound around a stem which is maintained at that point by a servo-controlled heater. As the moisture level changes, the mirror temperature is automatically adjusted to maintain a constant condensation layer on the mirror. As in a conventional hygrometer, dew/frost point is indicated by an imbedded temperature sensor, see Figure 3.28. The frost point reading can then be easily converted into ppm or ppb

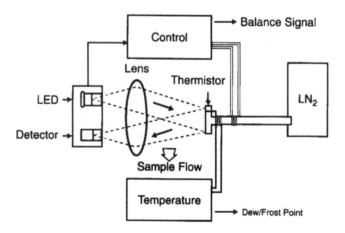

Figure 3.27   Block diagram of cryogenic dew/frost point hygrometer.

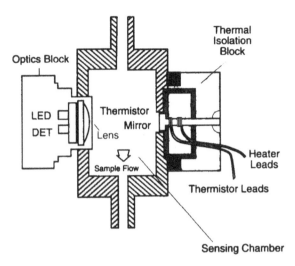

Figure 3.28   Cryogenic sensing chamber and mirror assembly.

using a micro-processor built into the instrument. Like the conventional chilled mirror instrument, the cryogenic hygrometer is a fundamental measurement device, offering very high accuracy, but at a much faster response. A lower operating limit of 2 ppb has been attained in the Buck Research instrument.

2. Cryogenic Dewar

The cryogenic hygrometer is often designed to operate with liquid nitrogen (LN_2) coolant or any other suitable cryogenic liquid, which is contained in an evacuated stainless steel dewar, normally with a capacity of two to five liters. This allows operation for four to six hours. For laboratory and other ground-based applications, the dewar may be modified to allow continuous replenishing of the cryogenic liquid, or a larger cryogen container may be used.

3. Cryo-Pump

In more advanced instruments, a continuously operating sterling cycle cryogenic refrigerator or cryo-pump is built into the instrument, allowing continuous operation at very low frost points. Cryo-pumps, which use helium as a working gas, require less than 35 watts input power to produce 1 watt of net usable cooling. This kind of instrument can be very useful in industrial process applications such as semi-conductor manufacturing. However, at this time the cost is very high and prohibitive for many applications. Future availability of lower-cost cryo-pumps would probably result in many interesting applications for this instrument.

4. Mirror Assembly

The mirror assembly is also shown in Figure 3.28. The mirror is made of thermally-conductive, rhodium-plated copper, with a small bead thermister imbedded in it to provide the dew/frost point measurement. All leads are bonded to the stem with conductive epoxy, and the mirror support structure uses a variety of structural and non-structural materials to maintain a precise alignment of the mirror with the detector optics and provide a combination of structural strength and low heat loss.

The mirror is thermally connected to the coolant through an attached rod. The coupling provides a carefully controlled thermal resistance and thus a cooling rate adequate for rapid acquisition of a frost layer, but not so great that excessive heating current is required. A flexible coupling is required because the large gradients encountered along the mirror stem produce expansion, contraction, and flexing.

5. Optics

The condensation layer is monitored by means of a light source and photo detector, positioned to measure mirror reflectance. As condensation forms, reflectance drops, and the photo detector signal changes accordingly. This signal drives the mirror temperature control circuit.

6. Sensing Chamber

The sensing chamber, shown in Figure 3.28, is designed for smooth airflow and easy access for inspection and cleaning. Surfaces exposed to the gas stream are made of electro-polished 316 L stainless steel. Flow and pressure sensing are located downstream from the sensing chamber. For aircraft use, it is equipped with a pressure sensor to correct for frost point errors due to pressure variations in the sensing chamber. For ground-based applications in which the pressure differentials are negligible, pressure monitoring is not necessary.

7. Thermodynamic Considerations

The principle heat flow paths in the system are from the mirror surface, through the insulation surrounding the mirror head assembly, and from the dewar and mirror stem/dewar interface. In general, good thermal insulators are poor structural materials. The mirror must be mounted in contact with the sample air, and well aligned with the optical system, but the sample air must also be blocked from any leakage paths into the space surrounding the mirror. The support material must be non-hygroscopic to prevent out-gassing of trapped moisture which would distort the readings.

8. Control Electronics

To provide optimal control over the required measurement range, a servo control system is used, which has no offset error under steady state conditions, regardless of the operating range. This translates into a fixed, stable frost layer, and thus allows a more accurate frost point measurement. Two rate feedback circuits are used: one for mirror temperature and the other for optical balance. These circuits sense any rapid change in either condensate level or dew/frost point temperature, and drive the control circuitry in "anticipation" of the changing conditions. This type of compensation circuit is effective for rapid accurate control under a wide range of conditions.

9. Air Sampling System

Inlet air can be introduced into the sensing chamber actively by pumping, or passively through an inlet configuration such as that used by Rosemount or by conventional dew point hygrometers. Active sampling, if it is done with an accumulator to smooth pump-induced fluctuations, has the advantage of more predictable pressure in the sensing chamber, and therefore decreases the need for pressure correction. In any system which reaches frost points below –60°C (–76°F), it is mandatory to use non-hygroscopic materials, i.e., the use of polished stainless steel throughout.

10. Response Time

Figure 3.29 shows the response time of a Buck Research cryogenic hygrometer, and Figure 3.30 the response time for a step change in frost point. These response times are comparable to the response times available in secondary measurement devices such as aluminum oxide, phosphorous pentoxide, and Piezoelectric hygrometers.

11. Accuracy

The cryogenic chilled mirror hygrometer is a fundamental measurement device offering excellent repeatability and an accuracy estimated at ± 0.5°C to ± 1.0°C (0.9°F to 1.8°F). These accuracies are lower than the typical ± 0.2°C (0.3°F) accuracies of conventional chilled mirror hygrometers operating in the –40°C (–40°F) to +100°C (212°F) range. However it is well known and generally accepted that below –40°C (–40°F) present state-of-the-art accuracies of conventional instruments also deteriorate to about ± 0.5°C (0.9°F) at –75°C (–103°F) and lower below –75°C (–103°F). The cryogenic hygrometer accuracy can therefore be considered to be about the best attainable and the instrument is quite suitable as a calibration standard for very low frost point and ppm/ppb ranges.

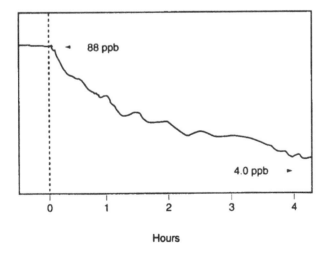

Hours

Figure 3.29   Response time of cryogenic hygrometer from 88 ppb to 4 ppb through 4 m of 30 mm (12 ft. 1/8 inch) line at a flow rate of 1 l/min. (60 cu inch/min.).

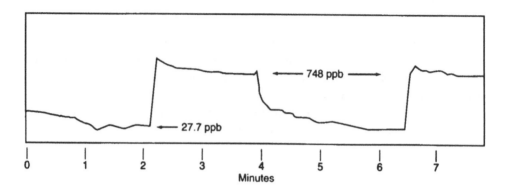

Figure 3.30 Response to step change.

Figure 3.31 Cryogenic hygrometer with dewar.
(Courtesy Buck Research)

12. Summary

The cryogenic dew/frost point hygrometer is a system, designed to operate with high stability and fast response over the dew/frost point range from below −100°C to 30°C (−148°F to 86°F). It provides a fundamental and accurate measurement with good thermal efficiency, durability, and reliability with several performance monitoring outputs. The only presently known commercial instrument of this type is offered by Buck Research, Inc. in Boulder, Colorado, and is shown in Figures 3.31 and 3.32.

The instrument with built-in cryo-pump can be operated for long periods of time without maintenance, but is more expensive than the instrument with built-in cryogenic dewar, which must be periodically replenished with liquid coolants (typically liquid nitrogen).

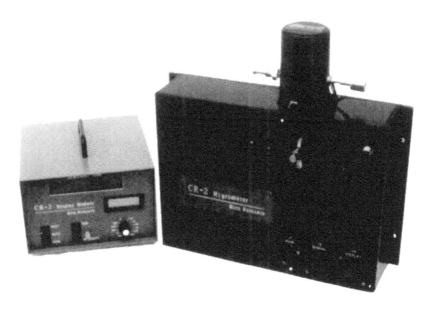

Figure 3.32   Cryogenic hygrometer with cryo-pump.
(Courtesy Buck Research)

4

RELATIVE HUMIDITY

I. General

Percent relative humidity is the best known and perhaps the most widely used method for expressing the water vapor content of air. Percent relative humidity is defined as the ratio of the prevailing water vapor pressure, to the water the vapor pressure if the air were saturated, multiplied by 100:

$$\% \text{ RH} = \frac{e_a \times 100}{e_s} \tag{4.1}$$

Basically, percent relative humidity is an indicator of water vapor saturation or deficit of the gas mixture. A measurement of RH without a corresponding measurement of dry bulb temperature is not of particular value, since water vapor content cannot be determined from percent RH alone.

It has become evident in recent years that the influence of humidity, or the effect of humidity control, is of paramount importance in many types of industries, such as in the manufacture of moisture sensitive products, storage areas, energy management, computer rooms, hospitals, museums, and libraries.

However, it is not so simple to measure humidity exactly. Many methods have been used to measure humidity, such as the method of using human hair or nylon film, the wet and dry bulb method, and the semiconductor sensor method. All of these methods have their advantages as well as disadvantages, and none of them can be labeled perfect. In addition, there are very few highly accurate calibration devices for humidity generation and measurement which makes it relatively difficult to precisely confirm the accuracy of a humidity measurement.

II. Bulk Polymer Humidity Sensor

The most advanced humidity sensors on the market today are the "Bulk Polymer Sensors," consisting of a miniature electrode base plate coated with a humidity sensitive hygroscopic macro polymer. An electrical grid structure is vapor deposited upon the element and an electrical measurement is made which is a function of the relative humidity.

The two most popular polymer sensors are the bulk resistance and the bulk capacitive sensors, depending on whether the sensor is designed for a change of resistance or change of capacitance measurement. The polymers used and the manufacturing techniques for resistive and capacitive sensors are different and the sensors have somewhat different characteristics with advantages and limitations for both.

Specifically, resistive sensors have a slower response, though they are generally fast enough for most applications. Most resistive polymer sensors on the market today are only suitable for a narrower temperature range, typically from –10°C to 80°C, whereas capacitive sensors are presently available for temperature ranges from –40°C to over 150°C. The resistive sensor offers good accuracy in the 15 to 99% RH range. Capacitive sensors can give good results down to 2% RH, but are sometimes less ideal at RH levels above 95%. Most capacitive sensors have a very fast response but tend to be slightly more expensive that the resistive types. The use of a resistive or capacitive sensor is often dictated by the application.

A. Resistive Polymer Sensor

In a typical resistive sensor, the sensitive material is prepared by polymerizing a solution of quaternary ammonium bases and the reaction of this functional base with a polymer resin. This produces a three-dimensional thermosetting resin which is characterized by its excellent stability in extreme conditions. A typical bulk polymer resistive sensor is shown in Figure 4.1.

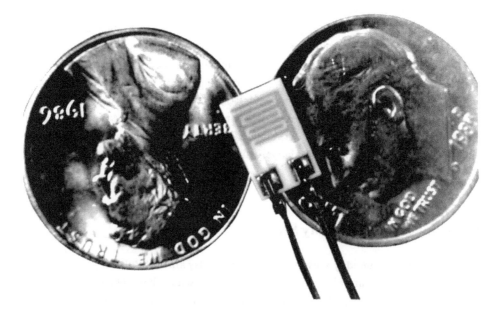

Figure 4.1   Resistive bulk polymer RH sensor.
                    (Courtesy General Eastern Instruments)

Humidity is measured by the change in resistance between anode and cathode. When the quaternary ammonium salt contained in the electrode shows ion conductivity caused by the presence of humidity, mobile ion changes are created. The more the humidity in the atmosphere increases, the greater the ionization becomes, and as a result, the concentration of mobile ions increases. Conversely, when humidity decreases, ionization is reduced and the concentration of mobile ions decreases. In this way, electrical resistance of the humidity sensitive film responds to changes in humidity absorption and desorption. The movement of such ions is measured using the variation in impedance caused by the cell as typically measured in a Wheatstone bridge. Contrary to the Pope Cell (see Section IV) which is a surface resistance element, the polymer sensor is a bulk effect sensor which means that the bulk resistance of the polymer is related to the relative humidity.

1. Features of Bulk Polymer Sensor

- 100% solid state construction

- Low power input

- Compact size

- Low cost

- Bulk polymer minimizes accuracy degradation resulting from contamination

- Broad humidity range (15% to 99%)

- Temperature range from $-10°C$ to $80°C$

- Fast response

- Excellent reproducibility (better than 0.5% RH)

- High Accuracy ($\pm$ 2% RH typical, $\pm$ 1% RH in narrow range)

Like all relative humidity sensors, when exposed to contaminants and/or extreme environmental conditions, accuracy degradation could result. Polymer-type sensors have proven to be more resistant to such errors or drift than the older Pope Cells and Dunmore Cells (see Sections III and IV). Typical results of extensive field tests performed on the General Eastern and Shinyei resistive polymer sensors are shown in Table 4.1.

2. Resistance to Contaminants

Because of the good chemical stability of polymer sensors, the sensors exhibit excellent resistance to most solvents and are therefore often used for industrial applications, except for those having high concentrations of corrosive chemicals. Many common substances such as petrochemicals have little effect.

Since different polymers are used, sensitivity to contaminants for resistive and capacitive sensors is different, i.e., some contaminants cause greater adverse effects on resistive sensors while others effect capacitive sensors more. Hence the choice of sensor is often dictated by the type of contaminants that are present.

Tables 4.1 and 4.2 show several examples of applications involving contaminants and different ambient temperatures. This table is by no means complete and applies to the General Eastern/Shinyei sensors only. But this information can often be helpful in determining likely effects of contaminants in the same or similar applications.

Table 4.1   Environmental tests.
(Courtesy General Eastern and Shinyei Corp.)

Test	Conditions	Time	Change in % RH
1. Low temperature (0°C)	At 0°C in moisture-proof bag	17 months	−1.4%RH
2. Low temperature resistance	At −20°C in moisture-proof bag	1000 hours	± 1% RH
3. Normal temperature (25°C)	In 25°C dryer, 1 Volt rms applied	24 months	2.1% RH
4. High temperature (50°C)	In 50°C dryer, 1 Volt rms applied	24 months	+3.0% RH
5. Resistance to heat (80°C)	In 80°C dryer, 1 Volt rms applied	5 months	+6.1% RH
6. Resistance to heat (100°C)	In 100°C dryer	20 days	+6.2% RH
7. Heat resistance	At 120°C in dessicator	1000 hours	± 2% RH
8. Temperature cycle test 50°C (30 min.)	0°C (30 min.)	4000 cycles	−0.4% RH
9. Temperature cycle (30 min.)	−20°C (30 min.) +100°C	100 times	± 2% RH
10. Resistance to humidity (25°C, 90% RH)	25°C, 90% RH in vessel 1 Volt rms applied	13 months	+3.0% RH
11. Humidity cycle	10% RH (30 min.) 90% RH (30 min.) Repeat cycle @ 25°C	1000 times	± 2% RH
12. High temperature and high humidity	50°C, 80% RH in vessel 1 Volt rms applied	12 months	+6.3% RH
13. Exposure to high temp. and high humidity	99.9% RH and 60°C	1000 hours	± 2% RH
14. Exposure to low humidity	20% RH or under and at room temperature	2000 hours	± 1% RH
15. Humidity cycle	25°C, 40% RH (30 min.) 90% RH (30 min.)	2000 cycles	+3.5% RH

Table 4.1   Environmental tests continued.

Test	Conditions	Time	Change in % RH
16. Condensation cycle	(10 min.) 100% RH (20 min.) Repeat @ 40C	1000 times	+2% RH
17. Storage test	Left in room in sealed (silica gel) bag	23 months	+2.1% RH
18. Resistance to gas (saturated acetone)	Saturated acetone atmosphere	6 months	+0.1% RH
19. Resistance to gas (saturated ammonia)	Saturated ammonia gas 1 Volt rms applied	9 days	−4.6% RH
20. Resistance to gas (Saturated formalin)	Formalin gas at normal temperature	2 months	+1.8% RH
21. Resistance to gas (Saturated NaCl)	NaCl saturated solution at normal temperature	5 months	+2.3% RH
22. Saturated water vapor	Water vapor at normal temperature	5 months	+2.5% RH
23. Resistance to oil and smoke	Atmosphere saturated with oil vapor and smoke	1 week	+1.6% RH
24. Resistance to oil	Apply salad oil by brush	1 time	± 1% RH
25. Resistance to oil	Impregnated with cooking oil	After 1 min.	+1.7% RH
26. Solder heat resistance	350°C; 5 sec.	2 times	± 1% RH
27. Water immersion	Room temperature	3 minutes	± 5% RH
28. Resistance to organic solvents	Stand in solution of benzene 30%, toluene 30%, and xylene 40%.	2000 hours	± 1% RH
29. Resistance to formalin	Stand in solution of 10% formalin	300 hours	± 1% RH
30. Resistance to alcohol	50% ethanol 50% methanol	300 hours 300 hours	± 2% RH ± 2% RH
31. Resistance to tobacco	Exposure to 3 smoking cigarettes	60 minutes	± 1% RH

Table 4.2   Environmental/field tests.
                 (Courtesy General Eastern and Shinyei Corp.)

Temperature:	Time	Calibration Change
–20°C in moisture-proof bag	17 months	–1.5% RH
0°C in moisture-proof bag	17 months	–1.4% RH
25°C in a dryer	24 months	2.1% RH
50°C in a dryer	24 months	3.0% RH
80°C in a dryer	5 months	6.1% RH
100°C in a dryer	20 days	6.2% RH
0°C (30 min.) 50°C (30 min.)	4000 cycles	2.0% RH
Humidity:		
90% RH @ 25°C	13 months	3.0% RH
20% RH @ 25°C	2000 hours	1.0% RH
100% RH 30 min. 99% RH @ 25°C	1000 cycles	2.2% RH
Gas Atmospheres:		
Methanol saturated @ 25°C	6 months	–0.5% RH
Acetone saturated @ 25°C	6 months	0.1% RH
Ammonia saturated @ 25°C	9 days	–4.6% RH
Formalin saturated @ 25°C	2 months	1.8% RH
NaCL saturated @ 25°C	5 months	2.3% RH
Oil mist, bug repellent	1 exposure	1.6% RH
Water, saturated @ 25°C	5 months	2.5% RH
Liquids @ 25°C:		
30% benzene, 30% toluene, 40% xylene	300 hours	1.0% RH
10% formalin	300 hours	1.0% RH
50% ethanol	300 hours	2.0% RH
50% methanol	300 hours	2.0% RH
Oil	1 dip	1.7% RH
Water	3 min.	4.0% RH

All humidity sensors are susceptible to contaminants that will cause loss of accuracy, response time and life expectancy. The above test results are provided as a guide to determine the suitability of the sensor for various applications. For exposure to other contaminants, it is recommended that the manufacturer be consulted, although most companies have little information available for this purpose. In many cases the user must perform the necessary tests.

3. Performance

The resistive polymer sensor has, for most applications, an adequately fast response, but is slower than a capacitive sensor. The typical sensor looses sensitivity below 15% RH and is therefore not an ideal choice for measurements at very low RH. The resistive sensor offers good performance at humidity levels close to 100% RH which is often not the case with capacitive sensors. Typical curves showing RH versus output for resistive polymer sensors are shown in Figures 4.2 and 4.3.

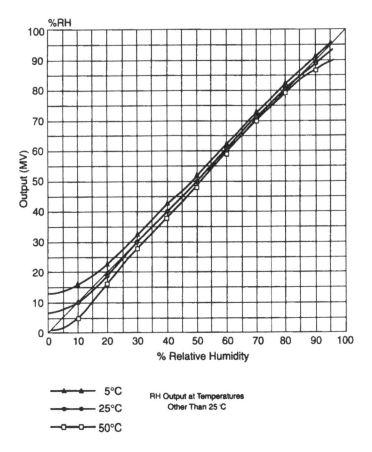

Figure 4.2   %RH versus output, sensor type A.

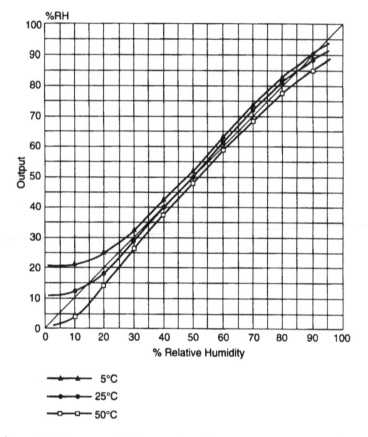

Figure 4.3   %RH versus output, sensor type B.

4. Advantages

Bulk resistive polymer sensors offer the following advantages:

- Fast response (but slower than capacitive sensors)
- Virtually no hysteresis or aging
- Low temperature constant
- Stable and repeatable operation in many applications
- Inexpensive
- Low power consumption

5. Limitations Include:

- Secondary measurement device, must be periodically re-calibrated
- Not suitable in the presence of certain contaminants
- Needs temperature compensation if used over a broad temperature range
- Modest temperature and humidity range

III. Dunmore Cell

The Dunmore cell, named after its inventor, Dr. Dunmore, is an older type of resistive sensor, developed in 1938, which has presently in many cases been superseded by the newer and more advanced polymer sensor.

The instrument operates on the principle that a lithium chloride solution immersed in a porous binder changes its ionic conductivity depending on changes in relative humidity. With a fixed concentration of lithium chloride, a linear change of resistance is observed for a narrow range of humidities. Because of the steep resistance to relative humidity change, it is necessary to vary the bifilar element spacing, or the concentration of lithium chloride, or both, to develop resistance curves for specific humidity ranges.

The sensor is comprised of a bifilar wound inert wire grid on an insulative substrate. This is then treated with a known concentration of lithium chloride in solution. Due to its hygroscopic nature, the lithium chloride will take on water vapor from the surrounding atmosphere. This modifies the surface resistivity (impedance) of the sensor because of the ionic nature of the salt ($LiCl$).

The Dunmore element is a good RH sensor but suffers one major drawback. It is only active over a small RH range due to the doping concentration of the salt. This can be remedied by placing multiple elements into an array and designing in cross over circuitry for range transition switching.

Like the Pope cell, the Dunmore array can be readily contaminated. However, if proper application techniques are employed, these sensors could provide reasonably good service.

IV. Pope Cell

The Pope cell is also named after its inventor, Dr. Pope. It is an ion exchange resin electric hygrometer. A conventional ion exchange resin consists of a high-molecular-weight polymer having polar groups of positive or negative charge in a cross-linked structure. Associated with these polar groups are ions of opposite charge, which are held to the fixed polar groups by electrostatic forces. In the presence of water or water vapor, the electro-statically held ions become mobile, and when a voltage is impressed across the resin, the ions are capable of electrolytic conduction. The Pope cell is an example of an ion exchange element. The cell is a wide-range sensor, typically covering 15% to 95% RH. Therefore, one sensor can be used where several Dunmore elements would be required to cover the same range. The Pope cell, however, has a very nonlinear response characteristic from approximately 1.000 ohms at 100% RH to several megohms at 10% RH. The element is quite sensitive to contaminants and exhibits a significant amount of hysteresis.

The ion exchange material used is sulphonated polystyrene. The Pope cell consists of a conductive grid over an insulative substrate. This substrate consists of polystyrene which has been treated with sulfuric acid. This acid treatment causes sulphonation of the polystyrene molecules. The sulfate radical, SO_4, becomes very mobile in the presence of hydrogen ions (from the water vapor), and readily detaches to take on the H⁺ ions. This alters the impedance of the sensor as a function of humidity.

The main advantage of the Pope cell is its wide operating range and well documented humidity response curves. This allows for simple mechanical and electrical designs. A disadvantage is that due to the extremely active surface of the sensor, the sensing surface can be easily washed off or contaminated to cause errors in readings. The sensor is much more sensitive to contaminants than the previously discussed resistive bulk polymer sensor which offers much better performance with virtually no hysteresis. Pope cells are therefore no longer widely used and have often been replaced with the more stable bulk polymer sensors.

V. Capacitive Polymer Sensor

The capacitive polymer sensors use either polyamide or cellulose acetate polymer thin film deposits between conductive electrodes. The film, acting as a capacitor dielectric with the interjected surface metal as electrodes, changes its dielectric constant as moisture is adsorbed or desorbed by the thin film. An alternative construction method uses a porous top metal layer that allows moisture transmittance.

The use of high temperature thermosetting polymers has resulted in the development of a new generation of capacitive RH sensors that can perform continuous measurements well above 100°C (212°F). The basic construction of a capacitive bulk polymer RH sensor is shown in Figure 4.4.

1. A substrate base, typically glass. Its main function is to support the other layers of the sensor.

2. One of the electrodes, made of conductive and corrosion resistant material.

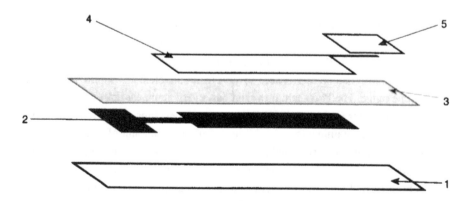

Figure 4.4   Schematic of capacitive polymer sensor.

3. A thin polymer layer. This is the heart of the sensor. The amount of sorbed water in the film varies as a function of the surrounding relative humidity. The thickness of this film is typically 1 to 10 μm.

4. The upper electrode, which also plays a role in determining the performance and characteristics of the sensor. For fast response it must have good permeability for water. It must also be electrically conductive and have strong corrosion resistance.

5. A contact pad for the upper electrode. Since there are many constraints on the design of the upper electrode, a separate metallization for making reliable contacts is often required.

A. Operation

The capacitance of the sensor is determined by the overlapping area of the upper and lower electrodes according to:

$$C = \frac{\varepsilon \cdot \varepsilon_0 \cdot A}{L}$$ (4.2)

where:

ε = Relative permittivity of the polymer film (typically in the range 2 ... 6)
ε_0 = Permittivity of vacuum (8.85×10^{-12} F m^{-1})
A = Overlapping area of the upper and lower electrodes
L = Thickness of the polymer film
C = Capacitance

The area A is constant and well defined, usually using lithographic techniques. Also the thickness of the polymer film can be considered constant (if not, very hygroscopic polymers are used). Thus only the relative permittivity of the film changes as water is absorbed into it.

As the water molecule is highly polar, even small amounts of water can change the sensor capacitance to a measurable extent. The relative permittivity of water is 80 as compared to 2 to 6 for the polymer material. This property of the water molecule makes capacitive measurement a natural choice for humidity measurement. It also makes the sensor less prone to interference from other atmospheric gases.

Polymer-based capacitive relative humidity (RH) sensors are proving increasingly effective for applications beyond the range of simple humidity measurements. Full-range capabilities have been demonstrated for some sensor types at moisture contents represented by –50°C (–58°F) to 100°C (212°F) dew point. This measurement range is equivalent to 24 ppm of water by weight at the dry end to atmospheric steam at the wet end. These sensors are suitable for both atmospheric and process measurements, and are capable of functioning down to a level below 2% RH. They can be used over a wide temperature range, in some instances without temperature compensation. High temperature thermosetting polymers have permitted the development of capacitive RH sensors that can perform continuous measurements at temperatures to 185°C (365°F), and with continuous exposure to 210°C (410°F). Because thin film sensor layer processing is performed at temperatures over 400°C (752°F), the maximum temperature is determined by the materials used in the sensor package. At atmospheric pressure, the maximum RH at 185°C (365°F) is 10%, but this is still within the operating range. Measurements in environments over 185°C (365°F) rapidly approach 0% RH. These measurements can be made by restricting the sensor temperature, thereby raising the RH at the sensor, and converting to absolute moisture (dew point) measurements. One advantage of the thermosetting polymer sensor is that it has a small temperature coefficient from –50°C (–58°F) to 100°C (212°F). This greatly simplifies accurate measurement at low to moderate temperatures and moisture levels. Any necessary temperature correction represents only a small percentage of the RH reading when external offset error sources are eliminated.

B. Temperature Dependence

All RH sensors are temperature sensitive and, when calibrated at one temperature, exhibit errors at other temperatures. One advantage of polymer sensors is that they generally exhibit far less temperature dependency, i.e., they have a lower temperature coefficient, than the older RH sensors such as the Pope cell and Dunmore cell. Hence, if the operating temperature is not too different from the temperature at which the sensor was calibrated, the error is small and can often be ignored. If the sensor is to be used in extreme temperature environments, or if optimum accuracy is desired, electronic temperature compensation must be incorporated. Temperature compensation can as a rule be easily accomplished in temperature spans of about 50°C, but is more difficult and less accurate over broader temperature ranges. Hence, when used in a very broad temperature range, polymer sensors lose some of their accuracy. Nevertheless, modern polymer sensors are available with accuracies of ± 1% RH in narrow ranges and at calibration temperature, to ± 3% RH when newly calibrated over a broad temperature and humidity range. When operated over a period of time, some accuracy deterioration should be expected, and when contaminated, more significant errors could occur, requiring re-calibration or sensor replacement.

C. Performance

Compared to the resistive polymer sensor, the capacitive sensor has the following advantages:

- Faster response, typically in the seconds range

- Better linearity and hence higher accuracy in the low RH range

- Can be used to higher ambient temperatures, to 185°C in some cases

- Low temperature dependence

- Can be used at RH levels as low as 2% RH

A typical curve of capacitance versus RH at 20°C (68°F) is presented in Figure 4.5. In this curve the capacitance is normalized to the dry capacitance, i.e., capacitance at close to 0% RH.

D. Advantages

Specific features of the capacitive sensor include:

- Extremely fast response
- Broad humidity range
- Broad temperature range
- Virtually no hystersis
- Good stability and repeatability
- Low temperature coefficient
- Low cost

E. Limitations

Limitations include:

- Sensitivity to certain contaminants
- Do not work well in corrosive atmospheres
- Temperature dependence (but low temperature coefficient)

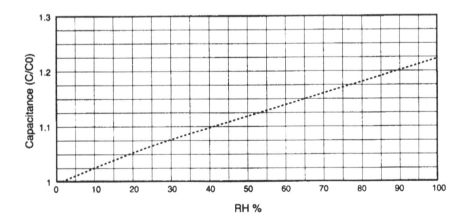

Figure 4.5   Typical capacitance change as a function of RH at 20°C. *(Courtesy Vaisala)*

VI. Displacement (Mechanical) Hygrometers

Perhaps the oldest type of RH sensor still in use is the displacement sensor. This device uses a strain gauge or other mechanism to measure expansion or contraction of a material in proportion to changes in relative humidity. The materials used for this purpose include: human hair, plastic polymers, cotton, wool fibers, wood, animal bones, and goldbeater's skin. The three main materials in use are hair, nylon and cellulose. They are coupled to pneumatic, mechanical linkages, or electrical transducers to form hygrometers. The inherent non-linearity and hysteresis must be compensated within the hygrometer. The advantages of this type of sensor are that they are inexpensive to manufacture and immune to most contaminants. Disadvantages include a tendency to drift over time, and to take a "set" if remaining at a given humidity for a protracted period of time, thereby losing sensitivity. These devices are generally unreliable below 0°C (32°F). In many cases the response time is inadequate for monitoring rapidly changing processes.

VII. Percent RH Transmitters and Instrumentation

In most cases %RH sensors are provided in the form of transmitters with a simple electrical circuit packaged in a small housing. The electrical circuit could be as simple as a Wheatstone bridge to measure resistance. Some electronics may be added to provide linearization and/or temperature compensation. The output may be 4–20 ma, or 0–5V. A display is usually not needed, but is often offered as an option. Portable instrument packages with digital display, often also available with battery supply, are also offered. Major applications for these sensors include air conditioning and energy management. Typical RH transmitters are shown in Figures 4.6 and 4.7.

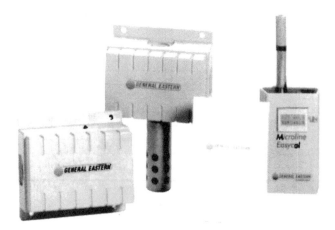

Figure 4.6   Typical %RH transmitters.
              (Courtesy General Eastern)

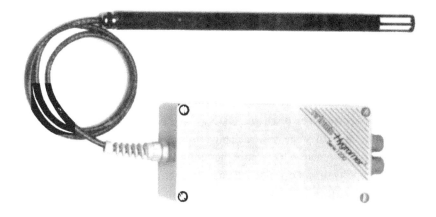

Figure 4.7   RH transmitter with remote probe.
                  (Courtesy Rotronic)

Rather sophisticated instruments using polymer type sensors are also available. Many of these instruments are offered with temperature probes as well and use micro-processors to provide various outputs and readings, such as relative humidity, temperature, calculated dew point, and sometimes mixing ratio and absolute humidity. These instruments are considerably less expensive than similar chilled mirror hygrometers. Although their initial accuracy may approach that of the chilled mirror hygrometer, it must be remembered that the measurement is of a secondary nature, i.e., resistance or capacitance measurement, and that long term accuracy cannot be guaranteed, unless the instrument is frequently recalibrated.

Figure 4.8 shows an advanced data processing instrument with multiple outputs and four sensors, and two convenient hand held instruments for general measurements.

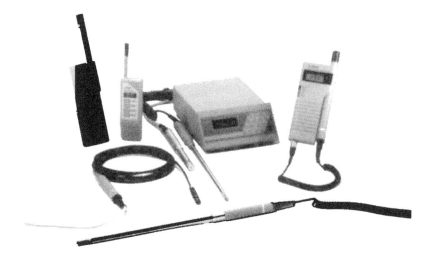

Figure 4.8   Typical polymer RH instrumentation.
                  (Courtesy Vaisala)

VIII. Summary

When choosing %RH sensors, it is important to understand:

- All instruments employ a sensor as well as varying amounts and types of supporting circuitry. The sensor itself is "transparent" as long as the instrument performs as required.

- All of the sensor types discussed above are empirical (secondary measurement devices); that is, they measure a change in some material, physical or electrical parameter as a function of relative humidity. They do not measure any fundamental property of water vapor. All empirical sensors are subject to drift and loss of calibration.

- Relative humidity is an expression of a ratio of vapor pressures. Difficult RH applications can sometimes be tackled with instruments that measure fundamental properties of water vapor. The results can be converted to RH either by formula or by using appropriate charts and tables as presented in Chapter 13. The technology that makes the most sense should be used for the application and the data can then be converted into the parameters desired.

5

TRACE MOISTURE INSTRUMENTATION

I. Aluminum Oxide Hygrometers

A. General

Aluminum oxide humidity analyzers are available in a variety of types. They range from low-cost, single-point systems, including portable battery-operated models, to multi-point, microprocessor-based systems with the capability to compute and display humidity information in different parameters, such as dew point, parts per million, and percent relative humidity.

A typical aluminum oxide sensor is in essence a capacitor, formed by depositing a layer of porous aluminum oxide on a conductive substrate, and then coating the oxide with a thin film of gold. The conductive base and the gold layer become the capacitor's electrodes, forming what is essentially an aluminum oxide capacitor. Water vapor penetrates the gold layer and is absorbed by the porous oxide layer. The number of water molecules absorbed determines the electrical impedance of the capacitor which, in turn, is proportional to water vapor pressure. Construction of an aluminum oxide sensor, as shown in Figure 5.1, is basically very simple. The complexity arises from the fact that such sensors cannot be manufactured with great consistency and interchangeability. They also exhibit varying degrees of aging and hysteresis, and require extensive calibration. Over many years the companies that offer aluminum oxide sensors, such as Panametrics (USA), Endress & Hauser (Germany), and Shaw (England) have developed elaborate calibration systems and per-aging routines, resulting in sensors that maintain acceptable accuracy for process applications over an extended period of time.

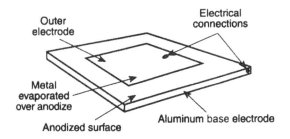

Figure 5.1   Construction of an aluminum oxide sensor.

Figure 5.2   Aluminum oxide sensor.
(Courtesy Panametrics)

     Aluminum oxide sensors are also suitable for measurements of water vapor in organic liquids. The pore wall openings are small compared to the size of organic molecules. Therefore admission into the pore cavity is limited to small molecules such as water. Parts per million by weight of water in organic liquids is given by Henry's law, which states that the mass of gas dissolved in a given volume of solvent at a constant temperature is proportional to the pressure of the gas with which it is in equilibrium. Hence:

$$\text{PPM water} = \text{Partial Pressure of Water} \times K$$

where K = Henry's law constant which must be computed for each liquid and is given by:

$$K = \frac{\text{Saturation Weight PPM of Water for that Solvent}}{\text{Saturation Vapor Pressure of Water}}$$

both at the temperature of measurement. Once the constant is known, water concentration in the organic liquid is computed by multiplying this constant by the vapor pressure of water as measured by the sensor probe. A typical aluminum oxide sensor is shown in Figure 5.2 and two typical aluminum oxide hygrometers in Figures 5.3 and 5.4.

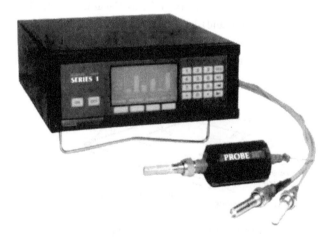

Figure 5.3   Microprocessor based aluminum oxide hygrometer and several sensors.
(Courtesy Panametrics)

Figure 5.4   Aluminum oxide hygrometer and sensor.
                   (Courtesy Endress & Hauser)

B. Aluminum Oxide Instrumentation

Aluminum oxide hygrometers respond to the vapor pressure of water over a very wide range of vapor pressures. The impedance of the aluminum oxide layer between gold and aluminum electrodes is sensitive to water vapor from saturation vapor pressure down to below 10^{-5} mb (10^{-3} Pascal) corresponding to $-110°C$ ($-166°F$) dew point. The strong affinity of water for this oxide, combined with the large dielectric constant of water, makes this device highly selective towards water. It does not respond to most other common gases or to numerous organic gases and liquids.

In situations where aluminum can be chemically attacked, silicon sensors with a tantalum metal base could be used as an alternative. Such sensors are the most inert, although their sensitivity is somewhat less. The oxide layer is generally described in the form of a mass of tubular pores running up from the metal base to the exposed surface. Change in the size of these tubules with time is presumed to be the cause of the slow shifts in calibration often experienced with these sensors. Water is absorbed in these tubules in amounts directly related to the moisture content of the gas in contact with it. The amount of water is sensed electrically by measuring the change in capacitance or admittance produced by this water. Because of the radius of the pores in the aluminum oxide, the sensor is virtually specific for water molecules.

The dew point range that is customarily covered by standard sensors of this kind is between $-110°C$ ($-166°F$) and $+20°C$ ($68°F$), which corresponds to a range from approximately 1 ppb to 0.2% by volume. Operation up to $100°C$ is theoretically possible but the shifts in calibration tend to be accentuated by higher temperature operation and very frequent calibration checks would be required. The sensor is therefore generally not recommended for operation above $70°C$ ($158°F$). Temperature coefficients are modest but need to be taken into account, especially in the higher gas temperature and dew point ranges. For accurate measurement in these ranges, it is essential that probes are calibrated at their expected operating temperature, though Panametrics uses a patented 100 Angstrom film thickness, as a result of which the sensor is claimed to be temperature independent. Response times for a 63% change are a few seconds at higher dew points, increasing as the dew point drops to a minute or more at $-50°C$ ($-58°F$) frost point. Commercially available sensors can operate at pressures up to 34×10^6 Pa (4.9×10^3 psi) and at vacuum down to 0.7 Pa (1.015×10^{-4} psi). However, it is important to note that the reading obtained is dependent on pressure. A sensor operating at 10^6 Pa (144 psi) will give the dew point of the gas at 10^6 Pa (144 psi), not at atmospheric pressure. If line pressure variations are common, it may be better to take a continuous sample of the gas and reduce its pressure (always ensuring that the cooling caused by the gas expansion does not bring its temperature to the region of the dew point) to atmospheric or other controlled value. Accuracies are often specified in the range of $\pm 1°C$ to $\pm 2°C$ ($\pm 1.8°F$ to $\pm 3.6°F$) at higher dew points, increasing to $\pm 2°C$ to $\pm 3°C$ (± 3.6 to $5.4°F$) at $-100°C$ ($-148°F$) frost point. However, these accuracies are obtained under laboratory and idealized conditions. In actual field use the accuracy will depend on the parameters of the application and the frequency of recalibration. These sensors can tolerate linear flow rates of up to 10 m (30 ft) per second at 10^5 Pa (14.4 psi). Calibrations do drift with time. The magnitude of these drifts for state-of-the-art sensors is modest. The stability varies from manufacturer to manufacturer.

Aluminum oxide sensors are free from interferences by gases, such as hydrocarbons, carbon dioxide, carbon monoxide, and chloro-fluoro-hydrocarbons, although the drift rate may vary for differing gases. Certain corrosive gases, such as ammonia, sulfur trioxide vapor, and chlorine, attack the sensing element and should be avoided. Where occasional dust is expected in the line a sintered metal filter should be used to cover the sensor. If heavier concentrations are encountered, it may be necessary to mount the sensor in a sample line with a cyclone or filter to remove the particulates. Performance of the sensor element is not affected by dust, but access of water vapor to the sensitive surface is hindered, resulting in a slower response time. The probes themselves can be made intrinsically safe if operated through Zener barriers or galvanic isolators, and can be used in all areas including zone 0.

C. Advantages

Aluminum oxide sensors offer the following advantages:

- Wide dynamic range from 1 ppm to 80% RH

- Probes can be installed remotely at distances of 300 m (1000 ft.) or more

- Probes can easily be used "in-situ"

- Aluminum oxide sensors are suitable for use in multi-sensor configurations

- Relatively stable, with low hysteresis and temperature coefficients

- Independent of flow rate changes

- Suitable for intrinsically safe applications

- High selectivity for moisture

- Operation over a wide range of temperature and pressure

- Low or modest maintenance requirements

- Available in small sizes

- Economical in multiple sensor arrangements

- Suitable for very low dew point levels without the need for cooling, with easy measurement of typical dew points down to −110°C (−166°F).

D. Limitations

- Slow drifts in calibration, which may accelerate at higher operating temperatures or in certain gases

- Probe affected by some corrosive gases. Analogous silicon or tantalum oxide sensors could be considered as alternatives.

- Temperature control often needed, especially at high temperature and dew points

- Offer only modest accuracy and repeatability

- Must be periodically calibrated to accommodate aging effects, hysteresis and contamination (not a fundamental measurement)

- Non-linear and non-uniform construction calls for separate calibration curves for each sensor.

Many improvements have been made over the last several years by various manufacturers of aluminum oxide sensors but it is important to remember that even though this sensor offers many advantages, it is a lower accuracy device than any of the fundamental measurement types. It is a secondary measurement and therefore can provide reliable data only if kept in calibration and if incompatible contaminants are avoided.

E. Calibration

Calibration of the aluminum oxide hygrometer probe is accomplished by measuring the probe impedance over a wide range of water vapor pressures. Calibration data is normally plotted as probe admittance (reciprocal of impedance) versus the dew point which correspond to the calibration water vapor pressure. At low gas pressures the calibration curve relates the measured probe admittance directly to the actual dew point in the gas. At higher pressures, a correction must be applied, due to the non-ideal behavior of the gas. In most practical cases the indicated dew point is up to several degrees higher than the actual dew point. Thus the use of the probe to warn of impending moisture condensation is not seriously affected. However, the actual dew point as well as the mole fraction of water in the gas can be obtained by the use of water enhancement factors.

There are several good reasons for making measurements at high pressure in-situ with aluminum oxide hygrometer probes. The foremost advantage is the increased water vapor concentration at high pressure, which increases speed and sensitivity of the measurements and reduces the probability of contamination from ambient moisture. The simplicity of a high pressure sampling system is another factor. Concerns about moisture freezing or condensing during gas decompression are eliminated. Finally, an advantage of increasing importance is the ease of returning sampled gas to the high pressure line rather than flaring or venting it.

Calibration is perhaps the most critical part of the manufacture of good aluminum oxide sensors. Due to production variations, it is not only necessary to calibrate each sensor, but each sensor must be calibrated several times after manufacture to make sure that sensor aging and calibration drifts, which take some time to stabilize, are down to a negligible, or at least acceptable, level. Some sensors may continue to drift indefinitely and must be discarded. Most manufacturers have installed rather elaborate calibration facilities in order to calibrate a multitude of sensors simultaneously, store the data in a computer, and repeat the calibration process several times over a period of 3 to 6 months. Sensors should only be installed when such calibration data have shown that the sensor is reasonably stable. Today's availability of computers has made it possible to greatly improve the quality of these sensors. A typical aluminum oxide sensor calibration facility is shown in Figure 5.5.

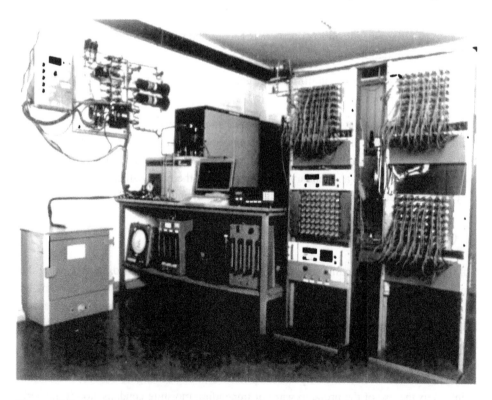

Figure 5.5   Aluminum oxide sensor calibration laboratory, located in Ireland.
(Courtesy Panametrics)

F. Applications

The range of applications of aluminum oxide sensors is very extensive: Ethylene feed gas in polyethylene manufacturing, dryer control, natural gas in pipelines, incandescent lamp filling, hydrogen recycle streams in catalytic reformers, and bottled gases, just to give an indication of the range of materials and uses. A more complete list is presented in Chapter 12, *Applications*. These sensors are used by a wide range of industries and have been the choice in a large number of applications, particularly where low dew and frost points have to be measured.

The aluminum oxide sensor is also used for moisture measurements in liquids (hydrocarbons) and, because of its low power usage, is suitable for intrinsically safe and explosion-proof installations. Aluminum oxide sensors are frequently used in petrochemical applications where low dew points are to be monitored "in line" and where reduced accuracies and other limitations are acceptable.

II. Silicon Oxide Hygrometers

Silicon oxide hygrometers use capacitive impedance type sensors that are in many ways similar to aluminum oxide sensors, but they employ silicon rather than aluminum as the humidity sensitive material. The sensors are fabricated using silicon technology and should not be confused with the silicon circuit-type detectors that employ an evaporated layer of aluminum, which is subsequently oxidized to form the sensing element.

Silicon oxide sensors have a typical operating range of –80°C to +80°C (–112°F to 176°F), but are not frequently used above +10°C (50°F) dew point. They offer modest stability and very fast response (faster than aluminum oxide). With suitable purging the sensor can be switched from measuring saturation to 1 ppm with a response time of about 15 seconds. They can operate at pressures from vacuum to 25×10^6 Pa (3.63×10^3 psi) and at gas temperatures of –40°C to +45°C (–40°F to 113°F) without greatly affecting performance.

A. Advantages

Silicon sensors have the following advantages:

- Wide dynamic range, from less than 1 ppm to saturation

- Operation independent of flow rate

- Operate over a wide temperature and pressure range

- High selectivity

- Very fast response (faster than aluminum oxide)

- Relatively inert.

B. Limitations

Silicon sensors have the following shortcomings:

- Require periodic calibration (not a fundamental measurement device)

- Subject to drift when used with certain gases, such as CO_2

- Very temperature dependent and difficult to compensate for temperature. For this reason commercial sensors are normally kept and controlled at a fixed temperature. Accurate measurements are only possible if the sensor is maintained at this precise temperature.

- More expensive than aluminum oxide systems

Because of the disadvantages listed, silicon oxide hygrometers have not been able to displace aluminum oxide sensors and are only used in selected applications, such as when their fast response and stability are of primary importance.

III. Piezoelectric Sensor

The Piezoelectric sensor measures moisture by monitoring the vibration frequency change of a hygroscopically-sensitive quartz crystal that is exposed alternatively to wet and dry gas. Sample gas is divided into two streams, "sample" and "reference," which are alternately passed across the measuring crystal. The reference gas is passed through a molecular sieve dryer, which removes virtually all its moisture.

As sample gas is passed over the measuring crystal, moisture is absorbed by the crystal's hygroscopic coating, thereby causing a vibration frequency change. At the end of a 30-second period, the microprocessor "reads" and stores the frequency difference ($\Delta F1$) between the crystal and a sealed local oscillator crystal. It then switches the gas flow to expose the crystal to dried reference gas for the next 30 seconds to dry the crystal. At the end of this 30 second period, the microprocessor again "reads" and stores the vibration frequency difference ($\Delta F2$) between the two crystals. The two different frequency values, $\Delta F1$ and $\Delta F2$, are then subtracted from each other to produce $\Delta F3$, which is proportional to the amount of moisture absorbed from the sample gas during that cycle. The moisture level is related to $\Delta F3$ through a polynomial expression with constants determined during the factory calibration of the cell. The microprocessor uses this expression to calculate the moisture level for each measured value of $\Delta F3$ and multiplies the results by a field-adjustable sensitivity factor to obtain the final output signal.

Both crystals are temperature controlled at 60°C (140°F) for optimum repeatability and accuracy. All electronics are solid state to obtain high reliability. The sensor head often also contains a calibrator that can be used to check the performance, although drift-free operation is usually possible for periods of a year or more. The instrument can cover a measurement range of about 2 ppb to 100,000 ppm by volume, though the sensor is normally only used for measurements up to 1,000 ppm. Typical attainable accuracies are 5% of reading in most ranges and ± 10 ppb in the lowest range. Sensors are usually available in explosion proof configurations and can be operated over a temperature range of –20°C to +55°C (–4°F to 131°F). A typical quartz crystal oscillator sensor with electronics is shown in Figure 5.6.

A. Operation of a Typical Piezoelectric Moisture Analyzer [42, 43]

1. Moisture Differential Sensing and Measurement

Analyzers of this type employ piezo oscillation frequency differential measurements for quantifying the gas moisture content. The differential measurement may be based on a known dry reference sample or a known wet sample. Both types of instruments condition a reference sample from which the differential is derived.

The analysis procedure sequence alternates the induction of reference and sample gases into a moisture sensitive sample chamber, or "cell." First, the test sample is introduced and characterized. Then the moisture reference sample is induced and measured. The relative change in the sample cell behavior is registered, and the absolute moisture content is derived by real-time computation.

2. The Piezo Oscillation Frequency Differential Moisture Sensor

The sample cell consists of a small chamber through which both the test sample and the reference sample gases alternately flow. Within the sampling chamber, a pair of electrodes support a thin, quartz crystal disc. A voltage potential is applied across the opposing faces of the disc, exciting a transverse oscillation within the crystal at a frequency in the 9 Mega Hertz range. Both faces of the quartz disc are partly coated with a hygroscopic polymer. As the moisture content of the polymer varies, the corresponding mass loading of the crystal significantly affects the oscillation frequency.

Figure 5.6   Sensing crystal oscillator unit.
           (Courtesy AMETEK)

Situated alongside the sample cell is a reference cell of identical design, except that it is hermetically sealed. The sample cell and reference cell crystals are excited by the same, constant voltage source. Thus, any change in their frequency differential is attributed solely to the moisture mediated variations in the polymer on the sample cell crystal. Utilizing this method, extremely fine measurements, down to a resolution of one part per billion, are possible.

3. Sample Flow and Reference Conditioning

An instrument of this type can be configured to internally condition the sample gas as a reference sample, or to use an externally supplied reference gas. The discussion below deals with the internal conditioning configuration.

The test gas may be brought into the instrument, through a 7 µ particulate filter to a "tee" connection. The first branch of the "tee" diverts part of the sample through a mini dryer molecular sieve. The mini dryer filters the gas output to less than a 5 ppb moisture level. This output is used as the dry reference in the differential measurement. After the mini dryer, the reference sample passes through a second "tee", a pressure regulator, and then a 200 ml/min. (12.2 in.³/min.) flow restriction orifice, which terminates at a solenoid valve at the inlet of the sample cell. The second "tee" branches to a calibrated moisture generator. This generator consists of a teflon jacketed water sample (a permeation tube) of precise configuration. The physical dimensions and stable mechanical properties of the generator components make it ideal as a calibration source for determining the measurement range or "span" of the sample cell. With the ppb moisture conditioned output from the mini dryer, the moisture generator typically yields an 800 ppb sample.

The system's measurement span is periodically calibrated, during which the 800 ppb sample gas is conveyed to the sample cell, and a reference measurement value is registered for use in the computation of process measurements.

The gas following the second branch of the inlet "tee" passes through a pressure regulator and a 200 ml/min. (12.2 in.³/min.) flow restriction orifice, terminating at a solenoid valve at the inlet of the sample cell. After sampling, all gases are fed to the instrument exhaust.

The process measurement cycle is shown in Figure 5.7 below:

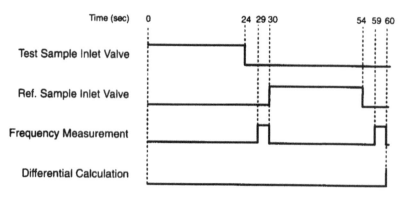

Figure 5.7   Piezoelectric process measurement cycle.
(*Courtesy AMETEK*)

Note that after 24 seconds of sample flow through the cell, there is a five second no-flow stabilization period, prior to frequency measurement.

4. Calibration

The Piezoelectric analyzer is not a fundamental instrument and must therefore be periodically calibrated. Such a calibration cannot be performed by NIST at levels below 1 ppm$_v$, and like with aluminum oxide and electrolytic hygrometers, special arrangements must be made. Most manufacturers have developed their own calibration methods. One such a system is discussed in Chapter 10, Section IV. Some instruments incorporate a built-in permeation-tube moisture generator, capable of making a rough field calibration check in about 30 minutes.

5. Construction

Materials used for building Piezoelectric sensors must be carefully selected. The use of non-hygroscopic materials is most important. Sample contacts are normally made of stainless steel, glass or teflon. To safeguard electronic components which are mounted within the explosion proof field unit, they are protected by installing a small air scrubbing unit inside the housing to remove any potentially corrosive gases that might enter the housing from sample system leaks or from the ambient air.

B. Response Time

Response time is very important in process analyzers since moisture can change rapidly. A slow response of the moisture sensor would prevent the sensor from coming into equilibrium and thus rendering the measurements useless to the operator. This is especially the case when low ppm or ppb level measurements are to be made.

The response time of a quartz crystal sensor depends on the difference between the partial water vapor pressure in the detector element and that in the sample gas. The higher this vapor pressure difference, the faster the sensor response to a moisture level change. The instrument is operated in such a way that the detector crystal is first exposed to the wet sample gas for a fixed period of time and then dried for another fixed period of time by a known dry reference gas. The difference in frequency between the wet and dry readings is proportional to the amount of water in the sample. By following this wet-dry cycle, the water partial pressure difference between the sample and detector is kept as great as possible. In this way, the quartz crystal moisture analyzer does not have to run in equilibrium with the moisture in the sample gas and can thus respond rapidly to small changes in moisture. Response times of 80% to a step change in less than 10 minutes at 10 ppb level have been obtained.

C. Sensitivity

The sensitivity of a Piezoelectric sensor can be affected by several variables including:

- *Cell pressure*
  This sensor is a moisture partial pressure detector. Hence any changes in total pressure in the cell will affect sensitivity.

- *Cell temperature*
  The rate of absorption and desorption is, among other things, dependent on temperature. At higher temperatures the sensitivity due to absorption is decreased. Likewise, the sensitivity increases at lower temperatures.

- *Coated polymer area*
  The more that surface is exposed to the wet gas, the higher the sensitivity.

- *Electronics*
  Sensitivity depends on the accuracy with which frequency changes can be measured.

- *Sample cycle*
  The detector is alternately exposed to the wet sample and then to a dry reference gas. To maximize sensitivity, it is desirable to make this period as long as possible. However, this results in a slower response time. The sample and reference cycles must therefore be optimized.

D. Advantages

The Piezoelectric moisture analyzer offers the following advantages:

- Wide operating range

- Low sensitivity to contamination

- Ability to measure very low frost points (down to ppb range)

- Auto-zeroing inherent in the design

- Built-in calibration checking

- Suitability for applications requiring explosion proofing

- Fast response time

- High accuracy at low frost point levels

E. Limitations

The main disadvantages are:

- More expensive than some other types
- Not suitable for in situ use
- Possible problems if the probe temperature is difficult to control

F. Applications

The quartz crystal sensor is especially useful in applications involving gases containing corrosive contaminants. Typical applications are:

- Catalytic reforming
- Moisture in olefins
- Natural gas (cryogenic extraction, transmission stations, storage, distribution, LNG, production)
- Petroleum refining (reforming, alkylation, LPG, hydrocarbons
- Petrochemicals (cracked gas, propylene, butadiene, ethylene)
- Chemicals (fluorocarbons gases, vinyl chloride, vinyl fluoride, refrigerants, reactor gas blankets, methyl chloride)
- Enhanced oil recovery (CO_2 pipelines)
- Electronics (doping gases, dielectric gases, soldering furnace atmospheres, blanketing gases)
- Metals (annealing furnace atmospheres)
- High purity gas production
- Semiconductor manufacturing
- Cylinder gases

G. Summary

The original concept of the quartz crystal (Piezoelectric) hygrometer was developed by Du Pont Company to meet the need for a reliable and accurate instrument to measure low ppm moisture levels in contaminated gases, such as gaseous petrochemical byproducts, gases containing chlorine, and natural gas containing glycols. Because of its high cost, the instrument is typically used in high value processes such as petroleum refining and catalytic recyclers. This product is now marketed by AMETEK. A typical commercial Piezoelectric instrument is shown in Figure 5.8. Crystal oscillator sensing cells are shown in Figure 5.6.

Figure 5.8   Piezoelectric moisture analyzer.
                   (Courtesy AMETEK)

IV. Electrolytic Hygrometer

The electrolytic hygrometer is desirable for dry gas measurements because it is one of the few methods that gives reliable performance for long periods of time in the low ppm and ppb regions.

An electrolytic hygrometer electrolyzes water vapor into its components, hydrogen and oxygen. The amount of electrical current required to dissociate water vapor at a particular flow rate is proportional to the water concentration in the sample. Electrolytic hygrometers require a relatively clean input gas which usually necessitates the use of a suitable filter. The electrolytic cells should not be used with gases which react with phosphorous pentoxide, such as NH_3.

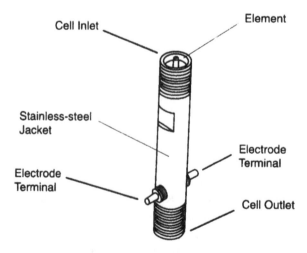

Figure 5.9   Electrolytic cell.
                   (Courtesy MEECO)

A. Introduction

The ability of an electrolytic hygrometer to measure water vapor content of a gas or gaseous mixture is based on the fact that a water molecule can be electrolyzed into molecular hydrogen and oxygen by the application of a voltage greater than two volts, the thermodynamic decomposition voltage of water. Since two electrons are required for the electrolysis of each water vapor molecule, the current in the electrolytic cell is proportional to the number of water molecules electrolyzed per second.

The electrolytic hygrometer has been commercially available for over 30 years, and is derived from the pioneering work done by F.A. Keidel[25] in the mid-1950s. It offers several advantages that make it well suited for trace moisture measurements in gases:

- It is a "fundamental" measurement technique, in that calibration of the electrolytic cell is not required. However, like chilled mirror hygrometers, annual recertification against a transfer standard is generally recommended.

- It is specific to water vapor.

- It provides good accuracy in the low ppm_v and ppb_v range at a reasonable cost.

In earlier years the application of the traditional electrolytic hygrometer has been somewhat limited because of problems associated with the design of the electrolytic cell:

- The instrument requires precise control of flow rate through the electrolytic cell. It is also highly sensitive to plugging and fouling from contaminants. The electrolytic cell sometimes requires factory rejuvenation after it is contaminated.

- Modern electrolytic cells are of epoxyless construction to minimize hygroscopic properties and improve response time. Their cost is not insignificant, but considerable progress and cost reductions have been accomplished in recent years resulting in many new applications.

B. Electrolytic Cell

A typical electrolytic hygrometer utilizes a sensing cell coated with a thin film of phosphorous pentoxide (P_2O_5), which absorbs water from the sample gas. A typical cell is shown in Figure 5.9. The cell has a double helix winding of inert electrodes on a glass capillary as shown in Figures 5.10 and 5.11. Direct current applied to the electrodes dissociates the water, absorbed by the P_2O_5, into hydrogen and oxygen. Two electrons are required to electrolyze each water molecule, and the current in the cell represents the number of molecules dissociated per second. A further calculation, based on flow rate, and current, yields the parts per million by volume concentration of water vapor (see Section C, Theory of Operation).

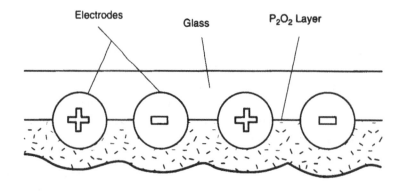

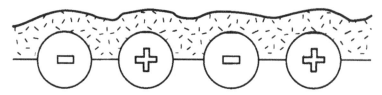

Figure 5.10   Electrolytic sensing cell.

A close-up view of the inner construction of the electrolytic cell is shown in Figure 5.11.

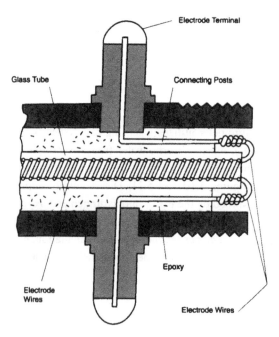

Figure 5.11   Inner construction of electrolytic cell.
                    (Courtesy MEECO)

To obtain accurate data, the flow rate of the sample gas through the cell must be known and properly controlled. The cell responds to the total amount of water vapor in the sample. It is most important that all components and sample line materials that come in contact with the sample gas prior to the sensor, be made of inert material, such as polished stainless steel. This is to minimize absorption/desorption of water. Most other materials are hygroscopic and act as a sponge at the low water vapor levels that are to be measured causing outgassing, errors, and a slow response time.

The electyrolytic cell was originally developed by the Du Pont Company and patented in 1958.

C. Theory of Operation

Operation of the electrolytic (phosphorous pentoxide) cell is based on Faraday's law as can best be understood using the example below:

Faraday's Law

In operation, the sample gas enters the electrolytic cell and the P_2O_5 absorbs all water vapor molecules present in the gas flow. A voltage, applied across the electrode terminals, electrolyzes the moisture in the film. Once equilibrium is reached, the rate at which water molecules enter the cell will exactly match the rate that molecules are electrolyzed. Each electrolyzed water molecule displaces two electrons from the anode to the cathode. The electrolysis current in Amperes defines the electrical charge in Coulombs that is displaced per second. Therefore, since the elementary charge of an electron is known, a current measurement will indicate the rate at which water molecules enter the cell. If the rate of water molecules entering the cell is known, the moisture reading is expressed in parts-per-million (ppm) by comparing the rate of water molecules to the rate of total gas molecules entering the cell. The rate of total gas molecules entering the cell can be determined by controlling the flow rate through the cell. The relationship between measured current and moisture concentration in the electrolytic cell is governed by the absolute principles of Faraday's law and no calibration versus a moisture standard is needed like in secondary measurement devices, such as aluminum sensors. The relationship can be derived numerically as shown in the following example.

It is assumed that the gas flow rate through the element is 100 cc/min. (6.1 in.^{3}/min.) which is standard for most electrolytic cells. A standard temperature of 25°C (298.16 K), pressure of 1.0 atm (14.7 psi), and a moisture concentration of 1 ppm are also assumed.

The total number of moles of gas that flow through the element per minute can be calculated by using the ideal gas law.

$$PV = nRT$$

where, P = Pressure (atmosphere)
V = Volume in liters
T = Temperature (Kelvin)
R = 0.08205 (atmosphere liter Kelvin^{-1} mole^{-1})
n = Number of moles of gas (mole)

Therefore, since:

$$n = \frac{P \times V}{(R \times T)}$$

$$n = \frac{1.0 \times 0.1}{(0.08205 \times 298.16)}$$

$$n = 4.0851 \times 10^{-3} \text{ moles}$$

Per definition, a mole of gas contains 6.022×10^{23} molecules of gas. Therefore, the total number of molecules of gas that flow through the element per minute is:

$$4.0851 \times 10^{-3} \text{ moles} \times 6.022 \times 10^{23} \text{ molecules} = 2.46 \times 10^{21} \text{ molecules}$$

Since the gas contains 1 ppm of moisture (one out of a million molecules is a water molecule), the number of water molecules that enter the element per minute is:

$$\frac{2.46 \times 10^{21}}{1 \times 10^6} = 2.46 \times 10^{15} \text{ water molecules}$$

As previously indicated, to electrolyze one water molecule, two electrons have to be transported from one electrode to the other. Also, an electron represents one elementary charge or 1.6022×10^{-19} Coulombs. Therefore, the total amount of charge due to electrolysis flowing through the electrodes per minute is:

$$2.46 \times 10^{15} \times 2 \text{ electrons} \times 1.6022 \times 10^{-19} \text{ Coulombs} = 7.85 \times 10^{-4} \text{ Coulombs}$$

Current is measured in Amps. One ampere is defined as one Coulomb per second. Since the total charge per minute in Coulombs is known, the charge flowing per second can be determined, which is:

$$\frac{7.85 \times 10^{-4} \text{ Coulombs}}{60} = 13.138 \times 10^{-6} \text{ Coulombs}$$

This corresponds to a current of 13.138 μAmp.

Thus, in accordance with Faraday's Law, a gas containing 1 ppm of moisture flowing through the element at 100 cc/minute (6.1 in.^{3}/min.), results in an electrolysis current of 13.138 μAmp.

For these equations to hold true in practice, the following conditions must be met:

• Accurate control of flow rate through the element

• Accurate measurement of the μAmpere current

• All water vapor molecules passing through the element must be absorbed by the P_2O_5.

D. Conventional Implementations

Most commercially available electrolytic hygrometers are variations on the design proposed by Keidel in 1956. Figure 5.11 shows a cross section of a conventional electrolytic cell. Two equally-spaced platinum wires are helically wound on a thin mandrel. A fluorocarbon extrusion over the windings holds them in place and forms a substrate for the hygroscopic material. Removal of the mandrel leaves a small tube of near capillary diameter. The inner surface of the tube is then wetted with a phosphoric acid solution. A DC current applied to the wires (electrodes) electrolyzes the solution and leaves the inner surface coated with phosphorous pentoxide (P_2O_5), a salt with a very high affinity for water. During operation, the P_2O_5 absorbs water vapor from the sample gas, producing a form of phosphoric acid. The water in the phosphoric acid is simultaneously electrolyzed to hydrogen and oxygen by the application of a DC potential across the electrodes, which, in turn, converts the phosphoric acid back to P_2O_5.

   In recent years, rhodium electrodes have in some cases replaced the original platinum wire electrodes. The use of rhodium electrodes virtually eliminates a problem known as recombination, which is the reverting to water of the products of electrolysis in a hydrogen-rich gas stream. This would produce relatively large errors in the instrument's readout as the recombined water is electrolyzed. It is theorized that platinum acts as a catalyst in a reaction between the oxygen product of electrolysis and hydrogen chemisorbed on the platinum to form water. Rhodium also minimizes another deleterious effect, the tendency of metallic ions to migrate under the influence of the applied potential between the electrodes. This results in the formation of metallic compounds and eventual bridging of the electrodes.

   If the efficiency of the electrolysis is 100% (all of the moisture in the sample is absorbed and electrolyzed), then the accuracy of the conventional electrolytic hygrometer is limited by its ability to control mass flow rate precisely, and the accuracy of the supporting electronics circuitry. Since, from the equation, the ppm_v calculation is based upon mass flow rate, an error in holding the mass flow rate to a constant value results in a direct error in the calculation of ppm_v. Thus the control of sample flow rate is of critical importance in the design of a conventional electrolytic hygrometer.

   Figure 5.12 shows the flow schematic for a typical electrolytic hygrometer. The sample gas is split into two streams at the inlet to the instrument. Constant volumetric flow is maintained in the electrolytic cell using a differential pressure regulator, adjustable restrictor, and rotameter. The remainder of the sample gas bypasses the electrolytic cell and is exhausted at the instrument outlet, along with the dry gas coming from the cell. If the rotameter is typically adjusted to exactly 100 cc/min. at 25°C (77°F) and 1 atm., then 13.2 µAmperes will be required to electrolyze one ppm_v of water. Other arrangements use a differential pressure regulator to maintain constant flow for small pressure fluctuations at the sample inlet. In some cases, a flow rate of 10 cc/min (0.61 in.³/min.). has been used, which has the advantage of a slower contamination rate in the cell.

   In commercially available electrolytic hygrometers, the flow rate is controlled on a volumetric rather than a mass basis. The flow meter supplied is usually calibrated for air at 20°C (68°F) and 1 atm. Thus, it is critical that the flow meter be re-calibrated to account

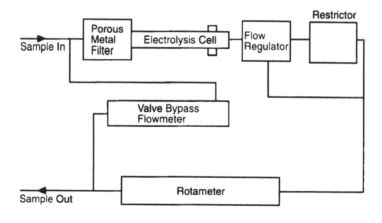

Figure 5.12   Schematic of a typical electrolytic hygrometer.

for gases other than air and temperatures or pressures other than "standard" conditions. This re-calibration can be done by a rough "correction factor" supplied by the manufacturer, or by an accurate but rather time-consuming procedure using the gas displacement or soap bubble method.

E. Application Problems

1. *High moisture levels*
   Early cell failure is often caused by high moisture levels resulting from a shorting of the windings caused by one electrode contacting the other. This is usually characterized by the formation of black deposits in the inter electrode space. When this occurs, the cell must be cleaned and re-coated with P_2O_5. Hence this sensor is generally not suitable for high moisture levels, i.e., higher than 2,000 ppm.

2. *Hydrogen recombination*
   When this happens, the electrically generated oxygen reacts with hydrogen to form additional water. This can occur in gas streams where the major constituent of the gas is hydrogen. In recent years leading manufacturers of electrolytic cells have introduced cells utilizing two steps to cancel the recombination effect, which is flow dependent, using a microprocessor to perform the calculations. Sensors have also been introduced that reduce the effect itself.

F. Applications

Typical minimum sample pressure which is required for this type of control is 68.95 kPa to 689.5 kPa (10 psi_g to 100 psi_g). If the sample pressure is more than 689.5 kPa (100 psi), an external upstream regulator must be employed. A filter is normally located upstream to prevent particulate contaminants from reaching the cell.

A typical electrolytic hygrometer can cover a span from a few ppb_v to 2,000 ppm_v, with an accuracy of ± 5% of reading or typically 1% of full scale. This is more than adequate for most industrial applications. The sensor is suitable for many applications involving clean, noncorrosive gases.

The electrolytic hygrometer can be used with most inert elemental gases, and many organic and inorganic compounds that do not react with P_2O_5. Suitable gases include air, nitrogen, hydrogen, oxygen, argon, helium, neon, carbon monoxide, carbon dioxide, sulfur hexafluoride, methane, ethane, propane, butane, natural gas, and certain freons. Unsuitable gases include corrosive gases and those which readily combine with P_2O_5 to form water (such as alcohols). Also to be avoided are certain acid gases, amines, and ammonia, which can react with the P_2O_5, and certain unsaturated hydrocarbons (alkynes, alkadienes, and alkenes higher than propylene) which can polymerize to a liquid or solid phase and clog the electrolytic cell.

Some typical applications include:

- Bottled gases, often used as a QC instrument

- Plant, instrument, and process air

- Semiconductor industry—process gases

- Glove boxes for manufacture and encapsulation of electronic components

- Light bulb manufacturing—inert gases

- Electric power industry—transformer gases (SF_6)

- Aerospace industry—argon and helium welding

When the electrolytic hygrometer is applied to clean gases in the low ppm_v region, the instrument offers long and reliable service. However, if misapplied, the electrolytic cell can be damaged. Although procedures are furnished by the manufacturers of these cells for unclogging the capillary and re-coating with phosphorous pentoxide, this is not a trivial procedure. In some cases, a failed electrolytic cell must be replaced. On the other hand, since the electrolytic cell is essentially a primary device, no calibration is required when replacing the unit. Compared to the aluminum oxide sensor, the electrolytic cell, since it utilizes a primary method of measurement, has better long-term accuracy, exhibits no hysteresis, and is easier to keep in calibration. However, a sampling system is required which must be designed using special nonhygroscopic materials to prevent errors or a very slow response time. The aluminum oxide sensor is more convenient to use when "in line" monitoring is desired, or when a multiple sensor system is preferred.

G. Advantages

The electrolytic hygrometer offers several advantages for the measurement of trace water vapor in gases:

- It is considered a "primary" measurement technique.

- It is specific to water vapor.

- It can provide good accuracy in the low ppm_v and ppb_v range if precise flow control and good sample conditioning are employed.

- It is lower in cost than other techniques of comparable accuracy, such as the chilled mirror hygrometer.

H. Limitations

The conventional electrolytic hygrometer has the following limitations:

- The cell cannot be exposed to high concentrations of water vapor for any length of time, since this results in high usage rates of the the P_2O_5. Cell life is considered to be roughly proportional to the parts-per-million levels measured and number of hours the cell is operated at these ppm levels.

- The instrument requires precise control of the mass flow rate of the sample gas, since an error in flow rate translates directly to an error in ppm_v.

- The geometry of construction of the electrolytic cell makes it susceptible to contaminants. If the cell becomes clogged or coated, or the electrodes short-circuit or break, it is not a trivial procedure to rejuvenate the cell. This often necessitates a return of the cell to the manufacturer for repair or replacement. Replacement costs are relatively high compared to the total cost of the instrument.

- Although the electrolytic cell can usually be reactivated after neutralization of the P_2O_5, the ease of performing this procedure varies significantly from manufacturer to manufacturer.

Typical commercial electrolytic hygrometers are shown in Figures 5.13 and 5.14.

Figure 5.13   Electrolytic hygrometer.
(*Courtesy Edgetech/EG&G*)

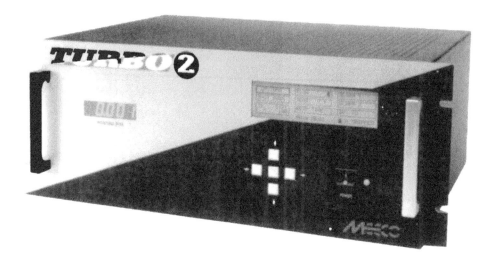

Figure 5.14   Electrolytic low ppm/ppb hygrometer.
(*Courtesy MEECO*)

6

OPTICAL ABSORPTION HYGROMETERS

This chapter discusses instrumentation based on water vapor absorption of radiation in certain optical wavelengths. The two most often used absorption-type optical hygrometers are the infrared and ultraviolet instruments. The latter is often referred to as the Lyman–Alpha hygrometer, indicating a specific absorption line in the ultraviolet spectrum.

Chilled mirror hygrometers, because of their use of optical detection techniques, are sometimes also referred to as optical hygrometers. However, they are not spectral absorption-type instruments and are discussed separately in Chapter 3.

I. Infrared Hygrometer

Transmission of infrared and ultraviolet radiation through a gaseous medium does not occur uniformly at all wavelengths. Selective absorption of radiation occurs in absorption bands which are characteristic of the particular gas through which the radiation is passing.

Operation of the infrared hygrometer is based on the absorption of infrared radiation at certain distinct frequency bands as a result of the presence of water vapor in the gas through which an infrared beam is passed. Water vapor strongly absorbs infrared radiation in bands centered at 2.7 μ and 6.3 μ. The feasibility of measuring water vapor using infrared absorption measurements was first demonstrated by Fowle in 1912,[1] but did not result in practical instrumentation until the 1980s.

A. Operation

Infrared optical hygrometers are noncontact instruments. The sensing element is protected with a sapphire window and never contacts the air or gas that is being sampled. It is therefore a method suitable for monitoring dew point in highly contaminated areas.

Operation of an infrared hygrometer is usually based on the dual-wavelength differential absorption technique. This requires identification of a primary wavelength within the most intense portion of the water vapor absorption band and a nearby reference wavelength at which water vapor absorption is negligible. The differential absorption technique involves computing the ratio of the transmission measured at the primary wavelength to that measured at the reference wavelength.

This ratio can be considered a normalized transmission which is a direct measure of the water vapor absorption, and is insensitive to drifts in electro-optical sensitivity, deposits on the windows and haze or fog in the sample volume. It is based on the assumption that transmission losses due to gases other than water vapor are the same for both the primary and reference wavelength bands.

The two strongest infrared water vapor absorption bands are those centered at 6.3 μ and 2.7 μ. The 2.7 μ band is used in most IR hygrometers. Figures 6.1 and 6.2 show the water vapor absorption spectrum and the absorption spectrum of the uniformly mixed atmospheric gases in the 2–3 μ region.

From the results shown in Figure 6.1, the absorption maxima at 2.67 μ and 2.60 μ are suitable as primary wavelengths while the region 2.30 μ to 2.50 μ is the obvious choice as the reference wavelength. Further analysis indicates that the 2.60 μ and 2.50 μ wavelengths are best suited for the IR hygrometer. The 2.67 μ line is close to the absorption line for CO_2. By selecting 2.60 μ instead, a heavily polluted environment containing CO_2 would cause no errors. The effects of natural aerosols, hazes and fog are quite uniform over this range and much smaller than the absorption from water vapor.

The water vapor absorption spectrum shown in Figure 6.1 is for a path length of 1 meter, at a pressure of 1 atm. The five curves represent humidity levels of –40°C, –20°C, 0°C, +20°C, and +40°C dew points. The resolution is 20 cm^{-1} or 0.015 μm at 2.7 μm.

The absorption spectrum shown in Figure 6.2 is for the uniformly mixed gases CO_2, N_2O, CO, CH_4, and O_2, for a path length of 1 meter, at a pressure of 1 atm. The only atmospheric absorption feature in this region is due to CO_2. Beer's law states that the amount of radiation that is absorbed by the gas element is proportional to the concentration of the absorbing gas element and to the path length. This then forms the basis for measuring and calculating humidity from IR spectral information. The sensitivity of the IR hygrometer is determined by the derivative of the transmission function. It decreases exponentially with decreasing absolute humidity.

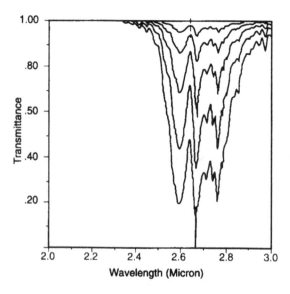

Figure 6.1   Water vapor absorption spectrum.

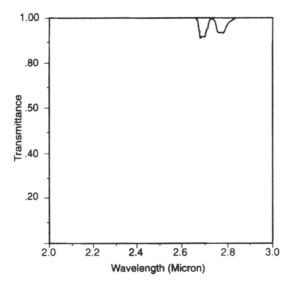

Figure 6.2   Absorption spectrum for uniformly mixed gases.

B. Advantages

The infrared hygrometer offers the following advantages:

- It is a noncontact instrument and is virtually unaffected by contaminants.

- The sensing element is protected by a sapphire window and never comes in contact with the gas which is being sampled.

- Sensitivity can be increased by increasing the path length. Low humidity level measurements require a long path length, sometimes more than one meter. To prevent an odd size for the instrument, the folded path version is customarily used. Using this technique, the beam will be reflected and bounced back and forth one or more times using reflectance mirrors. Hence the beam will travel the same long distance, but the size of the instrument will be reduced.

- It offers a fast response at high and low absolute humidities.

- There is virtually no drift and no hysteresis.

- If properly calibrated, it can provide good long term accuracy.

- Operation is available over a broad humidity range.

C. Limitations

Disadvantages of the infrared hygrometer are:

- The infrared hygrometer is more expensive than the chilled mirror hygrometer.

- As a secondary measurement device, it must be periodically calibrated.

- Optical windows can deteriorate over time or as a result of environmental conditions.

- The design is relatively complex.

- For low frost point measurements, the detector must be cryogenically cooled, adding further cost.

A typical infrared hygrometer is shown in Figure 6.3. In this instrument the infrared source and detector are isolated from the atmospheric sample. Infrared radiation leaves and re-enters the enclosure through sapphire windows. Sapphire is chosen because of its excellent stability and immunity to environmental attack.

D. Applications

Because of the limitations listed above, infrared hygrometers have not been used in a wide range of applications. Successful applications include outdoor meteorological measurements in areas with heavy contaminants, such as in the presence of smoke stacks, oil vapors and dust. The instrument has also been used in selective industrial applications where the sensor has to operate in heavily contaminated areas or gas streams.

Figure 6.3   Environmental infrared hygrometer.
(Courtesy Ophir Corporation)

II. Lyman–Alpha Hygrometer

A. Moisture Measurement

Like the infrared hygrometer, the Lyman–Alpha hygrometer operates on the principle of water vapor absorption of radiation in a narrow optical band, in this case the ultraviolet region.

The instrument is mostly used for atmospheric measurements. It provides a means of measuring atmospheric humidity over a wide range of values with high resolution in time and space. Response times of a few milliseconds and spatial resolution on the order of one centimeter can be achieved over a range of –80°C to +40°C (–112°F to 104°F) dew point.

The Lyman–Alpha hygrometer consists of a radiation source and detector, separated by intervening space, usually in the order of several millimeters to a few centimeters, windows on both the source and detector, along with an amplifier and supporting electronics. Lyman–Alpha technology covers the 120 nm spectral range, commonly referred to as the "vacuum ultraviolet" range. The Lyman–Alpha emission line of atomic hydrogen results from the decay of the electron of a hydrogen atom to its ground state, i.e., lowest energy orbit, from its next higher energy orbit. This radiation is often used to measure water vapor since, at its 121.56 nm wavelength, water vapor absorption is high. Oxygen absorption is uniquely low, and most other common gases are relatively transparent, e.g., nitrogen. A block diagram of a Lyman–Alpha hygrometer is shown in Figure 6.4.

A schematic drawing of the humidity sampling system is shown in Figure 6.5. The design allows for continuous variation of the path length in the range of, for example, 5 mm to 2 cm. Most of the pioneering development work on the Lyman–Alpha hygrometer has been done at NCAR and by Arden Buck of Buck Research, both in Boulder, CO (see list of references in Chapter 15).

B. Windows

The most common Lyman–Alpha radiation source is a DC-excited, cold-cathode glow discharge hydrogen lamp because it is simple and easy to drive. The detector is usually a photo ionization chamber containing ionizing gas and two electrodes. Incoming radiation

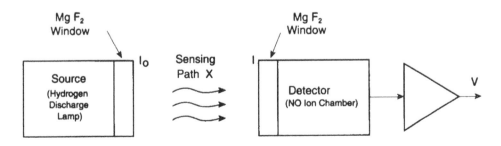

Figure 6.4 Block diagram of Lyman–Alpha absorption hygrometer.

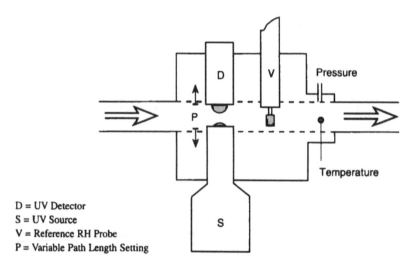

D = UV Detector
S = UV Source
V = Reference RH Probe
P = Variable Path Length Setting

Figure 6.5   Schematic of atmospheric Lyman–Alpha hygrometer.

ionizes the gas, and an electrical field is maintained between the electrodes to induce electron and ion drift. The resultant current is proportional to the incident light intensity.

With most materials opaque to Lyman–Alpha radiation, special window materials
must be used. The most common windows are about 1 mm thick crystals of magnesium
fluoride. Cut-off wavelengths for the windows are 132 nm and 115 nm which nicely sur-
round the Lyman–Alpha wavelength of 121.56 nm. They provide about 55% transmission,
deteriorating with time at the rate of roughly 0.5% to 5% per hour of operating time. This
is probably due to the reaction of atmospheric constituents with magnesium fluoride in the
presence of UV stimulation. Fortunately, the window degradation can be partially reversed,
since it occurs mostly on the outside of the source and detector. Window transmission can
be partially restored by washing with alcohol or acetone and lightly polishing with a fine
abrasive. One of the most difficult problems which limits accurate Lyman–Alpha measurements, is the contamination of windows by the environment, or by handling of the
window. The very large absorption coefficient of liquids and solids means that any oils,
greases, and dirt may cause serious problems. Even differential methods will fail unless
both paths are exposed to the same environmental absorption, such as atmospheric organics. Any invisible fingerprints will cause absolute errors of several tens of percentage. A
typical window transmission curve is shown in Figure 6.6.

C. Detectors

Both ionization chambers and photo multipliers can be used as detectors in this region of
the spectrum. Nitric oxide ionization chambers provide some wavelength discrimination.
Their low wavelength cutoff is determined by the window material, and the upper wavelength cutoff by the ionization potential of the gas. These devices are rather simple but
require careful construction and pure gases to prevent short lifetimes and sensitivity outside the desired wavelength region. The detector is usually a photo ionization chamber
containing ionization gas and two electrodes. Incoming radiation ionizes the gas, and an

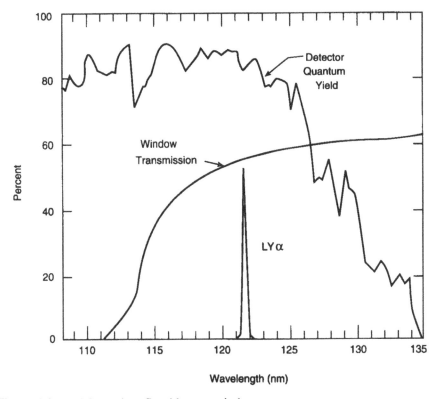

Figure 6.6   Magnesium fluoride transmission curve.

electrical field is maintained between the electrodes which induces electron and ion drift. The resultant current is proportional to the incident light intensity.

Photo multipliers can be used in this wavelength range to provide much higher output currents and noise levels equivalent to a few photons per second. The main problems are size, expense, a higher long wavelength cutoff than for ion detectors, and the requirement for a very highly-regulated high voltage power supply.

D. Sources

UV source design is a highly specialized field. Improper design can seriously affect the life expectancy of the device, and in time can cause changes in the spectrum. Poor window seals can result in leaks or outgassing of the sealing material, especially if it is an epoxy. The most common Lyman–Alpha radiation source is a DC-excited, cold-cathode glow discharge hydrogen lamp with a hydride reservoir because of its simplicity and ease of drive. When held at constant temperature, the hydride maintains a very predictable hydrogen pressure in the source. To provide reasonable lamp life in conventional sealed sources, moderately high hydrogen pressure in the order of about 1 kPa is required. However, low hydrogen pressures produce a spectrally cleaner Lyman–Alpha emission, since the resultant longer mean-free path increases the ratio of free atoms to molecules and gives a higher atomic emission relative to the extraneous molecular emission lines. A cutaway view of a photo ionization source is shown in Figure 6.7.

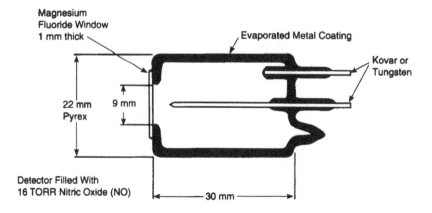

Magnesium
Fluoride Window
1 mm thick

Evaporated Metal Coating

Kovar or
Tungsten

22 mm
Pyrex

9 mm

Detector Filled With
16 TORR Nitric Oxide (NO)

30 mm

Figure 6.7   Cutaway view of ionization chamber.

A photograph of the ionization chamber is shown in Figure 6.8.

E. Calibration

Calibration of both the Lyman–Alpha hygrometer and the field reference probe are usually performed simultaneously in an enclosed airflow circuit. A chilled mirror hygrometer or psychrometer may be used as the calibration standard.

When a reference hygrometer is not available, a variable path length self-calibration technique can be used. However, this method is rather complex and not used very often.

F. Performance

The Lyman–Alpha hygrometer is a very fast-responding instrument and is therefore frequently used for fast-moving aircraft and cloud study measurements. A typical response time is 6 milliseconds. The instrument also offers a high resolution, typically 0.2% relative humidity and 0.03°C (0.054°F) dew point.

Figure 6.8   Photograph of an ionization chamber.
                   (*Courtesy Buck Research and NCAR*)

Typical accuracies are 4% RH and ± 0.6°C (± 1.08°F) dew point when calibrated against an accurate standard such as a chilled mirror hygrometer.

The system can be used over a dew/frost point range of –90°C to +50°C (–130°F to +122°F) at an environmental temperature of –25°C to +70°C (–13°F to +158°F) and at pressures of about 50 mb to 1200 mb (altitudes up to 6 km, or 20,000 ft).

The hygrometer has been used primarily on aircraft to obtain time series and spectral information. It has also been used on towers to obtain latent heat fluxes or measurements of refractive index structure. A drawback is that the Lyman–Alpha hygrometer is a secondary measurement device and must be periodically calibrated to provide accurate measurements. Instrumentation developed and used by NCAR (National Center for Atmospheric Research) is shown in Figure 6.9.

NCAR Lyman–Alpha instrumentation used on aircraft for atmospheric research is shown in Figure 6.10. An instrument designed for tower use is shown in Figure 6.11.

G. Summary

An ideal Lyman–Alpha instrument offers:

- Fast response time of less than 0.1 seconds

- Very short path length (mm to cm)

- Use of ion chamber detectors

- Use of simple, room temperature, low power sources and detectors for the hydrogen Lyman–Alpha emission line (as opposed to the frequent need for cryogenically-cooled devices in infrared systems)

- Detection specific to water vapor response at high and moderate humidities

- Small size and light weight.

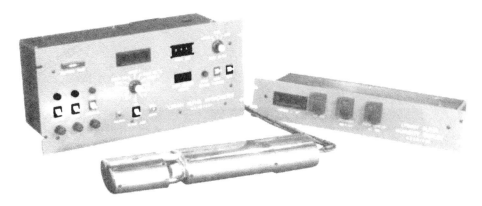

Figure 6.9   UV hygrometer.
            (Courtesy NCAR)

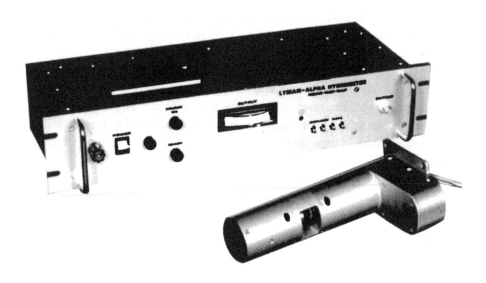

Figure 6.10   Lyman–Alpha instrument for aircraft use.
                      (*Courtesy NCAR*)

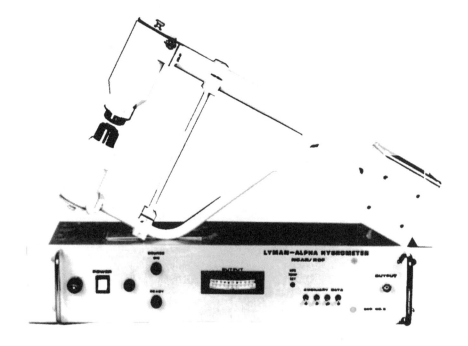

Figure 6.11   Lyman–Alpha hygrometer for use on a tower.
                      (*Courtesy NCAR*)

H. Advantages

The main advantages of the Lyman–Alpha hygrometer are:

- It offers a very fast response, even at very low frost point levels.
- It is useful for high altitude and cloud measurements when there is water in vapor as well as in droplet form.
- It can make accurate measurements at very low frost points and high altitudes (low pressure).

I. Disadvantages

Limitations of the Lyman–Alpha hygrometer are:

- Sensitivity to oxygen at low humidities
- Difficulty making the required magnesium fluoride windows which degrade with use
- Relatively high cost
- Secondary nature of measurement (dependent on calibration)

J. Applications

The Lyman–Alpha hygrometer is an important instrument for making certain highly-specialized environmental measurements, but has been used in few other applications, mostly because of its high cost and relative complexity, especially the source, detector and window designs. There is one known commercial manufacturer of this product in the USA (AIR, Inc. in Boulder Colorado). The instrument is shown in Figure 6.12.

Almost all Lyman–Alpha hygrometers have been built on a special basis for specific applications, by research or government laboratories.

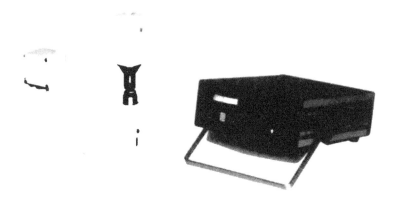

Figure 6.12   Commercial Lyman–Alpha hygrometer.
(*Courtesy AIR, Inc.*)

7

DRY/WET-BULB PSYCHROMETER

I. General Description

Psychrometry has long been a popular method for monitoring humidity, primarily due to its simplicity and inherent low cost. A typical industrial psychrometer consists of a pair of matched electrical thermometers, one of which is enclosed by a wick which is maintained in a wetted condition using distilled water. Air is passed over both thermometers, usually at a rate of about 200 m/min. (650 ft./min.) or more. The resultant evaporative cooling produces a wet-bulb temperature approximately equal to the thermodynamic wet-bulb temperature. The difference between the dry-bulb and wet-bulb temperature is called the wet-bulb depression which determines the humidity level. Charts converting wet bulb depression to relative humidity are shown in Chapter 13. Conversions can also easily be made using the computer disk in the back cover of this book.

In a well maintained psychrometer, with thermometer accuracies of ± 0.2°C (± 0.38°F) and within a temperature range of +5°C to + 80°C (41°F to 176°F), the relative humidity can be determined with an accuracy of about ± 3% RH.

In a so-called "sling psychrometer" as shown in Figure 7.1, the two thermometers are mounted side-by-side on a frame which is fitted with a handle for whirling the device through the air. The motion is stopped in order to take a reading of the thermometers. It is evident that such a measurement cannot be considered reliable since it is operator dependent. For low humidities, the wet-bulb depression is large and a high air velocity is needed to obtain the proper depression. In a ventilated or motor aspirated psychrometer, a stationary air flow is controlled and more accurate measurements can be made.

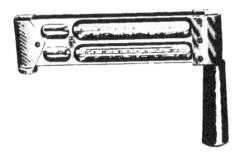

Figure 7.1   Sling psychrometer.[78]*

---

* Calibration section

Accuracy of a psychrometer is strongly dependent on the accuracy of the thermometers used. For most precise measurements platinum resistance thermometers are often chosen. Thermistors and thermocouples are also widely used. The absolute accuracy of the thermometers is of less importance than the ability to accurately measure the wet-bulb depression, i.e., the difference between dry and wet bulb measurements. This value is extremely small at very high humidities.

For air temperatures below 0°C (32°F), the water on the wet-bulb wick may either freeze or supercool. Since the wet-bulb temperature is different for ice than for water, its state must be known and the proper charts or tables used. An alcohol solution could be added to the water in the reservoir to avoid freezing. However this could affect the accuracy of the device.

Psychrometers can be used at temperatures above the boiling point of water (100°C or 212°F), but if the wet-bulb depression is large, the wick must nevertheless remain wet and water supplied to the wick must be cooled en route so as not to influence the wet-bulb temperature by carrying sensible heat to it. Such measurements are difficult to make with high accuracy.

Generally speaking the psychrometer is a fundamental measurement system and, if properly operated with accurately calibrated thermometers, a psychrometer such as the Assman laboratory type, can provide very accurate, reliable, and repeatable measurements. For this reason the psychrometer has frequently been used in the past as a fundamental calibration standard. However most operators, especially in industry, do not take the time and necessary precautions and obtain far less accurate and less reliable results. For this reason the psychrometer has often been replaced with more modern systems, though it is one of the least expensive instruments for industrial use. One of the original aspirated Assman psychrometers is shown in Figure 7.2.

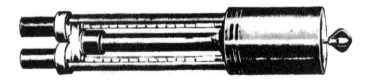

Figure 7.2   Assman psychrometer.[78*]

---

* Calibration section

II. Operation

A. Theory

Two temperatures are to be measured, i.e., the temperature of the sample gas, which is called the dry bulb or ambient temperature, and the temperature of a thermometer covered with a porous substance (wick) which is saturated with pure water and in equilibrium with the sample gas which is passed over it at an adequate velocity. The depression of temperature of the wet-bulb versus the dry-bulb can be correlated with water vapor pressure using the equation:

$$e = e_w - 66 \times 10^{-5} P(T_a - T_w)(1 + 115 \times 10^{-5} T_w)$$

where:

e = vapor pressure of water in the sampled gas

e_w = is the saturation vapor pressure of water at temperature t_w

P = is the atmospheric pressure

T_a = dry bulb temperature in °C

T_w = wet bulb temperature, in °C

This equation is only valid if the thermometers are in a gas stream of sufficient velocity (about 200 meters/min. or 650 ft/min.). At low velocities, errors in measurement could easily occur. The sensors have in the past often been used in meteorology, with the thermometers placed in a hand-held vane which can be rapidly rotated to achieve the necessary air velocity for the measurement, i.e., a simple, inexpensive, manually operated measurement device. In industrial versions, platinum resistance thermometers are used for their high accuracy, with the "wick" being fed from a reservoir of distilled water. This reduces the frequency of routine maintenance. Humidity values can be obtained from a psychrometric chart (see Chapter 13). In some cases the simple depression of temperature can be used effectively to control dryers. In a properly designed psychrometer, both sensors are thermally shielded to minimize errors from thermal radiation.

Because of its earlier extensive use as a standard, a considerable amount of literature has been generated on this subject and many theories have been developed to explain the operation of the psychrometer. For further study, reference is made to the bibliography in Chapter 15.

B. Advantages

The psychrometer does have certain inherent advantages, i.e.:

- The psychrometer attains its highest accuracy near 100% RH. Since the dry bulb and wet bulb sensors can be connected in a differential mode, this allows the wet-bulb depression (which approaches zero as the relative humidity approaches 100%) to be measured with high accuracy. Although large errors can occur if the wet bulb becomes contaminated or improperly fitted, the simplicity of the device affords easy repair at minimum cost.

- The psychrometer can be used at ambient temperatures above 100°C (212°F), and the wet-bulb measurement is usable up to a few degrees below the boiling point.

- It is a fundamental measurement service

- Its cost is low

C. Limitations

Major shortcomings of the psychrometer are:

- As the relative humidity drops to values below about 15% RH, the problem of cooling the wet-bulb to its full depression becomes difficult. The result could be impaired accuracy below 15% RH, and few psychrometers work well below 5% to 10% RH.

- Wet-bulb measurements at temperatures below 0°C (32°F) are difficult to obtain with a high degree of confidence. Automatic water feeds are not feasible due to freeze-up.

- Because a wet-bulb psychrometer is a source of moisture, it can only be used in environments where added water vapor from the psychrometer exhaust is not a significant component of the total volume. Generally speaking, psychrometers cannot be used in small, closed volumes.

- Most physical sources of error, such as dirt, oil or contamination on the wick, insufficient water flow, etc., tend to increase the apparent wet-bulb temperature. This results in the derived relative humidity being higher than the actual relative humidity.

- Errors in psychrometric practice and derived %RH

III. Error Analysis

A. Temperature Errors

Errors in humidity occur when the wet- or dry-bulb have a calibration error. The most serious errors happen when these errors combine to cause an appreciable error in the depression. Errors of this kind can be minimized by matching the thermometers or applying calibration corrections to the indicated temperature values. As an example, a depression error of 0.1°C (0.18°F) results in an error of about 1% RH at 20°C (68°F) ambient. A 0.2°C (0.38°F) depression error results in a 2% RH error.

B. Pressure Errors

Large errors will arise if pressure corrections are not applied or if inappropriate pressure correction tables are used. For example, in the case of a table valid for 1000 mb, a correction of 10% must be applied to the depression values for 900 mb.

C. Radiation Errors

Errors due to radiational heat exchange between the surroundings and the wet-bulb have been discussed by various authors (see list of references). Errors from this cause are generally independent of the ventilation speed. In a normal psychrometer, precautions are taken to shield the wet-bulb from external sources of radiation, but the shielding is at or near the ambient temperature. The radiation correction will therefore be a function of the depression and will be at a maximum in dry warm conditions when the depressions are large. The effect of radiation must be determined by theoretical empirical considerations for a specific psychrometer.

D. Errors Arising From Other Sources

The errors mentioned above are the most important in normal psychrometric practice. However, conduction of heat into the wet-bulb along the sensing element may also cause large errors when metallic sensing elements are used, and particularly in the case of thermocouple elements, unless precautions are taken to insulate the sensing element from its supports, or cool the apparatus to the wet-bulb temperature.

E. Accuracy

The overall accuracy that may be expected from a ventilated, properly designed psychrometer, when careful attention is given to the various error sources, is in the order of ± 2% when temperatures are above the frost point. Below frost point, the errors increase with decreasing temperature.

Where psychrometers are used under laboratory conditions and extreme care is given to all factors that could contribute to errors, a higher accuracy can be obtained. Where psychrometers are used under industrial conditions, the errors experienced are generally much higher than indicated above.

IV. Applications

A properly designed and utilized psychrometer, such as the Assman laboratory type, is capable of providing accurate data. However, very few industrial psychrometers meet these criteria. They are therefore limited to applications where low cost and moderate accuracy are the main requirements.

V. Summary

Cooling by the evaporation of water into the air has been known and applied for more than 200 years. The specific use of the wet bulb depression as a measure of humidity in the atmosphere was first mentioned in the mid 18th century and subsequently gave rise to the well known psychrometer. This instrument, even today, is a very practical and frequently used system. The psychrometer, if properly designed and operated, is a fundamental measurement device and is still often used as a laboratory calibration standard, although through the years it has been replaced to a large extent by the easier to operate, but more expensive, chilled mirror hygrometer.

8

OTHER HUMIDITY INSTRUMENTS

I. Saturated Salt (Lithium Chloride) Sensor

A. General Discussion

The Saturated Salt, also called Lithium Chloride, sensor, trademarked by the Foxboro Company as the "Dew Cell," has in the past been one of the most widely used dew point sensors. In earlier years it was a popular choice for air conditioning systems, due to its inherent simplicity, ruggedness, low cost, and ability to be "reactivated." The sensor also found many applications in outdoor meteorological applications and was for many years the standard for the National Weather Bureau. In the 1980s and 90s saturated salt sensors have been more and more replaced with bulk polymer %RH sensors and also by chilled mirror hygrometers. However the sensor still has many uses and a brief discussion in this book is warranted.

B. Theory of Operation

The sensor measures the dew point or partial pressure of water vapor, but not the relative humidity, though its operation is based on the equilibrium relative humidity of a salt. Humidity determination results from the fact that, for every water vapor pressure in contact with a saturated salt solution, there is an equilibrium temperature, which is always higher than the ambient temperature, and at which the solution neither absorbs nor gives up moisture. This equilibrium humidity is characteristic for the salt used (see saturated salt tables in Chapter 10). Below the equilibrium temperature, the salt solution absorbs water, and above this temperature, it evaporates water vapor.

The sensor consists of a thin walled tube (bobbin) covered with an absorbent fabric and a bifilar winding of inert electrodes which is coated with a dilute solution of lithium chloride. A conventional temperature sensing element (RTD, thermistor, or thermocouple) is placed inside the bobbin. An alternating current is passed through the winding and hence through the salt solution causing resistive heating. As the bobbin heats, water evaporates into the surrounding atmosphere from the diluted lithium chloride solution; the rate of evaporation is controlled by the vapor pressure of water in the surrounding air. As the bobbin begins to dry out, due to evaporation, the resistance of the salt solution increases causing less current to flow through the winding, and this allows the bobbin to cool. In this manner, the bobbin alternately heats and cools until an equilibrium is reached where the bobbin neither takes on, nor gives off, any water. This equilibrium temperature of the bobbin is a function of the prevailing water vapor or dew point of the surrounding air. A simple, offset calculation is needed to arrive at the correct dew point. Lithium Chloride is the only salt that is widely used for this type of sensor, because it reaches its natural equilibrium humidity at approximately 11% RH, which gives it a broad range (11% RH to 100% RH).

This measurement method is known by the following names: Lithium Chloride Sensor, Saturated Salt Sensor, and Dew Cell.

Some attention must be paid to flow rates when applying the saturated salt dew point sensor. High flow rates tend to extract heat from the sensor, causing it to give a low reading and this results in difficulties in maintaining the bobbin at the correct temperature. Low flow rates encourage stratification of water vapor in the humidity sensor, causing high readings. A flow rate in the range of 0.05 m/sec to 1 m/sec (10 ft/min to 200 ft/min.) will result in proper performance of the sensor. In certain applications, an adjustable sleeve may be used over the sensor to permit the flow rate to be brought within the desired range. An example of a saturated salt dew point sensor is shown in Figure 8.1.

Properly used, a saturated salt sensor is accurate to ± 1°C (± 2°F) between dew point temperatures of –12°C (10°F) and 38°C (100°F). Outside these limits, errors may occur due to multiple hydration characteristics of lithium chloride which produce ambiguous results at 41°C (106°F), –13°C (9°F), and –33°C (–29°F) dew points. Maximum errors at these ambiguity points are 1.4°C (2.5°F), 1.7°C (3°F), and 3°C (6°F), respectively, but actual errors encountered in typical applications are usually less.

Figure 8.1   Saturated salt sensor.
(*Courtesy General Eastern Instruments*)

C. Advantages

Advantages of the lithium chloride sensor include:

- Low cost

- Low sensitivity to contaminants

D. Limitations

Limitations of the sensor are:

- Low to modest accuracy

- Cannot be used in small contained areas (generates heat)

- Needs frequent service (sensor cleaning and sensitizing)

- Very slow response

- Limited range (cannot measure below 11% RH)

- Errors due to hydration at certain temperatures

E. Applications

The saturated salt sensor has certain advantages over other electrical humidity sensors, such as electrical percent RH instruments. Because the sensor operates as a current carrier saturated with lithium (Li^+) and chloride (Cl) ions, addition of contaminating ions has little effect on its behavior compared to a typical percent RH sensor, which operates "starved" of ions and is easily contaminated. A properly designed saturated salt sensor is not easily contaminated since, from an ionic standpoint, it can be considered pre-contaminated.

If a saturated salt sensor does become contaminated, it can be washed with an ordinary sudsy ammonia solution, rinsed and recharged with lithium chloride. It is seldom necessary to discard a saturated salt sensor if proper maintenance procedures are observed.

Limitations of saturated salt sensors include a relatively slow response time and a lower limit to the measurement range, imposed by the nature of lithium chloride. Also, once the cell is coated with lithium chloride, electric power must be applied continuously, or it must be kept in an environment with less than 11% relative humidity. Left unpowered in a humid environment, a lithium chloride cell can absorb enough water to dissolve the salt which will drip off the sensor.

The sensor cannot be used to measure dew points when the vapor pressure of water is below the saturation vapor pressure of lithium chloride, which occurs at about 11% RH. In certain cases ambient temperatures can be reduced, increasing the RH to above 11 percent, but the extra effort needed to cool the gas usually warrants selection of a different type of sensor. Fortunately, a large number of scientific and industrial measurements fall above this limitation and are readily handled by the sensor.

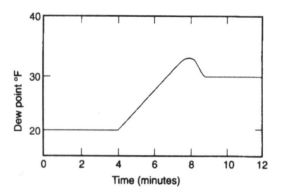

Figure 8.2   Response of lithium chloride sensor.

Slow response of the sensor is due to the bobbin control process, which is dependent on the thermal mass of the bobbin, amount of electrical current, and flow rate of the surrounding air. The typical response time of the sensor is shown in Figure 8.2. Although the response of a saturated salt sensor is slow, it is faster than some industrial processes. The sensor is sometimes used for monitoring dry air systems, chambers, curing ovens, dryers, etc., where the mass of the system changes slowly and slow response of the sensor is not a prohibitive characteristic.

Saturated salt sensors are desirable when a low cost, rugged, slow responding and moderately accurate sensor is needed. For applications requiring greater accuracies, or for humidities lower than 11% RH, condensation-type, electrolytic, or aluminum oxide hygrometers should be considered. The sensor has lost much of its original popularity due to its slow and sluggish response, modest accuracy, and need for sensor rejuvenation at certain time intervals.

II. Fog Chamber

The fog chamber, like the dew cup discussed in Chapter 3, is another type of manual instrument for measuring dew point. It is used throughout industry, mostly by heat treaters, because of its portability, simplicity, and ability to cover a broad range of dew points without the need for cooling. The fog chamber, an example of which is shown in Fig 8.3, operates on the principle that a rapidly expanding gas which is cooled adiabatically will produce a fog only when specific requirements of pressure drop, ambient temperature, and moisture content in the gas sample are satisfied. The gas sample to be tested is drawn into the apparatus and held under pressure in an observation or fog chamber by a small hand pump. A pressure ratio gage indicates the relationship between the pressure of the furnace atmosphere sample (when used for heat treating) and the ambient atmospheric pressure. The temperature is indicated by a high-grade mercury thermometer that extends into the observation chamber. The atmosphere sample is held in the observation chamber for

Figure 8.3   Fog chamber, as often used for heat treat applications.
               (*Courtesy Alnor Instrument Company*)

several seconds to stabilize the temperature, after which the quick opening valve is depressed, releasing the pressure and creating adiabatic cooling, which causes visible condensation or fog to be suspended in the chamber. The fog is easily observed with the lens system that provides a beam of light in the fog chamber when the quick opening valve is depressed. The procedure is repeated to find the end point—the point at which the fog disappears. The dew point is then determined by referring to a chart based on the initial temperature reading of the thermometer and the pressure ratio gage reading at the point where the fog disappeared. Though simple in construction and easy to operate, the method has the same drawback as the dew cup (see Chapter 3) in that it is subject to human interpretation and skill of the operator. Measurement errors of 5°C are not uncommon. It is also a one time measurement and not suitable for continuous control applications.

III. Impedance-Based Ceramic Sensors

Utilizing the adsorption characteristics of oxides, humidity-sensitive ceramic oxide devices have been developed. These devices employ either ionic or electronic measurement techniques to relate adsorbed water to relative humidity. Ionic conduction is produced by dissociation of water molecules forming surface hydroxides. This dissociation causes migration of protons such that the device impedance decreases with increasing water content. The ceramic oxide is sandwiched between porous metal electrodes, which connect the device to an impedance measuring circuit for linearizing and signal conditioning. These sensors have good sensitivity, are resistant to contamination and high temperatures up to 200°C (392°F), and may become fully wet without sensor degradation. These sensors are accurate to approximately ± 1.5% RH or to ± 1% RH when they are temperature compensated.

Ceramic sensors have some advantages and limitations. A large number of ceramic materials can be used for humidity sensing. One primary limitation, is the requirement of regeneration and cleaning. For example, porous $MgCr_2O_4$-TiO_2 solid-solution spinel (35 mol % TiO_2) sensors must undergo periodic regeneration and cleaning. The permittivity change with humidity is similar to polymer type RH sensors and the porous Al_2O_3 humidity sensors. Some ceramic sensors, such as the porous ZnO-Cr_2O_3 sensors, are somewhat less sensitive to contaminants and thus may require less frequent recalibration.

IV. Fiber Optic Humidity Analyzer

A new fiber optic humidity analyzer was developed during the last few years and put on the market in Europe. The use of optical fibers for humidity sensing offers some advantages, i.e., immunity to electromagnetic interference, safety in explosive environments, the potential of remote monitoring, and compatibility to fiber optic networks.

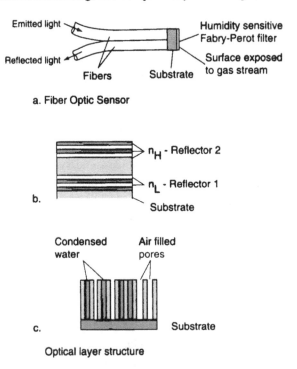

a. Fiber Optic Sensor

b.

c.

Optical layer structure

Figure 8.4   Fiber optic sensor.

The instrument provides high sensitivity and stability even over a long transmission path length. A wavelength encoding system is used as the transmission independent measuring system. The instrument includes a sensor probe consisting of a humidity sensitive optical thin-film reflection filter on fiber ends, and an optical polychromator unit for the detection of the humidity-dependent reflection spectrum of this filter.

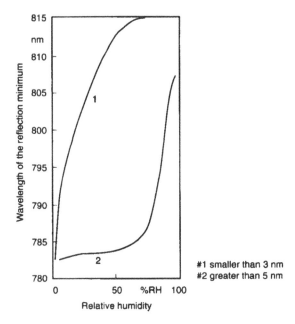

Figure 8.5   Relative humidity versus wavelength.

The principle of the sensor is shown in Figure 8.4. The moisture sensitive probe consists of a series of alternately high and low refracting, microporous dielectric layers representing a Fabry–Perot filter. Because of their porosity the layers adsorb or desorb water depending on the water vapor partial pressure of the gas. This causes the refractive index of the layers to increase or decrease, and the peak wavelength of the filter spectrum is shifted.

In order to monitor these shifts with high accuracy, a compact polychromator is used as an optical wavelength decoding device with fiber optic connectors for sensor coupling. The wavelength resolution is better than 0.1 nm in the humidity measuring regime. The detected reflection spectra of the thin film probe are mathematically treated and compared with the calibration characteristic of the sensor.

Figure 8.5 shows typical wavelength shifts of two different reflection filter types as a function of the relative humidity at 21°C. The humidity dependence is reversible. Sensitive ranges and sensitivities depend on the number and sizes of the micropores in the optical layers. Both of these can be controlled during the processing of the microporous layer structures. In this way, it has been possible to adjust the sensor sensitivity to different humidity ranges from close to 0% to 98% RH. The low moisture detection capability in terms of dew point measurements is about −70°C (−94°F). A curve showing dew point versus wavelength is shown in Figure 8.6.

Advantageous applications include the measurement of residual humidities in industrial processing, in microwave and induction furnaces, in vacuum chambers, in gas pipes, and others. The measurement principle involves a fiber optic signal transfer over distances of several hundred meters and more. An additional advantage is the small size of the sensing element which is about 1 mm in diameter or less.

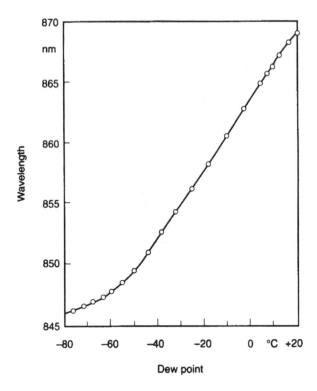

Figure 8.6   Dew point versus wavelength.

Applications

This type of instrumentation has only recently been put on the market and has not yet been used on a wide scale. Potential applications include:

- Heat treating
- Natural gas
- Kilns
- Measurements in contaminated areas

V. Other Types

There are many other, less well known humidity instruments and sensors which have their uses in particular applications including:

- *Color change*

  These are humidity indicators which show changes in humidity as a color change on a paper strip or other material. The sensing material is impregnated with cobalt chloride. The color change takes place as a result of a moisture reaction with this chemical. Other color change measurements involve pumping gas through a vial filled with crystals which change color according to the humidity of the gas.

- *Acoustic*

  Transmission of sound in air or other materials that can be used to indicate humidity.

- *Thermal conductivity*

  Heat loss from a hot wire is affected by water vapor as well as other constituent gases and could be used to measure humidity.

9

METEOROLOGICAL SYSTEMS

I. General

Water vapor measurements are indispensable for weather forecasting, upper atmosphere studies, air and noise pollution measurements, toxic and nuclear power plant emissions, etc. Several measurement techniques are used depending on the application. The different areas of meteorological measurements are discussed below:

II. Weather Stations

Through the last six decades, several types of meteorological humidity instruments and sensors have been used, starting in the 1930s with the old and well known psychrometer and presently utilizing the modern polymer RH sensors and precision chilled mirror hygrometers.

A. Psychrometers

During the 1930s and '40s most weather stations were outfitted with dry/wet bulb psychrometers. These systems are simple, inexpensive, and fairly simple to operate. However, as discussed in Chapter 7, psychrometers have several pitfalls:

- When improperly maintained, they give cause for large errors. Good and frequent maintenance can rarely be provided in meteorological stations or towers in the field, especially when located in remote areas.

- The water reservoir must be replenished at frequent intervals. Unless clean, distilled water is used, errors will result.

- In areas where winter conditions exist, (i.e., frost) or could occur, water freeze-up would prevent the psychrometer from operating. This can be remedied by adding alcohol to the water, though this could also lead to errors.

- When used in very dry areas, like deserts, very low relative humidities are normally encountered. The psychrometer would have to operate at a large temperature depression (drop of wet bulb temperature below the dry bulb measurement). This requires many precautions and providing the proper air flow over the temperature bulbs. Below a certain relative humidity level of about 10%, the psychrometer is generally not very useful.

B. Saturated Salt Sensor (Dew Cell)

When the "Dew Cell" came on the market in the early 1940s, it appeared to have some significant advantages over the psychrometer and many weather stations converted to this sensor, which is a saturated salt device, normally designed to operate with lithium chloride salt.

Advantages of the Dew Cell are:

- Simple, rugged construction

- The lithium chloride does not have to be replenished as often as the water in a psychrometer

- It is much less sensitive to contaminants

- Low cost (though higher than the psychrometer)

Disadvantages are:

- Response time is very slow, i.e., several minutes for a small step change in relative humidity.

- Limited range. Because of the properties of lithium chloride, the sensor can theoretically not operate below 11% RH. As a practical matter, humidity measurements below 12% are not practical.

- For proper and accurate operation, the sensors must be periodically cleaned and resensitized, which cannot easily be done in the filed. Hence replacement sensors are needed so that a new sensor can be installed while the previously installed sensor is taken to a service station for service. This process adds much labor.

- Accuracy is low at certain hydration points (see Chapter 7).

C. Electrical RH Sensors

The use of the much faster electrical relative humidity sensors was considered in the past but few of the earlier model RH sensors have been used in meteorological applications for the following reasons:

- High sensitivity to contaminants

- Inaccuracies resulting from the relatively high hysteresis exhibited by these sensors

- Frequent need for recalibration

- Need for replacement with new sensing elements when heavily contaminated or when no longer recalibrateable

- Errors due to temperature changes

D. Polymer RH Sensors

When the newer and more stable polymer sensors came on the market in the 1980s, more and more of these found their way into meteorological applications. Overall, the capacitive sensor is more suitable for this application than the resistive types, because of their ability to operated over a broader humidity and temperature range. Advantages are:

- The capacitive polymer sensor has a very fast response.

- Compared to earlier RH sensors, the polymer sensor has virtually no hysteresis.

- The new polymer sensors are much less sensitive to contaminants than the earlier Pope and Dunmore cells.

- Because of the above, they require less frequent recalibration or replacement.

- Polymer sensors have a low temperature coefficient. Hence errors resulting from temperature fluctuations are limited. If necessary, these sensors can be provided with electronic temperature compensation circuitry.

- Modern capacitive polymer sensors can be operated over a broad range, typically from 2% to 100% RH.

- Their cost is low.

- Calibration of sensors can often be performed in the field.

Disadvantages of the polymer sensors are:

- They are secondary measurement devices. Unless properly calibrated, their measurements are meaningless. Periodic calibration is therefore essential.

- If a sensor is heavily contaminated, its readings are in substantial error. This is often not known until the sensor is recalibrated and previous measurements are then in doubt.

A typical modern humidity/temperature probe for meteorological applications is shown in Figure 9.1 and a solar shield for such a sensor in Figure 9.2.

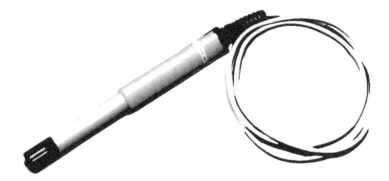

Figure 9.1   Meteorological RH/temperature sensor probe.
                 (Courtesy Vaisala)

E. Aluminum Oxide

Aluminum oxide sensors were considered for meteorological applications in the 1960s but were for the most part found to be unsuitable for operation under typical meteorological conditions.

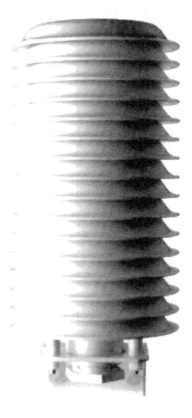

Figure 9.2   Meteorological solar shield for sensors.
                 (Courtesy Vaisala)

F. Chilled Mirror Hygrometer

In cases where high accuracies and fundamental measurements were desired, the chilled mirror hygrometer came into the picture after the mid 1960s. Advantages of the condensation hygrometer for meteorological applications are:

- Very high accuracy, to ± 0.2°C dew or frost point

- Fundamental measurement device, i.e., it does not require periodic recalibration. It will maintain its accuracy for long periods of time and can be made traceable to national standards (NIST, NPL, etc.).

- No hysteresis and excellent repeatability

- Can cover a very broad range, down to frost points of –40°C and up to dew points of 50°C

- Accuracy is not dependent on the rate of aspiration. Natural aspiration, i.e. wind flow, is adequate. If desired a motorized aspirator may be used to obtain a more stable response time.

Disadvantages of the chilled mirror hygrometer are:

- Much higher cost than any of the other types of instruments

- Requirement for periodic mirror cleaning

- Potential problems with mirror "flooding" during fog conditions. This problem can be alleviated by heating the mirror at all times to a temperature slightly above the ambient temperature, but this reduces the measurement range.

- Under conditions of heavy wind and rapidly changing dew points, the sensor could get flooded and go out of control in which case no useful information will be obtained. The sensor can easily be returned to its original condition by cleaning the mirror, but this requires field maintenance.

Chilled mirror meteorological hygrometers have been successfully used in critical areas, such as nuclear power plants, and for critical applications such as plant site locations for nuclear power plants, and factories producing toxic fumes and emissions. They have also been widely used by research establishments such as NCAR (National Center for Atmospheric Research), NOAA (National Oceanic and Atmospheric Administration), DLR (German Aerospace Research Establishment), KMI (Netherlands Meteorological Institute), for windsheer research, and several other special studies and applications. The chilled mirror hygrometer was chosen by the US Airforce for use at all of its military airports across the world. A typical chilled mirror meteorological hygrometer is shown in Figure 9.3.

The US Weather Bureau and FAA have, during the late 1980s and early 1990s, installed a large number of chilled mirror hygrometers which were designed to be produced at much lower cost than conventional systems. However, excessive maintenance requirements of these systems have been a major drawback.

The use of CCM type hygrometers could offer a more satisfactory solution, but has not yet been field tested.

Figure 9.3   Meteorological chilled mirror hygrometer.
                   (*Courtesy Edgetech*)

G. Infrared Hygrometer

In some meteorological applications, such as on board of aircraft carriers where planes are continuously landing and taking off, and where the sensor is subjected to hydrocarbon contamination, none of the previously mentioned humidity measurement systems are suitable. This is also the case when meteorological measurements are required in heavily contaminated areas, such as sand storms, pollutants, and heavy chimney smoke.

To find a solution to these problems, the US government sponsored a research project in the 1980s to develop alternative technologies and this has resulted in the development of an infrared meteorological hygrometer by Ophir Corporation in Denver, CO. This hygrometer, which is discussed in Chapter 6, uses optics which are shielded from the contaminants by using appropriate windows. A limited number of these infrared meteorological hygrometers have been successfully used in such applications.

The advantages are:

- None of the optical or other components used need to be exposed to the contaminated environment. They can all be shielded by windows which are transparent to the optical wavelength used.

- Maintenance, i.e. cleaning of the optical windows, is rather minimal.

Disadvantages are:

- This instrumentation is more expensive than the chilled mirror hygrometer and has for this reason found only limited applications.

- It is not a fundamental measurement method.

- The instrument must be carefully calibrated, which is usually done against a chilled mirror hygrometer.

- In order to obtain good sensitivity at low frost points, a very long optical path length must be used. This could result in an awkward size for the system, but which can be avoided by using so called "folded path" techniques as is discussed in Chapter 6.

- In order to make accurate measurements at low frost points, the infrared detector must be cryogenically cooled, adding further cost to the instrument.

An infrared meteorological hygrometer is shown in Chapter 6.

III. Noise Pollution Measurements

The EPA (Environmental Protection Agency) in the USA has established stringent limits on the volume of sound that can be generated by aircraft when approaching populated areas and airports. Aircraft must be subjected to such tests when in flight. Noise levels are known to be significantly affected by the relative humidity of the environment in which the aircraft operates. The EPA specifications are based on a certain relative humidity, usually 50%. It is impractical to perform such tests only when these exact conditions exist. Hence the measurements are made at any level of humidity and corrections in the end result are made to correct for the proper relative humidity. For this reason there is a the need for accurate, and verifiable humidity measurements. This is generally accomplished using chilled mirror hygrometers. Typically some hygrometers are mounted on airport towers and others on aircraft flying at various altitudes. From a large number of such measurements, a computer generated humidity profile can be obtained from which the corrections can then be made with a good degree of accuracy.

As previously stated, the chilled mirror hygrometer, because of its fundamental nature is generally the chosen method.

IV. Communications and Cloud Studies

For high altitude meteorological measurements and cloud studies, both chilled mirror and Lyman–Alpha hygrometers are used. Both are operated from airplanes flying at high altitudes and high speeds. Sophisticated and very expensive instruments were developed for this purpose, including specially designed aircraft chilled mirror hygrometers developed by EG&G (now Edgetech) and General Eastern. Lyman–Alpha and infrared hygrometers have also been used, for example to study radio propagation through the atmosphere, aircraft and missile detection, and monitoring of surface and near-surface properties of the earth.

The chilled mirror hygrometer which is less expensive than the Lyman–Alpha systems, is sometimes used for low frost point measurements and to calibrate the often needed much faster response Lyman–Alpha system. Some UV hygrometers are shown in Chapter 6.

V. Upper Atmosphere Studies [8, 9]

A research program was started by NRL (Naval Research Laboratory) in Washington, DC, in the 1960s and is now conducted by NOAA in Boulder, CO, to study effects of air pollution in the upper atmosphere, specifically as it may affect the ozone layer. Sounding rockets and balloons are launched from different locations around the earth at different time intervals. From continuing studies and comparisons with previous measurements, conclusions can be drawn as to the effects of pollutants on the upper atmosphere. As expected, there is considerable evidence that ozone is being depleted and that some so called "ozone holes" already exist. This causes considerable concern because of its effect on plant growth and health, specifically the onset of skin cancer.

The frost points to be measured are extremely low at these very high altitudes (about –80°C) and the atmospheric pressure is also very low (near vacuum). For this reason a different technique was developed, which is the cryogenic hygrometer discussed in Chapter 3. In fact, today's cryogenic hygrometer is an outgrowth of the original NRL balloon borne hygrometer.

In the balloon borne hygrometer, the mirror is cooled, usually using dry ice, to a temperature well below the frost point to be measured. A heater imbedded in the mirror is used to heat the mirror until the frost layer disappears. The temperature at which this happens is by definition the frost point. This technique results in a much faster response than would be possible if the mirror was cooled to the frost point like in a conventional chilled mirror hygrometer.

10

CALIBRATION

I. Importance of Calibration

An essential part of humidity and moisture measurement is the calibration against a standard. Without calibration few instruments provide accurate or reliable information.

The most fundamental standard that is employed by national standards laboratories is the so called "gravimetric hygrometer." Using this method, a certain amount of bone dry gas is weighed and compared with the weight of the test gas in exactly the same volume. From this, the amount of water is determined and vapor pressure calculated. This method is capable of providing the most accurate measurements possible, but such a system is very cumbersome, expensive, and time consuming to use. Some national laboratories, such as NIST (National Institute for Standards Testing) in the United States, NPL (National Physical Laboratory) in the United Kingdom, NRLM (National Research Laboratory of Metrology) in Japan, and PTB (Physicalish-Technische Bundesanstalt) in Germany have the availability of a gravimetric hygrometer. However, these laboratories only use the system to calibrate other standards which are easier and faster to use for day-to-day calibrations, such as the two-pressure humidity generator, a precision chilled mirror hygrometer, or a carefully designed psychrometer.

In this chapter a rather detailed description will be given of calibration techniques and procedures employed by NIST and NPL. Calibration services at other national laboratories will be discussed in somewhat less detail and mostly to the extent they differ from NIST and NPL. Portions of the section on NIST were abstracted from Reference 41 (see Chapter 15). Sections on NPL were derived from References 51 to 56. Much of the information on CETIAT was obtained from Reference 67 and information on NRLM from References 76 and 78.

A. Traceability to National Standards

Most commercial humidity measurement instruments are supplied with a calibration report showing the accuracy at the time of manufacture or shipment from the factory. In most cases, this does not truly reflect the way the instrument will perform in the field, nor does it need to in many of those cases. The user should know what to expect from the instrument in terms of performance in the field. Traceability means that the instrument has been calibrated against a primary or transfer standard.

145

B. Calibration Standards

With regard to humidity measurements, it is commonly accepted that a standard is a system
or device which can either produce a gas stream of known humidity by reference to fundamental base units, such as temperature, mass, and pressure, or an instrument that can
measure humidity in a gas in a fundamental way, using similar base units. There are established standards for humidity in many countries, operating on various principles, such as
gravimetric systems, two pressure generators, and other devices. Some of these national
standards, other than the gravimetric method, and the approximate ranges they are capable
to cover, are shown in Table 10.1.

Table 10.1   Comparison of non-gravimetric national standards.

Country	Authority	Type	D.P. Range	Uncertainty
UK	NPL	Two Temperature	−75°C to +90°C	± 0.03°C to 0.10°C
USA	NIST	Two Pressure	−75°C to +40°C	± 0.04°C
ITALY	IMGC	Two Temperature	−15°C to +90°C	± 0.05°C
GERMANY	PTB	Two Pressure	−50°C to +75°C	± 0.05°C
		Coulometric	−75°C to −10°C	± 0.1°C to 0.5°C
FRANCE	CETIAT	Recirculation Generator	−60°C to +70°C	± 0.05°C to 0.10°C
JAPAN	NRLM	Two Pressure Two Temperature	−10°C to +20°C	± 0.2% RH

C. Uncertainty Versus Accuracy

In the field of humidity measurements, as in many other areas, there are different ways of
expressing the level of performance of an instrument. Two terms that are often used are
"uncertainty" and "accuracy." Uncertainty is usually expressed by those who are making
more precise measurements and where a full explanation of the meaning of the calibration
is in order. Accuracy is more commonly used as a general statement of performance of an
instrument and is often the type of information provided by the manufacturer in product
information sheets.

   As a rule, uncertainty is quoted at a 95% confidence level and represents a statistically-
proven method of assessing the likely error from a true value. For example, if the
uncertainty of a measurement is 1°C (1.8°F), 95 out of every 100 measurements made at
that dew point level will lie within ± 1°C (1.8°F) of the true value.

   A statement of accuracy is quite often produced on a much less scientific basis, and
as such, is not necessarily as significant to the scientific user. If the supplier of the statement

provides an explanation of the basis of that statement, the user can make a sensible assessment of its significance in comparison with the application requirements and/or the claims of other suppliers. Few suppliers use a standard convention and therefore inter-comparisons are often difficult.

D. Types Of Standards Used

Standards used to calibrate humidity instruments fall into three classifications: primary standards, transfer standards, and secondary devices. Below is a brief description of examples of the three classes of instruments.

1. Primary Standards

These systems rely on fundamental principles and base units of measurement. A gravimetric hygrometer is such a device. It measures humidity in a gas stream by physically separating the moisture from the carrier gas and collecting it in a storage vessel. This is subsequently weighed to give the mixing ratio of the sample, and from this information other hygrometric parameters can be calculated. This method is extremely accurate but is cumbersome and time consuming to use. A gravimetric hygrometer is very expensive to build, and, at low humidity levels, can require many hours of operation to obtain a large enough sample. It is therefore not an attractive system for day-to-day use. At a lower level, and at somewhat lower accuracies, the two-pressure, and two-temperature generators, and some other systems, are customarily used as primary standards.

2. Transfer Standards

A number of instruments fall into this category. They operate on fundamental principles. These standards are capable of providing good, stable, and repeatable results but, if not properly used, can give incorrect results. Examples of commonly-used instruments are:

a. Chilled Mirror Hygrometer

This is probably the most widely used transfer standard. As discussed in Chapter 3, a mirrored surface in contact with the gas stream to be monitored is cooled until condensation is formed. The temperature at which condensation is formed is known as the dew or frost point of the gas and directly relates to the saturation water vapor pressure of the sample. From this data any hygrometric equivalent parameter can be calculated, provided that other information such as gas temperature and pressure are also known.

b. Electrolytic Hygrometer

This instrument, discussed in Chapter 5, operates on the principle of Faraday's law of electrolysis to determine the amount of moisture in a gas stream. The water vapor in the gas stream is passed through the instrument's measurement cell which electrolyzes the water molecules into their component parts (H_2 and O_2). The current consumed in this process is directly related to the amount of electrolyzed water. Provided the cell converts all the water in the gas stream into its component parts, the measurement of current represents an absolute measure of the moisture content.

c. Psychrometer

In a dry/wet bulb psychrometer, pure water is evaporated from a wick surrounding a temperature probe placed in the gas stream which passes over the wick at a proper velocity. The evaporation process causes a lowering of the temperature of the wet bulb, which, under ideal conditions, is directly related to the relative humidity of the gas at the prevailing gas temperature, which is measured with the dry bulb thermometer. If all measurement parameters are known, the temperature difference (depression) is related fundamentally to the heat of evaporation of water. Hence an absolute determination of the water vapor pressure of the gas can be made. This technique has been widely used in the past, but is presently considered to be the least desirable of the three methods discussed because of the large number of variables that must be controlled and which can affect the measurement results. The psychrometer also requires skilled operators in order to make accurate measurements.

3. Secondary Devices

Secondary devices are nonfundamental and must be calibrated against a transfer standard or other fundamental system. To provide accurate data, they require recalibration at certain time intervals. These secondary systems are rarely used for laboratory calibration, but have many applications in industry and in commercial buildings, for example, air conditioning. Examples of secondary humidity analyzers are:

a. Impedance Hygrometer

The ceramic, aluminum oxide, and silicon oxide-based sensors rely upon a change in electrical properties (capacitance, resistance, or impedance) of a porous layer which is then processed by simple electronics to give an output calibrated in a suitable hygrometric unit. Properly manufactured, calibrated, and operated, these devices can provide good, on-line service for many years. However, calibration is required on a regular basis, and adjustments are often needed. Accuracies are modest and much lower than for transfer standard type instruments.

b. Polymer Film RH Sensors

Polymer film RH sensors are similar in principle to impedance hygrometer sensors discussed above. These sensors are constructed from polymer material with a hygroscopic dielectric and designed to provide an electrical response in terms of relative humidity. Significant improvements have been made in recent years and these devices can now provide excellent low-cost service, particularly in normal ambients, and sometimes also at high temperatures. A summary of instrument classifications is given in Table 10.2.

Table 10.2   Instrument classifications.

Type	Suitable for use as a standard	Class	Typical Range	Typical measurement accuracy
Gravimetric	Yes	Primary	+50/–50°C dew point	0.1°C dew point
Chilled Mirror Hygrometer	Yes	Fundamental (Transfer)	+90/–90°C dew point	0.2°C dew point
Electrolytic Hygrometer	Yes	Fundamental	1 to 2000 ppm$_v$	5% of reading ppm$_v$
Psychrometer	Yes	Fundamental	10 to 100% RH 0 to 100°C ambient	2% RH
Impedance Hygrometer	No	Secondary	–100 to +30°C dew point	2 to 4°C dew point
Polymer RH Sensor	No	Secondary	5 to 95% RH 0 to 100°C ambient	2 to 5% RH

E. Field Applications

In real-time applications of humidity measurement instruments, specifications of a manufacturer and calibration data of a standards laboratory data lose some of their significance. Operation of instrumentation under non-ideal conditions, which is rarely taken into account by manufacturers and calibration laboratories, introduces a large number of variables which are likely to affect even the most reliable and accurate systems in some way. Some of these variables are:

1. Temperature Effects

Most hygrometers are calibrated at a fixed ambient temperature. This may vary from manufacturer to manufacturer, but is usually around 25°C ± 1°C or 75°F ± 2°F. Variations from this ambient temperature can easily affect the measurement results and accuracy.

Sensor temperature coefficients in impedance-type sensors, like the capacitors and resistors they are derived from, exhibit some level of temperature dependency. Many systems compensate for this either by an electronic compensation method or by controlling the sensor temperature to negate the effect. Temperature coefficients of 0.1 to 0.2 are not uncommon, even in compensated systems, and result from the nonuniformity of sensors and electronic components used in the compensation systems. Sensor temperature effects in chilled mirror hygrometers are usually negligible. However, temperature changes could cause changes in measurement range as noted in Chapter 3 on chilled mirror hygrometers.

2. Electronics

Modern electronic instrumentation is usually rather temperature insensitive over a normal daytime ambient temperature range. However, large temperature swings in certain exposed locations can give rise to performance errors of many electronic components. Instrument suppliers usually report the temperature range over which electronic performance is unaffected, and it is fair to assume that if these limits are adhered to, temperature effects can be ignored. However, if the operating temperature exceeds these limits, the user can expect errors and in certain circumstances, especially at low temperatures, complete failure.

3. Pressure

Pressure effects are much easier to quantify and therefore easier to correct than temperature effects. Variations in pressure only have an effect on the actual water vapor pressure of the gas being measured. Therefore knowledge of the pressure at the measurement point will allow the effect to be fully compensated, provided that the nature of the gas and its behavior under pressure are understood.

4. Flow

In theory, flow rate should have no direct effect on the measured moisture level in a gas system, but in practice, the flow rate of a gas can affect the accuracy. Excessive flow rates in piping systems can introduce pressure gradients. Care should be taken to ensure that the sampling system can accommodate the required flow rate for the measurement instrument. An inadequate flow rate can result in errors due to a number of side effects:

a. Back Diffusion

If an open-ended sampling system is used, ambient air can flow back into the system.

b. Ineffective Purging of the Sampling System

In a complex system, inadequate flow can allow pockets of undisturbed wet gas to remain in the sampling system or sensor, which will gradually be released into the sample flow. In worst case conditions, liquid water may exist for long periods in "moisture traps" in the sample handling system. This could happen after hydraulic testing, or after a high moisture fault condition.

c. Low Flow Rate

This will accentuate adsorption and desorption effects in the total volume of gas passing through the sampling system. This will become more significant as flow rates are lowered.

5. Contaminants

Humidity measurement instruments are susceptible to errors caused by contamination. The full list is almost endless, but typical examples are as follows:

a. Particulates

Particulates eventually cause a chilled mirror hygrometer to go out of control, clog the porous structure of impedance and polymer film sensors, and can potentially inflict physical damage by impingement. Particulates can also reduce the evaporation rate of a psychrometer wet wick and they will clog an electrolytic cell.

b. Inorganic Salts

Inorganic salts can affect the accuracy of chilled mirror systems and psychrometers by modifying the saturation vapor pressure of liquid water on the surface of each type of device. Most salts are to some degree hygroscopic and, when deposited within the porous structure of an impedance or RH device, can modify its characteristics. In fact, older types of such sensors relied on a salt dopant to give the required electrical response to moisture changes. Salts deposited on an electrolytic cell can change the coulombmetric relationship and cause damage to the cell.

c. Organic Compounds

Although organic compounds tend not to have a direct interference with water vapor, they can condense on a mirror surface at a higher temperature than water, or evaporate from a psychrometer wick, causing extra evaporative cooling. It is possible for organics to damage some secondary sensors if glues or epoxies are used in their manufacture. To eradicate, or at least minimize, these problems, conditions should be evaluated and steps taken to properly remove all or most contaminants from the gas stream by a suitable filtration method.

F. Methods of Calibration

Calibration of instruments used in the field can be carried out at any level from a full evaluation against a transfer standard down to a one-point check against an assumed humidity level. Broadly speaking, four categories can be established, each of which has its merits and disadvantages. Before exploring these, the calibration hierarchy will be reviewed.

1. Primary Calibration

Calibration carried out directly against a standard, usually a national standard, represents the highest level of calibration. The national standard used may be the gravimetric hygrometer, or an easier-to-use standard method like a two-temperature or two-pressure system.

2. Transfer Standard Calibration

In the USA, many laboratories perform such calibrations using NIST-certified transfer standards, in most cases the chilled mirror hygrometer. In the UK, a number of laboratories are accredited under the national accreditation of measurement and sampling (NAMAS) to perform calibrations on behalf of the National Physical Laboratory. These NAMAS laboratories are strictly monitored and are required to provide not only copies of all calibration certificates provided to NAMAS, but also confirmation of their abilities by periodically calibrating test instruments as an audit procedure. Measurement capabilities of these laboratories are not as high as those of the national standards laboratories, but provide the next best level of capability. Under the NAMAS system, instrument errors will not usually be corrected, but a certified error analysis with an uncertainty statement will be provided. Although less expensive than calibrations against national standards, the use of transfer standards still follow similar procedures and hence can be time consuming.

NAMAS has recently combined with NACCB to form UKAS, the United Kingdom Accreditation Service.

3. Field Calibration

It is possible to provide calibration information on field instruments in-situ either by reference against a gas stream of known humidity (generator or certified compressed gas cylinder), or by comparison against a portable transfer standard device. This can be very useful in terms of assessing the performance of an instrument in its normal operating conditions, but because of the departure from ideal laboratory conditions, they may not corroborate directly with formal calibration data of the instrument in question. Generally this type of evaluation is used to give a close estimate of the performance of an instrument and to assess whether it needs to be removed from service for a more rigorous examination.

4. Summary

At the highest international level an excellent network of national standards provides an extremely valuable precision calibration reference to hundredths of a degree. Transfer standard laboratories can spread the availability of this service at uncertainties of tenths of a degree. The application of an instrument in the field can give rise to errors of several degrees, or even tens of degrees.

While the issues of calibration accuracy and traceability are extremely important in all fields of instrumentation, this is particularly important in humidity measurement. Accurate knowledge of the application of the instrument is most important. A user can be impressed by a calibration certificate from NIST or NPL showing errors of less than 0.2°C (0.38°F). However, if that instrument is then installed in a gas line with an outlet restriction causing a pressure build-up and a badly constructed elbow giving a dead space (moisture trap) prior to the measurement point, the end result is going to be much less accurate. The burden is therefore on the user to take account of all factors likely to affect the measurement being undertaken and to eliminate problems where possible. The user must ultimately make a value judgment as to the true measurement capability of the system, which could be orders of magnitude worse than the data contained on the instrument's calibration certificate.

Table 10.3 represents a rough estimate of the levels of measurement capability obtainable in different applications with various types of instruments, assuming reasonable precautions are taken to minimize application errors. The table is only intended as a rough guide and is based entirely on experience gained in the field. This information is therefore subject to wide variations in different situations.

Table 10.3   Typical application accuracies.

Application	Dew Point	Accuracy Level Achievable				
		Cooled Mirror	Electrolytic Hygrometer	Psychrometer	Impedance Instrument	RH
Laboratory	−75°C/+90°C	±0.2°C	±5% of reading	± 2% RH	± 2 °C	± 2% RH
Instrument air	−40°C	± 0.5°C	± 8% of reading	n/a	± 4°C	n/a
High purity gases	−75°C	± 0.5°C	±8% of reading	n/a	±5°C	n/a
Ambient monitoring	−20°C/+20°C	± 0.5°C	n/a	± 5% RH	± 4°C	±7% RH
Process plant	−75°C/−40°C	± 0.7°C	± 10% of reading	n/a	± 4°C	n/a
Drying processes	+90°C/+20°C	± 0.7°C	n/a	± 5% RH	n/a	± 7 to 10% RH

G. Accuracy

Considerable misunderstandings can exist when defining accuracy of humidity sensors. Each manufacturer has its own way of measuring and specifying accuracy, and the user is often left confused. While most manufacturers are not deliberately misstating accuracy claims, competition forces them to specify accuracy of their instruments in the most favorable way, sometimes only over a narrow range, at a fixed temperature, and under clean, ideal laboratory conditions. The following considerations are important and should be considered by any user of humidity instrumentation:

- *What calibration standard was used to calibrate the instrument against?*

  Obviously, its accuracy is an important determining factor.

- *Over what range was the instrument calibrated?*

  It is much easier to obtain a high accuracy at one point, or over a narrow range, than, for example, from 5%–100% RH or from −60° to +90°C dew point (−76°F to +194°F).

- *At what ambient temperature was the instrument calibrated?*

Most sensors are temperature dependent. If the calibration is defined at one temperature only (for example, 25°C or 75°F), the accuracy will be much better than when the accuracy is defined over a range of temperatures, for example, from –20°C to +75°C (–4°F to +167°F).

- *How long ago was the instrument calibrated?*

Some sensors exhibit aging and are less accurate if calibrated many months before their use.

- *What contaminants exist?*

This is perhaps the most important consideration. All known humidity sensors are affected by chemical and/or particulate contamination. If calibrated in a clean laboratory environment and subsequently used in a contaminated factory environment, or outdoors in the presence of air pollutants, the sensor will perform at a much lower accuracy, especially over time.

The various sensors discussed in this book have different levels of sensitivity to contaminants and such contamination effects vary depending on the type of sensor used and on the type of contaminant to which the sensor is exposed. Unfortunately, manufacturers generally have limited information on how their sensors perform when exposed to various contaminants, and most often it is the user who must find this out by field tests.

It is important to realize that the sensor with the best factory specified accuracy is not always the most accurate in field use. For example, in heating, ventilating, air conditioning, and cooling (HVAC) applications, some sensors specified to be ±3% RH will still be ± 6% RH after a year's operation in a lightly contaminated environment, while other sensors specified at ±1% RH by the manufacturer, may be ±10% RH or worse after a few months. It is therefore important to look beyond the accuracy claims and try to determine how accurate the sensor's performance will be in actual use. The best attainable accuracies for frequently used dew point sensors and the dew point ranges they are able to cover are shown schematically in Figure 10.1. Typical accuracies for a variety of humidity measurement methods is shown in Figure 10.2 These accuracies can be considered laboratory accuracies rather than accuracies obtainable in field use. There are also wide variations between instruments and manufacturing techniques. Hence these curves serve as a general indication only.

II. National Standards Laboratories

National standards laboratories in different countries offer similar services and use similar equipment for calibration. Most of the larger national laboratories such as NIST, NPL, and NRLM, employ a gravimetric system and one or more transfer standards.

Precise comparisons between different international calibration laboratories are unavailable. Therefore, reference is often made to "NIST traceable" or "NPL traceable

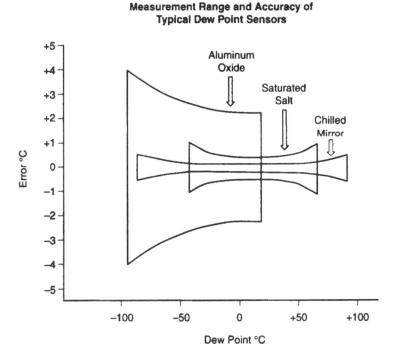

Figure 10.1   Dew point sensor accuracies.

standards." Traceability certificates are not necessarily equivalent, however NAMAS now recognizes a NIST certificate as being equal to NPL.

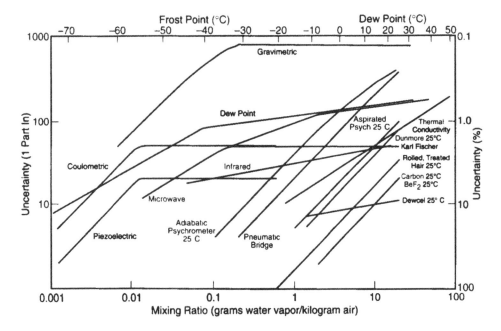

Figure 10.2   Typical accuracies for various measurement methods.

A. NIST (USA)[41]

1. Objectives of NIST

The primary responsibility of NIST (previously the National Bureau of Standards, also known as NBS) is to provide the central basis for the national measurement system, to coordinate that system nationally and internationally, and to furnish essential services leading to accurate and uniform measurements throughout the USA. This section summarizes the activities of NIST and some of the humidity measurement systems used. The hierarchy of humidity standards used by NIST is shown in Figure 10.3. There are two routes by which the units of humidity are propagated. Both originate with the national standards for the base units. One route leads directly to the primary standard for humidity and then through precision generators, secondary standards, and fixed points, to working instruments and controls. The second route leads to the national standard for derived units, such as pressure, which in turn, is used for measuring water vapor pressure over fixed points and for characterizing pure water substances. With the enhancement factor or correction for the real gas behavior of water vapor and gas mixture, water is used as the reference material in precision generators to produce known levels of humidity for calibrating secondary and working standards.

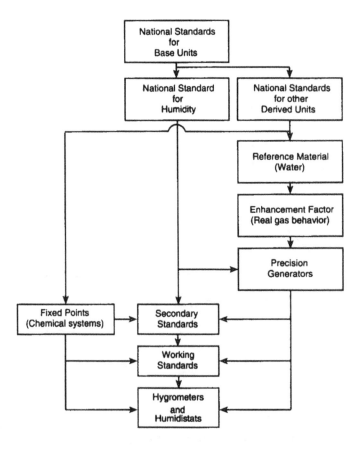

Figure 10.3   Hierarchy of humidity standards.

2. NIST Gravimetric Hygrometer

The NIST standard hygrometer is based on the well-known gravimetric method of water vapor measurement. In the hygrometer, the mass of water vapor mixed with a volume of gas is absorbed by a chemical desiccant and weighed. The volume of dry gas is measured directly. Since mass and volume are fundamental quantities, this method yields an absolute measurement of humidity. The unit of measurement is mixing ratio, that is, mass water/ mass dry gas. The NIST gravimetric hygrometer is shown schematically in Figure 10.4.

The apparatus basically consists of a drying train for absorbing the water vapor from a moist gas, usually CO_2-free air, and a gas volume measurement system for measuring the volume of the dry gas. The main drying train consists of three removable absorption U-tubes. The first U-tube is filled with anhydrous Mg $(ClO_4)_2$ and backed with a plug of anhydrous P_2O_5. If the desiccant in the first U-tube is near exhaustion or if the air flow is too fast, some water vapor may remain in the effluent. The second tube removes the trace amount that is still left in the gas stream. If water vapor is inadvertently present downstream of the main drying train, the third tube serves as a guard to prevent it from diffusing back to the second tube. At the termination of a run, the mass of water vapor removed by the drying train is determined by precision weighing. The total volume of dry gas that was associated with collected water vapor is measured by counting the number of times two cylinders of known volume, each of about 0.03 m³ (1 ft³) capacity, are alternately evacuated and filled.

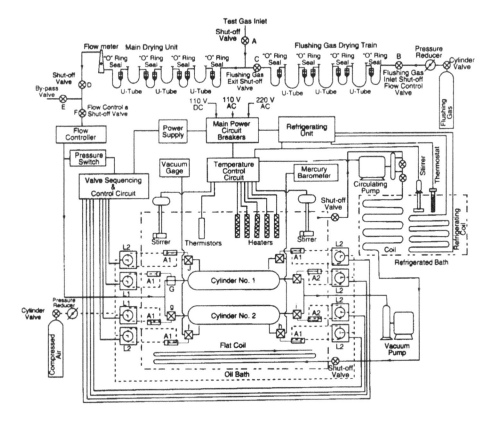

Figure 10.4   Schematic diagram of NIST gravimetric hygrometer.

By means of a pressure switch, a vacuum pump, and associated automatic controls, each of the two cylinders is evacuated and brought to a preset pressure, producing a continuous flow of test gas. The calibrated internal volume of each cylinder, together with the pressure and temperature of the gas therein, provide a highly accurate measurement of the gas volume. The mass of dry gas is obtained by multiplying the volume by the gas density, which is, in turn, computed from the pressure and temperature in the cylinder. The total gas mass is the sum of the individual fillings of each cylinder. In theory, the gravimetric method has great inherent precision and accuracy. In practice, great care must be exercised in making a mixing ratio determination, and careful attention must be given to a multitude of details. Measurements are time-consuming and results are average values. Hence, it is common practice to use the gravimetric hygrometer with a humidity generator that produces gas of constant humidity. The gravimetric method is principally used for research work or for making fundamental calibrations.

The NIST (NBS) gravimetric hygrometer has a nominal measurement range of 27 mg/g to 0.19 mg/g mixing ratio. The highest value is determined by consideration of operator comfort. An ambient room temperature slightly above 30°C (86°F) is tolerable for a short period and, to prevent condensation in the lines, the room temperature must be above the dew point temperature. This limits the upper dew point of the test gas to 30°C (86°F), corresponding to a mixing ratio of nominally 27 mg/g, although the hygrometer itself is capable of measuring a higher value. The lower limit, 0.19 mg/g, is determined primarily by error considerations, e.g., systematic errors from leakage and incomplete absorption. The estimated maximum error magnitude encountered over the operational range of the instrument is 12.7 parts in 10^4, or 0.13% of the measured value.

3. Saturation Vapor Pressure Over Liquid Water

An accurate equation for the vapor pressure of water is essential to establish and maintain humidity standards, to calibrate hygrometers, and to make precision humidity measurements. Until the late 1970s, it was the practice at NIST to use the Goff and Gratch vapor pressure formulation[13] in all work pertaining to hygrometry. In 1976, Wexler[30] developed a vapor pressure equation, valid from 273.16 K to 373.15 K, which is to within 43 ppm or better with the precise measurements of the vapor pressure of water made by Stimson[22] and Guildner, and Johnson and Jones[32]. The agreement of the vapor pressure formulations between Wexler and Goff-Gratch is better than 100 ppm over the temperature range 25°C to 100°C (77°F to 212°F). However, below 25°C (77°F), the differences between the two formulations increase from 100 ppm at 25°C (77°F) to 900 ppm at 0°C (32°F). This corresponds to a temperature difference of less than 2 mK to 12 mK.

Recently Wexler and Hyland[36, 37] extended the vapor pressure formulation for temperatures to 473.15 K. Their vapor pressure formulations are also summarized in Reference 38. The maximum difference between Wexler (1976) and Wexler-Hyland (1983) formulations is less than 50 ppm over the temperature range of 273.16 K to 373.16 K. Therefore, either of the two formulations, Wexler (1976) or Wexler-Hyland (1983), may be used for temperatures to 373.16 K. However, the Wexler-Hyland (1983) formulation must be used for the temperature range 373.16 K to 473.16 K.[36]

4. Vapor Pressure Formulation For Ice

Using similar methods as those used in the formation for water vapor, Wexler[35] derived a formulation for the vapor pressure of ice from the triple point down to −100°C (−148°F). The agreement with the Goff-Gratch equation is better than 0.08% at 0°C (32°F) and 0.29% at −100°C (−148°F). In 1983, Wexler and Hyland[36] derived an updated equation for the saturation vapor pressure over ice. The agreement of the vapor pressures derived from the Wexler-Hyland[36] and Wexler[35] equations is within 5 parts in 10^4 from −100°C (−148°F) to the triple point.

5. Enhancement Factor

The saturation vapor pressure over a flat surface of its pure condensed phase is an accurately known function of temperature. However, if a second gas is introduced over the surface of the water, the saturation vapor pressure is increased. This increased vapor pressure (the effective vapor pressure) differs from the saturation vapor pressure of the pure phase and is expressed by:

$$f = \frac{X_w P}{e_s} \qquad (10.1)$$

where:

f = the enhancement factor
X_w = the mole fraction of water vapor in the saturated mixture
P = total pressure above the surface of the condensed phase (liquid or solid)
e_s = the pure phase saturation vapor pressure

Hyland[24] obtained an accurate value for water vapor and CO_2-free air mixtures for temperatures between −50°C to 90°C (−58°F to +194°F) and for pressures between −0.25 × 10^5 to 100 × 10^5 Pa.

6. Precision Humidity Generators

There are several practical methods for producing atmospheres of known humidity with sufficient precision and accuracy, eliminating the need to use an auxiliary hygrometer for direct measurement. Equipment incorporating these methods are called precision humidity generators. Precision generators based on three different principles were developed at NBS. The methods used are sometimes referred to as the two-flow, two-temperature, and two-pressure methods.

a. Two-Flow Method

In this method, there are two streams of gas. One stream is dry gas and the other is saturated with water or ice. The two streams are combined in the test chamber. The humidity in the test chamber can be calculated from the known flow rates of the two streams of gas. In this device, the stream of dry air is divided into two parts using a proportioning value. One part is maintained dry and the other part is saturated with respect to ice. The two parts are recombined in the test chamber and then exhausted into the room. The relative humidity in the test chamber is given by the ratio of the flow division. This method is rather simple and is also widely used in university and corporate metrology laboratories.

b. Two-Temperature Method

The principle of the two-temperature method is to saturate a stream of air with water vapor at a given temperature and then to raise the temperature of the air to a specified higher value.

Figure 10.5 is a simplified schematic diagram of a two-temperature recirculating humidity generator.

Wexler[30,31] designed a two-temperature recirculating humidity generator consisting of four saturators and four test chambers. The arrangement permits rapid interchanges of the test chambers with saturators operating at different temperatures, so that operation need not be delayed while waiting for a new temperature equilibrium to be established. Using the two-temperature principle, a low frost point humidity generator[25] was built for producing gases with very low levels of water vapor content. The system was designed to operate at pressures from 500 Pa to 200,000 Pa (0.005 atm to 2 atm). Special provisions were provided to dry the gas flow system. The components also used in the generator were selected to reduce all sources and sinks of moisture downstream of the saturator. The generator was designed to produce frost points from –30°C to –100°C (–22°F to –148°F), with an uncertainty estimated at less than 0.09°C (0.162°F).

c. Two-Pressure Method

In this type of generator, a stream of air at an elevated pressure is saturated at a fixed temperature and then iso-thermally expanded into a test chamber at a lower pressure, usually atmospheric pressure. Measurements of the temperature and pressure within the saturator and the test chamber or other test space establish the humidity of the test air.

The principle of operation is illustrated in Figure 10.6. The air enters at A, through a pressure regulator B, saturator C, expansion valve E, and test chamber F. A two-pressure generator[7, 34] was designed and built to encompass the relative humidity range, 20% to 98%,

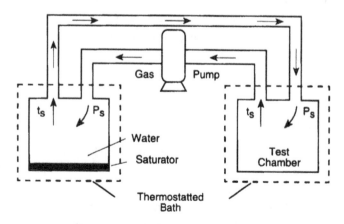

Figure 10.5   Schematic of two-temperature method.

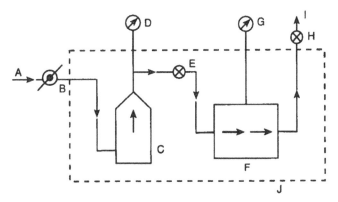

Figure 10.6   Block diagram of NIST two-pressure humidity generator.

and ambient temperature –40° to +40°C (–40°F to +104°F). After careful evaluation of the performance and operational characteristics of the two-flow, two-temperature, and two-pressure humidity generators, it was determined that the two-pressure method was the most suitable generator for providing an improved humidity calibration and testing facility at NIST.

Most humidity generators for use as a calibration standard are built by national calibration laboratories or by the research testing facilities themselves, though some commercial systems are also available (see Section III).

d. The NIST Two-Pressure Generator[34]

The two-pressure method for generating or producing air of known humidity involves saturating the air or other gas at a high pressure with water vapor and then expanding the air to a lower pressure. The NIST generator is designed for a maximum saturator operating pressure of 3.3 MPa (478.5 psi) which, in turn, controls the minimum relative humidity. Figure 10.7 is a simplified flow diagram which illustrates the principle of operation and the basic components.

Compressed air from the house pressure line is first cleaned by using commercially available filters and air dryers, D. Two pressure regulators, R_1 and R_2, are used to control the pressure in the humidifying system and a flow meter, F, is used to monitor the flow rate.

Saturation of the air at high pressure is accomplished in two saturators, S_1 and S_2. Pre-saturator, S_1, is immersed in a bath, B_1, which is maintained at a temperature of 10°C to 20°C (18°F to 36°F), warmer than the desired saturation temperature. The pre-saturator utilizes a centrifugal flow pattern. Air enters the pre-saturator tangentially to the inner wall and is directed slightly downward into the water. A liquid level controller automatically maintains a fixed water level in the pre-saturator by controlling the solenoid valve, V_1. Supply water is maintained in a container, C, which is at the same pressure as the pre-saturator. The air enters the pre-saturator through a coil of 1.27 cm (0.5 in.) outside diameter tubing and exits at the top of the saturator through a 5.08 cm (2 in.) outside diameter tubing. Types 316 or 304 stainless steel tubing and fittings are used throughout the system.

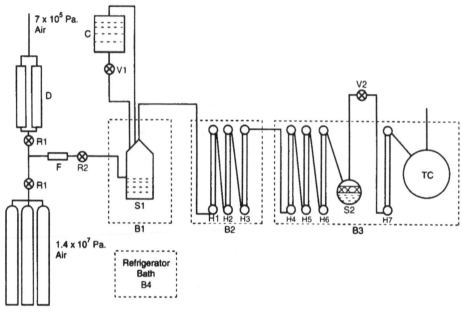

D	Air Dryer and Filters
R_1, R_2	Pressure Regulators
F	Flow Meter
S_1	Pre-Saturator
B_1	Pre-Saturator Bath
V_1	Water Supply Control Valve for Pre-Saturator
C	Water Reservoir for Pre-Saturator
B_4	Refrigeration Bath for Cooling Baths B1, B2, and B3
B_2	Conditioning Bath
H_1 - H_6	Radiator Type Heat Exchangers
B_3	Final Bath
S_2	Final Saturator
V_2	Digital Expansion Valve
H_7	Heat Exchanger Coil
TC	Test Chamber

Figure 10.7   Schematic diagram of two-pressure humidity generator.

The air then passes through the three heat exchangers, H_1, H_2 and H_3, located in the bath, B_2, which is operated at a temperature of 0.5°C to 1.0°C (0.9°F to 1.8 °F) warmer than the final saturation temperature. The air which enters the heat exchangers in bath B_2 is warmer than, and also supersaturated with respect to the temperature of bath B_2. The purpose of bath B_2 is to make it easier to achieve temperature control in the final bath, B_3, by tempering the air so that it is close to, but still slightly above, the final saturation temperature.

The six radiator-type heat exchangers, H_1 through H_6, located in baths B_2 and B_3 are designed for minimum pressure drop. The air enters the bottom of each heat exchanger which is a horizontal tube measuring 5.08 cm (2 in.) in outside diameter and 0.5 m (20 in.) in length. The top of the heat exchanger, which is similar in design and dimensions to the bottom is connected to the bottom with 16 parallel tubes measuring 1.9 cm (0.75 in.) in outside diameter and 0.5 m (20 in.) in length as shown in Figure 10.7.

After the air emerges from the conditioning bath, B_2, it is brought to the final saturation value by flowing through the three heat exchangers, H_4-H_6, and the final saturator, S_2, which are located in bath B_3. Under the worst operating conditions, i.e., at a flow rate of 0.005 m³/sec (0.18 ft³/sec) and at a pressure of one atmosphere, the maximum pressure drop between the pre-saturator, S_1, and the outlet of the final saturator, S_1, is 300 Pa (0.044 psi). Although the pressure drop in the three heat exchangers in the final bath, B_3, is small, there is still a finite pressure drop and if this pressure drop occurs after the air reaches the final saturation temperature, it will cause the air to be unsaturated. To make up this deficit, a final saturator, S_2, is installed downstream of the heat exchanger, H_6. In addition to resaturating the air, the final saturator provides additional surface air to assure that any water droplets, which may be entrained in the air stream, precipitate out.

The final saturator, S_2, is a tube having an outside diameter of 11.43 cm (4.5 in.) with a wall thickness of 0.95 cm (0.375 in.) and a length of 66 cm (26.0 in.). The tube is placed in a horizontal position and is half filled with water. A segment from the top of the tube was removed and replaced with a flat plate so that the height of the air space above the water is 2.5 cm (1 in.). In addition, the air stream is divided into three channels above the water surface and in each channel there is a maze of air deflectors to force the air into the water surface.

Upon emerging from the final saturator, S_2, the air is expanded to a lower pressure through an expansion valve, V_2, which is placed external to the bath, B_3.

After expansion, the air passes through the heat exchanger, H_7, which is a coil of tubing measuring 3.8 cm (1.5 in.) in outside diameter and approximately 11 m (36.1 ft.)in length, to bring the air back to the bath temperature before it enters the test chamber, TC. The air from the test chamber discharges into the laboratory or into a vacuum source. In the former case, the test chamber will remain at or near atmospheric pressure; in the latter, the test chamber can be controlled at sub-atmospheric pressure through the use of a suitable back-pressure regulator. A length of flexible stainless steel tubing between the test chamber and the final heat exchanger permits the test chamber to be taken out of the bath. The test chamber is a cylinder 32.4 cm (12.75 in.) in length with 0.95 cm (0.374 in.) wall thickness and 40.6 cm (16 in.) in length. Tubular outlets extend from the chamber to allow mechanical controls and electrical leads to be brought in and out of the working space and for measuring the temperature and pressure inside the chamber.

The temperature of the final bath, B_3, can be maintained over the range –60°C to +80°C (–76°F to +176°F). The temperatures of the baths, B_1, B_2 and B_3, are each maintained by balancing a small amount of constant cooling with controlled heating. Cooling is induced by pumping the liquid bath fluid from each bath through its associated heat exchange coils located in the refrigeration bath, B_4, which in turn is cooled with liquid carbon dioxide. The thermocouple indicator and temperature controller for bath B_4 is an on-off type which controls the opening and closing of a solenoid valve on a liquid carbon dioxide line. The temperature controllers for baths B_1 and B_2 are of the power proportional type with a resistance thermometer sensor and for bath B_3, the time-proportional with reset using a platinum resistance thermometer sensor.

7. Derived Humidity Values

a. Saturation Mixing Ratio

The saturation mixing ratio r_w of the moist gas emerging from the generator is:

$$r_w = \frac{M_w f(P_s \cdot T_s) e_w(T_s)}{M_g[P_s - f(P_s \cdot T_s) e_w(T_s)]} \tag{10.2}$$

where:

M_w = molecular weight of water vapor

M_g = molecular weight of the carrier gas

$e_w(T_s)$ = saturation vapor pressure over a plane surface of the pure phase of liquid or solid water at the saturator temperature T_s

P_s = saturator pressure

f = enhancement factor

b. Enhancement Factor

The enhancement factor f at the saturator pressure P_s and temperature T_s is expressed by:

$$f(P_s \cdot T_s) = \frac{x_w P_s}{e_w(T_s)} = \frac{(1 - x_g) P_s}{e_w(T_s)} \tag{10.3}$$

where:

x_g, x_w = the mole fractions of gas and water vapor in the saturated mixture, respectively

c. Relative Humidity

The definition of the relative humidity RH in the test chamber of the generator is:

$$\%RH = (x_v/x_w) P_c T_c \times 100 \tag{10.4}$$

where:

x_v = the mole fraction of water vapor in a given sample of moist air characterized by pressure P_c and temperature T_c

x_w = the mole fraction of water vapor in the saturated mixture at the same values of pressure P_c and temperature T_c

Substituting appropriate expressions for the mole fractions yields:

$$RH = \frac{f(P_s \cdot T_s)}{f(P_c \cdot T_c)} \times \frac{e_w(T_s)}{e_w(T_c)} \times \frac{P_c}{P_s} \times 100 \qquad (10.5)$$

where:

$f(P_c \cdot T_c)$ = enhancement factor at test chamber pressure P_c and temperature T_c

$e_w(T_c)$ = saturation vapor pressure over a plane surface of the pure phase of liquid or solid water at the test chamber temperature T_c

P_c = test chamber pressure.

d. Dew/Frost Point

The thermodynamic dew point (or frost point) temperature T_d of a moist gas at absolute total pressure P is defined as that temperature at which the moist gas is saturated with respect to a plane surface of pure liquid (or solid) water. The dew point T_d of the moist gas of the two-pressure generator is obtained by the iterative solution of:

$$f(P_c \cdot T_d)e_w(T_d) = f(P_s \cdot T_s)e_w(T_s) \frac{P_c}{P_s} \qquad (10.6)$$

where:

e_w and f = values obtained from suitable tables

P_c = is the absolute pressure in any space or volume that is filled with the moist gas, e.g., the mirror chamber of a dew point hygrometer

e. Volume Ratio

The volume ratio V of the moist gas of the two-pressure generator is:

$$V = \frac{f(P_s \cdot T_s)e_w(T_s)}{P_s - f(P_s) \cdot T_s)e_w(T_s)} \qquad (10.7)$$

Hence, with the establishment of constant temperature and pressures in the saturator and test chamber, the various units of humidity can be calculated for the moist gas produced by the two-pressure humidity generator. The formulations of Wexler and Greenspan[23] and Goff[13] are used for obtaining the saturation vapor pressure of water and ice respectively. When air is used as the carrier gas, the values for the enhancement factor f or air, given by Hyland[26] over the temperature range of −50°C to +90°C (−58°F to +194°F) and pressures from 0.25×10^5 to 10^7 Pa are used in the above equations. Greenspan[29] has obtained a simplified equation for f which can be easily programmed for a computer or can be calculated with the aid of a programmable pocket calculator.

The RH generated by this two-pressure principle is dependent only on the pressure and temperature of saturation and on the temperature and pressure after expansion. Precise control of the generated humidity is obtained by measuring and controlling these values. Since the generated humidity is based entirely on fundamental measurements of temperature and pressure, no standard humidity sensors need to be used.

B. National Physical Laboratory (NPL) United Kingdom

This laboratory, located in Teddington, near London, holds the British standard for humidity, based on a primary gravimetric hygrometer (PGH) and a standard humidity generator (SHG). This facility was built in the 1980s in response to an increasing demand for calibration of humidity instruments in Britain and throughout Europe. This gravimetric hygrometer contains many improvements over the much older system employed at NIST in the USA, although both are considered to offer about the same degree of accuracy. Photographs of the PGH hygrometer and NPL calibration facility generator are shown in Figures 10.8 and 10.9 respectively. The NPL system can cover calibrations in the range of –75°C to +90°C (–102°F to +196°F) with an accuracy of ± 0.05°C (± 0.09°F) or better.

Figure 10.8   PGH hygrometer.
                       (*Courtesy Sira, Ltd.*)

Figure 10.9   NPL humidity calibration facility.
                       (*Courtesy National Physical Laboratory*)*

---

* Crown Copyright 1995, Reproduced by permission of the controller of HMSO.

1. NPL Gravimetric Hygrometer[51, 52]

The NPL gravimetric hygrometer was developed by Sira under a contract with the U.K. Department of Trade and Industry, and was later installed and commissioned at NPL. The NPL humidity generator was constructed on location.

There are national standards of humidity in other countries, such as Italy, which have been established on the basis of a dew point generator alone. Elsewhere, standard facilities have used generators or transfer standard hygrometers like that of the French standard for nominal values of humidity, while placing ultimate reliance on a gravimetric primary standard such as at the NIST facility in the USA and NPL in the U.K. The performance of the NPL humidity generator is known to have a level of uncertainty comparable to, or better than, that quoted by other national standards laboratories.

Perhaps the most fundamental expression of humidity is the establishment of saturation vapor pressure in equilibrium with a surface of water or ice at a particular temperature. However, in practical situations the humid atmosphere is isolated and static. This makes it an impractical method for calibrating most transfer standards, which are designed to sample a flow of gas and then exhaust it to the atmosphere. A practical alternative is to return the measured gas to the saturation chamber, first conditioning it in a temperature such that the equilibrium is not disturbed. This is the essence of a recirculation humidity generator with ultimate reference to the base unit of temperature.

The most accurate determination of humidity is by gravimetric measurement of humid gas, i.e., by separating the water from the carrier gas and weighing the two components with reference to the base unit of mass. However, the humid gas measured in this way is consumed and cannot be returned to a process. This method also has the disadvantage of being time consuming so that it would be an impractical basis for day-to-day calibration service.

The NPL facility combines both approaches by using a recirculation humidity generator which permits the consumption of a moderate flow of gas by a gravimetric hygrometer or other instrument. The lost gas is replaced with gas pre-conditioned to a virtually identical humidity which is then carefully equilibrated in the recirculation system. Thus the generator can be compared against a gravimetric hygrometer with only a small penalty in departing from an ideal recirculation design.

Figure 10.10 shows a schematic of the NPL gravimetric hygrometer. Dry gas is passed over a surface of water or ice within a saturator controlled at the desired dew point temperature in the range of −75°C to +80°C (−103°F to +176°F). The gas is recirculated a number of times and achieves an equilibrium saturation at the temperature indicated by a platinum resistance thermometer (PRT) situated at the exit of the saturator, subject to efficiency and draw-off rates. Gas from the system is bled off at a rate of typically 1 liter/min. (61 in³/min.), and supplied to the instruments under test.

Filtered air is supplied to the system via an air dryer which normally reduces the inlet gas dew point to below approximately −80°C (−112°F). This dry gas then enters the system's piping through a pressure controller which maintains an operating pressure of 105 kPa. For all except the lowest dew points, the gas passes through an initial stage of moisture conditioning (pre-saturator) to bring it close to the desired humidity. The gas

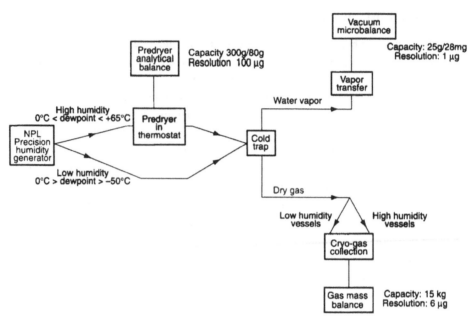

Figure 10.10   Schematic of NPL gravimetric hygrometer.

then achieves final equilibrium within the main saturator. Following this, the gas proceeds into the recirculation path where flow is maintained by a centrifugal impeller.

The saturator consists of a flat coiled 19 mm (3/4 in.) diameter steel tube, half filled with water or ice, with a spiral inlet pipe of sufficient length to equilibrate the temperature of the gas with that of the bath, before humidification takes place. The saturator is held in a bath controlled to a stability of 0.03°C (0.054°F) or better.

On leaving the main saturator, the gas immediately passes to a manifold where it is sampled by the instrument under test, or by the gravimetric hygrometer. The effects of leaks and desorption within the piping are minimized by the small distance between the saturator and manifold, and by the relatively fast recirculation flow. This flow also maxi-mizes the heat transfer between the gas and the PRT at the saturator exit, where the defin-itive temperature measurement takes place. Flow around the system is monitored with a mass-flow meter and laminar flow element with low pressure gradient. All components in the flow path are made of internally polished stainless steel except for teflon O-rings and ball valves. The use of at least 12.7 mm (1/2 in.) internal diameter tubing throughout ensures that pressure gradients are minimal.

For measurements at dew points above room temperature, a thermostatically-controlled enclosure and trace heated lines are employed to ensure that no moisture is lost by condensation in the flow path. In the case of low dew points of –50°C (–58°F) and below, the pre-saturator is omitted. Prior to all low and medium range measurements, the piping between the saturator and the instrument is back purged with dry gas until just before the test gas of the required humidity is allowed to flow. This conditions the piping surfaces much faster than if the test gas itself is used to dry the system to the required level.

All temperature measurements are made using a precision-resistant bridge. Automatic data logging is carried out using a microcomputer. LED, and LCD displays of instruments under calibration and can be monitored by a computer using a CCD video camera with a real-time digitizer. Additional monitoring of the generator's operation is provided using a transfer standard quality hygrometer. Like the NIST system, the PGH operates by separating a humid gas stream into its components—water and dry gas—and weighing the two to give a result in terms of mixing ration. The humid gas is supplied by the standard humidity generator.

For measurements at humidities above about 1°C (34°F) dew point, the bulk of the water is separated out by passing the gas through a pre-dryer, a stainless steel vessel containing a molecular sieve held in a bath controlled at 1°C (34°F). For lower humidities, the pre-dryer is omitted. The remaining moisture is then extracted by passing it through a specially designed cold trap. The water is retained as ice on the surfaces of the trap which is held at a low temperature approaching that of liquid nitrogen. This method of water extraction is very efficient. The water collected in the cold trap is later evaporated onto a desiccant and weighed on a micro balance in vacuum. A diagram showing the operation of the NPL standard gravimetric hygrometer is shown in Figure 10.10. The dry gas which emerges is collected by cooling and liquefying it in an aluminum cylinder. This enables the collection of up to 1 kg (2.2 lb.) of gas, with a corresponding amount of water collected varying between 10 mg (0.35×10^{-3} oz.) and 80 g (2.8 oz.) across the operating range.

The gas and water containers are each weighed before and after the collection process against an identical reference vessel, thus compensating for the effects of influences, such as air buoyancy. The end result is a measurement in terms of mixing ratio which can be converted into other units such as dew point. As a result, the humid gas can be provided with a calibration traceable to the base unit of mass.

2. Performance Limits

Operating range in grams of water per kilogram of dry gas	0.007 g/kg to 170 g/kg mixing ratio
Corresponding dew point range	–60°C to +60°C (–76°F to +140°F)
Accuracy above 0.08 g/kg (above –40°C or –40°F)	± 0.2% mixing ratio (± 0.03°C dew point) (± 0.054°F)
Accuracy below 0.08 g/kg (below –40°C or –40°F)	± 0.2% to ± 1.5% mixing ratio (± 0.02°C to ± 0.1°C frost point) (± 0.036°F to ± 0.18°F)

3. Standard Humidity Generator (SHG)[53]

The NPL standard humidity generator is the primary standard for humidity in terms of dew point temperature. It is used to provide calibrations for transfer standard hygrometers traceable to NPL. The humidity generator is also a secondary standard for mixing ratio through calibration against the gravimetric hygrometer. Figure 10.9 shows a photograph of the NPL humidity calibration facility, and Figure 10.11, a schematic of the NPL humidity generator.

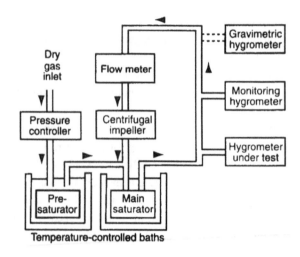

Figure 10.11   Schematic of NPL humidity generator.

The system consists of a recirculation system in which gas passes over a surface of water or ice controlled at the desired saturation temperature. After a number of recirculation passes, the gas reaches an equilibrium of saturation at the temperature indicated by a thermometer situated at the exit of the saturator. The process is controlled at a pressure of 105 kPa (15.23 psi) and humidity gas is bled off to instruments under test at a rate of up to 1.5 liters/min. (346 in.^{3}/min.).

4. Performance of the Standard Humidity Generator [56]

Dew Point Range	–75°C to +82°C (–103°F to +180°F)
Accuracy	± 0.10°C @ –75°C (± 0.27°F @ –103°F)
	0.03°C @ +20°C (± 0.09°F @ 68°F)
	± 0.04°C @ +82°C (± 0.144°F @ 180°F)
Generator Pressure	105 kPa ± 0.02 kPa absolute
	(15.23 ± 0.003 psi)
Gas Withdrawal Rate to External Hygrometer	Up to 1.5 liters/min. (346 cu. in./min.)

5. Acknowledgement

Information and data on the NPL gravimetric hygrometer and the NPL standard humidity generator were abstracted from publications by M. Stevens, S.A. Bell, and D. Armitage of NPL, References 54, 55, and 56, and from the NPL publication titled, "Humidity measurement standards for the UK" under Crown copyright January 1994. This information is published by permission of the Controller of HMSO.

6. Transfer Standard Calibration Facilities in the UK (NAMAS)

Where the highest accuracy calibrations are required, the calibration should be performed by NPL. However, for day-to-day calibrations, traceable to NPL, users in the UK can turn to one of a number of humidity calibration facilities which are approved as competent and traceable to NPL. Such approvals are provided by the national accreditation of measurement and sampling (NAMAS). There are at present several NAMAS approved and certified humidity calibration facilities in the U.K.

Recently NAMAS combined with NACCB to form UKAS, the "United Kingdom Accreditation Service." The NAMAS accreditation service is now provided by UKAS.

7. Sira Test and Certification Ltd.

Sira Test & Certification Ltd., (ST&C) is an internationally recognized organization which serves both manufacturers and users of industrial and scientific equipment. It offers a comprehensive range of calibration, test, quality assurance, consulting, training, and certification services.

a. Sira Calibration

ST&C operates one of the UK's leading calibration centers, offering one of the most comprehensive ranges of NAMAS accredited facilities. Measurements covered by Sira include, but are not limited to, temperature, pressure, optics, flow, and humidity.

The NAMAS accredited humidity measurement laboratory is capable of calibrating all commonly available hygrometers. Humidity, temperature, pressure, and composition of the test gas can be varied over a wide range to allow calibrations to be carried out under most actual conditions of use.

Humidity calibration is traceable to the UK National Humidity Standard at the National Physical Laboratory (NPL). The primary gravimetric hygrometer that is part of this standard was designed and built by Sira. The Sira Test & Calibration Facility has a dew point accuracy of 0.15°C (0.27°F) and a dew point temperature range of –75°F to +82°C (–103°F to 180°F). This represents the best accuracy and range of any metrology laboratory in the UK. Each hygrometer calibrated by Sira is provided with a NAMAS certificate which states the test conditions, instrument outputs, and measurement uncertainties.

Automation of the calibration process allows routine calibrations, for example calibrating relative humidity hygrometers at five points at 20°C (68°F), which can be performed within a short time and at low cost without sacrificing range or accuracy.

Standard calibrations are carried out with electronics/display units at 20°C ± 2°C (68°F ± 3.8°F) and less than 60% relative humidity, but other conditions can be provided. Temperature measurement is included in the accreditation, so the temperature outputs of hygrometers are calibrated at the same time. Sira also offers the service of calibration and characterization of environmental test chambers, using portable equipment, and consulting services for particular humidity measurement problems.

Instruments that are routinely calibrated by Sira include condensation dew point hygrometers, dew point probes, relative humidity instruments, psychrometers, mechanical hygrometers, and humidity generation systems, such as saturated salt standards, and calibrators.

8. SIREP International Instrument Users Association

SIREP is an independent non-profit company owned by its members, all of whom are instrument-using organizations. SIREP was founded in the UK in 1961, and has more than 30 members internationally. These members, through SIREP, and in cooperation with individual instrument manufacturers, sponsor the evaluation of industrial measurement and control systems. Reports are exchanged with WIB, a similar international group of instrument users whose operations are based in The Netherlands, and with EXERA in France. A SIREP member receives about 40–70 comprehensive reports per year, all written in English, and in a standard format. The reports are impartial and objective, prepared by independent test laboratories.

a. Instrument Users

For instrument users, the independent laboratory evaluations provide much of the detailed performance information needed, particularly on new models, to make intelligent selection and purchasing decisions. These reports are also often an important part of quality assurance programs. SIREP member users typically receive copies of about 60 different evaluation or survey reports each year at a cost of less than the cost of one individual project. In addition, members are periodically updated on instrument trends and field experience through surveys and regular technical meetings in Europe and North America.

b. Instrument Manufacturers

Instrument manufacturers can also participate in, and initiate, SIREP evaluations. In recent years up to 40% of the SIREP evaluations were sponsored by equipment manufacturers. The internationally recognized authority of independent SIREP evaluation reports provides an effective means of promoting a product world-wide. In addition, the feedback of unbiased evaluation data gives valuable information to the manufacturer to guide its product improvement programs.

c. Membership

Each SIREP member company pays a minimum basic annual subscription which is used to fund the evaluation of instruments in which the company is interested. At the annual general meeting, members elect a Board of Directors. The Board is assisted by a management staff for carrying out the day-to-day activities and monitoring progress of evaluations at the laboratories.

Frequent regular SIREP technical panel meetings are held in Europe and North America where decisions are made on instruments to be evaluated, and on significant test results which are reported verbally to members. The meetings also provide a forum for the discussion of mutual problems and experience between senior instrumentation personnel in major user companies. The meetings are often organized around visits to members' plants or manufacturers' facilities. The combined membership is drawn from 16 different countries, indicating the extent of international cooperation in instrument evaluation.

d. Evaluation Procedure

The purpose of an evaluation is to demonstrate whether or not the manufacturer's performance specifications for an instrument, are met. The evaluation projects are conducted at
independent laboratories, for the most part at Sira. The evaluation laboratory prepares a
test program, based on international and national standards, and provides a fixed price
quotation for approval by SIREP and the manufacturer.

The test program defines a series of tests which simulate, under controlled conditions, various working environments likely to be encountered in industry. The manufacturer lends the instrument to the independent laboratory for the evaluation. The defined
tests are undertaken by qualified professional personnel using test equipment with measurement uncertainties traceable to national standards. If the instrument does not perform
as expected, the manufacturer is informed and asked to service or repair its product.

e. Reports

The laboratory produces an objective draft report on the evaluation for comment by the
sponsors and the manufacturer. The manufacturer is allowed to provide its formal comments on the report together with the manufacturer's technical specification sheet.
Abridged reports giving the introduction, conclusions, and manufacturer's comments, are
also printed. The final reports are mailed on a regular basis to all SIREP members.

C. Calibration Facilities at CETIAT, France[67, 68]

The calibration system, developed and put into use at CETIAT, is a recirculation humid
air generator. The system is useful for calibrating dew point hygrometers and also for
secondary change of impedance type %RH sensors.

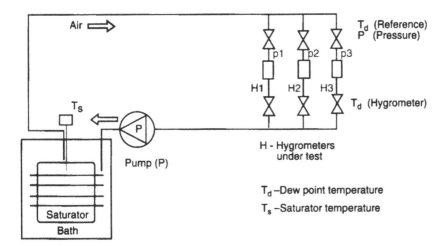

Figure 10.12   Recirculation humid air generator for dew point calibration.

The main part of this calibration system is a saturator filled with distilled water. When equilibrium is reached in the saturator, the vapor pressure saturates at the water surface temperature or at the temperature of the ice contained in the saturator as shown in Figure 10.12.

At the discharge end of the saturator, the humid air is fed into the measurement chambers of the hygrometers which are to be calibrated (H_1, H_2, and H_3). Humid air is then returned to the saturator by means of a circulating pump (P). When equilibrium is reached in the saturator, the dew point measured by the instrument under calibration is compared to the temperature read on probe (T_s) in the saturator. The saturator vapor pressure, corresponding to the dew/frost point temperature, is corrected, if necessary, to account for any difference in pressure in the saturator versus the hygrometers being calibrated. In practice, this correction is often negligible. Hygrometers can be tested at various dew points by equilibrating and operating saturators at selected dew points.

For calibrating relative humidity sensors, the air leaving the saturator is fed into a temperature controlled test chamber as shown in Figure 10.13.

By varying and equilibrating the temperature of the RH test chamber, the %RH can be accurately determined and changed to obtain as many calibration points as are desired. Typically, an RH instrument is calibrated at a minimum of 3, but usually at 5 or 6 points.

CETIAT's transfer standard (chilled mirror hygrometer) was checked against NIST. Factors that could contribute to measurement errors are:

1. *The imperfect nature of the saturator.*

    The air may not be 100% saturated as it exits the saturator. This is especially critical for very low dew point measurements.

2. *Temperature measurement.*

    - Proper installation of the temperature probes

    - Thermal conductivity between the temperature sensing element and the point where the probe is attached to the saturator

    - Calibration of the temperature probes and their electronics

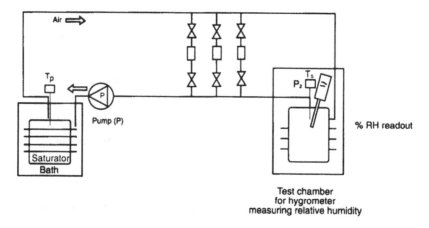

Test chamber
for hygrometer
measuring relative humidity

Figure 10.13   Recirculation humid air generator for %RH calibration.

3. *Proper equilibrium of the dew point temperature at the point where the hygro-meter is calibrated.*

4. *Proper air flow in the circulator.*

   Insufficient flow would cause problems when calibrating several hygrometers at a the same time. Excessive flow could result in water deposits in the sample lines and on the mirror surface if chilled mirror hygrometers are calibrated.

5. *Errors in pressure correction between the saturator and sensors being calibrated.*

One feature of the CETIAT calibration system is its capability to perform, not only as a humid air generator, but also as a dew point calibration reference. Reportedly, the satura-tor can provide humid air in the dew point range from –60°C (–76°F) to +85°C (+185°F) with an uncertainty known to be within ± 0.05°C to ± 0.1°C (±0.09°F to ± 0.18°F).

D. Ecole Polytechnique, Two-Temperature Calibration System, France [63]

1. General

This system was built in 1970 at the Laboratoire de Meteorologie Dynamique du CNRS in France, to perform calibration of atmospheric humidity sensors for balloon-borne experiments. It later proved to provide good enough accuracy for calibration of other humidity instruments as well. In this system moist air circulates in a closed circuit through a saturator, which is temperature controlled to determine the dew point, followed by a sensor chamber.

The saturator and sensor chamber temperatures can be easily controlled between –80°C and +30°C (–112°F and +86°F) so that the dew point can be set at any value and the sensor temperature can be chosen to simulate high altitude atmospheric conditions. This system can also generate relative humidities over a wide temperature range. The pressure can be varied from 100 kPa to 10 kPa (14.5 psi_a to 1.45 psi_a). Dew points and temperatures are measured with platinum resistance sensors. A standard chilled mirror hygrometer is used to monitor the sample air.

The life expectancy of scientific sounding balloons is a few days to several months, depending on the type of balloon used. Among the basic atmospheric parameters, humidity is the most difficult to measure. At high altitudes the water vapor content is extremely low and instruments also have to work at very low pressures and temperatures. Outgassing of the balloon and instrument components has to be carefully considered. For long duration flights, sensors must be sufficiently stable since no calibrations are possible during the flight.

Hygrometers must be tested in an accurate calibration facility which is capable of simulating atmospheric conditions at different altitudes. For this reason the two-temperature method was chosen.

Although this facility was primarily built for calibrating high altitude humidity sensors, its design is also of significant value for general use to complement metrology laboratories like NPL and CETIAT.

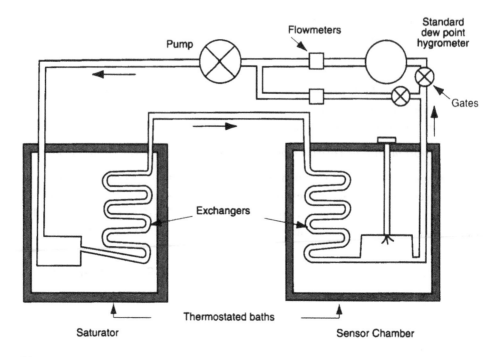

Figure 10.14   Schematic diagram of calibration system.

2. Description

A schematic diagram of the system is shown on Figure 10.14. Moist air is circulated in a closed circuit through the saturator at temperature T_1, followed by the sensor test chamber at temperature T_2 which is higher than T_1. A dew point hygrometer is used to maintain equilibrium and detect any discrepancies. At equilibrium, water vapor is saturated at T_1 and the dew point is given by the measurement of T_1. The relative humidity follows from the well known equation:

$$RH = 100 \ \frac{e_{S1}}{e_{S2}} \qquad (10.8)$$

where:

e_{S1} = Saturation vapor pressure at temperature T_1

e_{S2} = Saturation vapor pressure at temperature T_2

The saturator and test chamber are placed inside the thermostatically controlled liquid baths, which are temperature regulated. All sample lines and components that are in direct contact with the humid air must be made of non-hygroscopic materials. The chambers and sample lines are therefore made of special polished stainless steel and the pump uses a teflon diaphragm. The volumes are reduced to improve the response of the system at low frost points, especially when measuring below –60°C (–76°F).

The calibration system is designed for "low" or "high" humidity levels. Two interchangeable saturators are used, one for vapor pressures form –10°C to +30°C (+14°F to +86°F) and the other for –80°C to +10°C (–112°F to +50°F).

3. Low Humidity

Figure 10.15 shows the low humidity configuration of the system with pressure adjust-ment capability. Dew point and sensor temperatures can be controlled independently between +10°C and –80°C (+50°F to –112°F) and the pressures between 100 kPa and 10 kPa (14.5 psi$_a$ to 1.45 psi$_a$). The pressure is controlled by a manometer using a Piezoelectric sensor. This manometer is calibrated using a quartz-Bourbon pressure gage. For calibrating relative humidity sensors, the sensors are placed in the sensor chamber. Knowledge of the dew point, temperature, and pressure, allows calculation of any other humidity parameter. If dew point hygrometers are to be calibrated, they can be placed in the sampling system in series with the standard chilled mirror hygrometer.

4. High Humidity

The high humidity configuration uses the same design, except that the sample air and components that come in contact with the sample are mounted inside a thermostatically controlled enclosure to avoid water condensation when the dew point is higher than the ambient air temperature. The saturator design is also quite different since the air passes first through the exchanger and is then forced to pass over the water surface in the chamber.

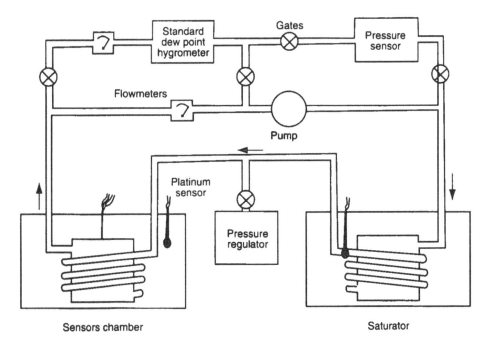

Figure 10.15   Low humidity calibration system.

E. National Research Laboratory for Metrology (NRLM), Japan

The National Research Laboratory for Metrology (NRLM) started to develop humidity standards and offer humidity calibration services in the 1960s. This laboratory has also developed and built a standard gravimetric hygrometer which is used with precision two-pressure and two-temperature systems. Figure 10.16 shows a flow chart of the humidity standard, standard generator, and dew point hygrometer as a transfer standard.

1. NRLM Gravimetric Hygrometer [76]

NRLM employs standards similar to those used at NIST, as previously discussed. A flow chart of the NRLM hygrometer is shown schematically in Figure 10.17.

In the NRLM gravimetric hygrometer, the mixing ratio of moist air is determined by an absolute measurement. The mass of water vapor is determined by weighing the amount absorbed by the desiccant in glass U-tubes. The quantity of dry air is determined by measuring its pressure and temperature after it is fed into a cylinder of known volume.

Unique features of the NRLM hygrometer are:

- The use of a four-way valve for sampling

- Provision of a by-pass of the U-tube for air flushing

- Control of the sample air flow rate by a critical (sonic) nozzle

- Computer control of valve operation, manometric measurements and calculations

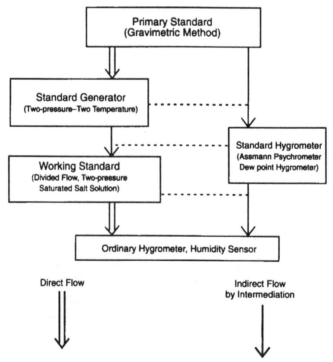

Figure 10.16   Flow chart of NRLM humidity standard.[70]

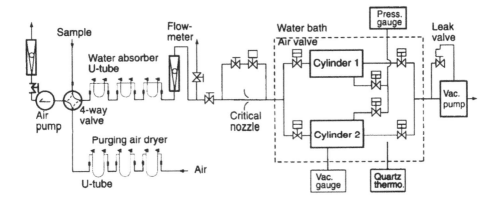

Figure 10.17   Schematic diagram of NRLM gravimetric hygrometer.

Sample air is passed through the U-tubes, which are filled with desiccant, into a cylinder of known volume. The mass of water m_w in the sample air is determined by weighing the U-tubes with an electrical balance before and after absorption. The mass of air m_a in the cylinder can be calculated from:

$$m_a = \frac{M_a PV}{zRT} \qquad (10.9)$$

Where:
M_a = Molar mass of dry air (2.89635 × 10^{-2} kg/mol)
m_a = Mass of air in the cylinder
P = Pressure in the cylinder
V = Volume of the cylinder
z = Compressibility factor
R = Gas constant (8.31441 J. K^{-1} mol^{-1})
T = Temperature in the cylinder

The mixing ratio of the sample air is the ratio between the mass of water, m_w, and m_a (m_w/m_a).

As shown in Figure 10.17, incoming air is fed to a 4-way valve, which supplies the sampling air, purges the air dryer, and feeds air through the water absorber. The system also contains a thermostatically controlled water bath and a control panel for automatic valves. Sample air is introduced through a fluorocarbon polymer tube. An air pump, flow meter, and flow control valve, are provided to purge the tube with sample air before each run. For purging, the 4-way valve is switched to the "broken line" position shown in Figure 10.17. During measurement the valve is in the "solid line" position and sample air is passed into the water absorber.

The purging air dryer consists of three glass U-tubes, filled with desiccant, and connected in series. Air from this dryer is used to purge the water absorber line. The water absorber also contains three U-tubes. The flow controller contains a "critical nozzle" and a bypass line with a solenoid and needle valve. The bypass line is only used for purging and drying the lines in the instruments. Although the pressure of the cylinder changes from near vacuum to atmospheric pressure, air flow can be controlled by the "critical nozzle." The flow rate is not affected by the downstream pressure when the pressure ratio of the up and down stream is larger than a constant value.

The hygrometer has two cylinders which are immersed in a thermally controlled water bath. The temperature of the water bath is controlled within 0.01°C (0.018°F).

2. Precision NRLM Humidity Generator[75, 78]

The NRLM precision humidity generator was designed to establish a humidity standard and to calibrate humidity sensors. The generator provides air of constant humidity, in the range of 10% to 100% RH with an uncertainty of 0.2%. It was calibrated against the NRLM gravimetric hygrometer. Principal features of the generator are:

- The use of two-pressure and two-temperature methods at the same time to extend the humidity range

- Addition of a pre-saturator to complete the saturation and to stabilize the temperature of the saturator

- A test chamber immersed in a thermostatically-controlled water bath of which the temperature is controlled within 0.01°C (0.018°F)

- Pressure measurement using a fused quartz Bourdon gauge and temperature measurement with a quartz thermometer

- Use of special non-porous stainless steel in the test chamber and sample lines

There are several methods to generate the atmosphere of known humidity. Nowadays, measurements of pressure and temperature are more accurate and precise than those of volume, flow rate, etc. Therefore the two-pressure and two-temperature methods were chosen for the standard humidity generator at NRLM. The relative humidity in the test chamber is obtained from the following equation:

$$\%RH = \frac{f(P_s, t_s) \cdot e_s}{f(P_t, t_t) \cdot e_t} \times \frac{P_t}{P_s} \times 100 \tag{10.10}$$

Where:
f = Enhancement factor
P_s = Pressure in the saturator
t_s = Temperature in the saturator
e_s = Saturated water vapor pressure at temperature t_s
P_t = Pressure in test chamber
t_t = Temperature in test chamber
e_t = Saturated water vapor pressure at temperature t_t

A schematic of the humidity generator is shown in Figure 10.18. Compressed air is used as a carrier gas. The maximum pressure is 700 kPa (49.8 psi), but 500 kPa (35.6 psi) is normally used and obtained with a pressure regulator. A pre-saturator is used to obtain

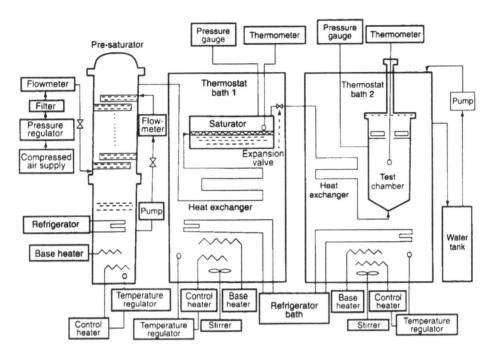

Figure 10.18   Schematic of NRLM humidity generator.

water vapor at a temperature higher than in the saturator. Two temperature controlled water baths are used, one for temperature control in the saturator, and the other for the test chamber. Both are temperature controlled to within 0.01°C (0.018°F).

NRLM employs primary standards similar to those used at NIST. A detailed description would be repetitive. Reference made is to the list of references and bibliography in Chapter 15.

F. Physicalish-Technische Bundesanstalt (PTB)

The PTB laboratory operates a gravimetric hygrometer as shown schematically in Figure 10.19, but provides most calibration services using a two-pressure generator as shown in Figure 10.20. In addition PTB operates a coulometric trace moisture generator for calibrations in the low ppm range. This system is shown in Figure 10.21.

1. Two-Pressure System

Compressed air or other non-reactive gas is at saturated at a certain pressure and temperature with respect to water or ice. A pressure reducing valve drops the pressure to atmospheric or any other desired pressure. The basic generator consists of the saturator inside the thermostatic bath. This is in principle a heat exchanger.[80] Wet air is taken from the saturator at a temperature a few degrees higher than the thermostatic bath and flows into the saturator through a number of plates. The vapor pressure of the air coming out of the saturator

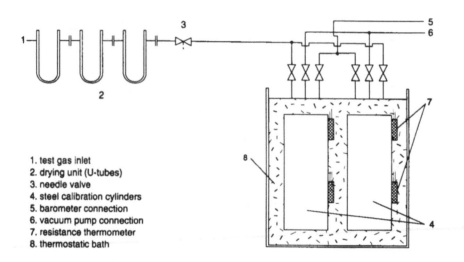

1. test gas inlet
2. drying unit (U-tubes)
3. needle valve
4. steel calibration cylinders
5. barometer connection
6. vacuum pump connection
7. resistance thermometer
8. thermostatic bath

Figure 10.19   PTB gravimetric hygrometer.

equals the saturation pressure with respect to the temperature of the bath. From this, the mixing ratio can be calculated from the temperature and pressure of the saturator as follows:

$$r = \frac{0.62198 \, e_w{}'}{P - e_w{}'} \qquad\qquad (10.11)$$

Where:
   r = Mixing ratio
   P = Pressure
   $e_w{}' = f e_w$
   f = Enhancement factor
   e_w = Saturation vapor pressure at saturation temperature

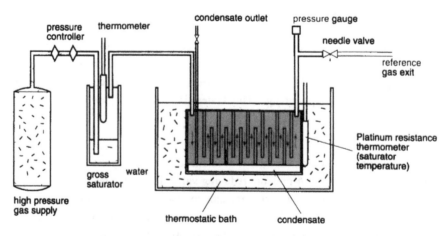

Figure 10.20   Schematic of PTB two-pressure generator.

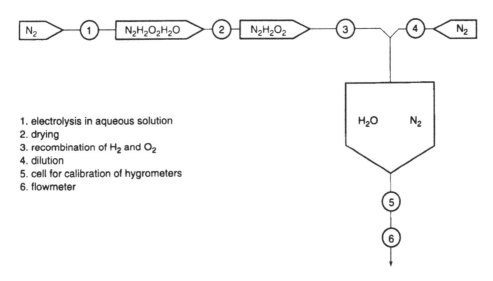

1. electrolysis in aqueous solution
2. drying
3. recombination of H_2 and O_2
4. dilution
5. cell for calibration of hygrometers
6. flowmeter

Figure 10.21   Flow sheet of the PTB coulometric trace humidity generator.

If there is no evaporation or condensation in or behind the pressure reducing valve, the mixing ratio will remain constant, and at every point downstream, the humidity parameters can be calculated from the mixing ration and values of temperature and pressure. Conversion to different humidity parameters can be performed using the known formulas. To be able to measure at dew points above the prevailing ambient temperature, all parts of the system that come in contact with the sample gas must be heated to a temperature above the dew point to be measured.

   Establishing relative humidity values requires the temperatures to be adjusted to the appropriate values. This is done in a thermostatically controlled chamber capable of controlling the temperature to better than 0.01 K. In this chamber the reference air is passed through various lines and cells for the calibration of the different types of humidity sensors.

- Available Ranges are:

   Flow rate of the reference air   = 0 to 15 liters/min. (0 to 915 in.^{3})
   Air Temperature   = –20°C to + 75°C (–4°F to +167°F)
   Dew Point Temperature   = –50°C to + 75°C (–58°F to +167°F)

- Uncertainties are:

   Air Temperature   = 0.02 K
   Dew Point Temperature   = 0.05 K
   Relative Humidity   = 0.05 %

2. Gravimetric Hygrometer

Figure 10.19 shows the PTB Gravimetric Hygrometer. This system is very similar to the NIST (NBS) system. The working range in terms of mixing ration is 1 g/kg to 30 g/kg. The relative uncertainty is 0.2%.

3. Coulometric Humidity Generator

This system is shown schematically in Figure 10.21. In an aqueous solution an electrolytic current causes the formation of H_2 and O_2 which are transported by a carrier gas running through the cell. A after drying by means of molecular sieves, the two components are recombined to water on a palladium contact. From the flow rate of the reference gas and electrolytic current, the humidity may be calculated using the equation:

$$W = k.I/v \qquad\qquad (10.12)$$

Where:

W = Parts of water vapor per volume in PPM
k = 456.4 liters/hour $\times$ mA
I = Electrolytic current in mA
v = Flow rate in liters/hour

Dilution with extremely dry nitrogen extends the humidity range towards lower dew points. This type of humidity generator is used for calibrations in the range of –75°C to –10°C (–103°F to +14°F) dew point temperature or 1 ppm_v to 3,000 ppm_v respectively. The uncertainty is estimated at about 0.1 K to 0.3 K dew point temperature.

G. Other National Standards Laboratories

Many other national standards laboratories are available for calibration certification, but do not all have a gravimetric hygrometer like NIST and NPL. They offer calibration services using primary standards such as the two-pressure, two-temperature, and chilled mirror systems. These international laboratories include:

- University of Wageningen, The Netherlands

  This university is primarily engaged in agricultural sciences and technology. The humidity calibration facility employs a two-pressure generator and a chilled mirror hygrometer. Calibration services are provided on a broad scale, but predominantly for agricultural related instruments. Much calibration work was performed by this laboratory on %RH sensors for plant growth areas.

- National Laboratories in Italy, Korea, Taiwan, China, and Singapore, Australia and others. For a more complete list and addresses, see Chapter 14.

Most of these organizations maintain calibration correlation with one of the large national laboratories that own a gravimetric system, such as NPL and NIST.

III. Commercially Available Primary Standards

A. Dew Point Calibration Chambers[44]

Primary standards like two-pressure and two-temperature systems that are used by calibration laboratories, government installations, and private corporations, are often custom built. However, there is a company in the USA, Thunder Scientific in Albuquerque, NM which offers a variety of two-pressure generators and chambers which are tailored after the NIST two-pressure system. Some of these, like the one shown in Figure 10.22, are small enough to be moved around on push carts and can thus be used at various test sites. These systems are often computer controlled and can be very effective and convenient to use. Since these calibration chambers have proven to be accurate, repeatable, and reliable, several metrology laboratories have obtained these commercial humidity generators rather than built their own.

An automated commercial two-pressure, two-temperature generator for very low frost point measurements is shown in Figure 10.23.

The instrument is capable of measuring humidity in the 0.05 ppm$_v$ to 12,000 ppm$_v$ range and frost points from −95°C (−139°F) to +10°C (+50°F) with an accuracy of ± 0.1°C (± 0.18°F). Such a unit is suitable for verifying calibration data of electrolytic, aluminum oxide, and chilled mirror sensors.

B. Relative Humidity Calibration Chambers

Another recently developed commercial humidity generator designed for relative humidity calibration, is shown in Figure 10.24. This generator has a very large temperature controlled chamber and utilizes an integral traceable chilled mirror hygrometer allowing calibrations to be performed with a high degree of accuracy and repeatability.

Figure 10.22   Self contained commercial two-pressure generator.
(Courtesy Thunder Scientific)

Figure 10.23   Two-pressure, two-temperature humidity generator.
(Courtesy Thunder Scientific)

The relative humidity chamber shown is based on the proven divided flow technology. The divided flow method is recognized as a reliable method for accurately controlling relative humidity at a fixed temperature. The user supplies a source of dry gas to the generator and provides distilled water as necessary. The desired RH is selected on the front panel. Relative humidity is automatically controlled by time proportioning a fraction of a constant flow dry air stream through a saturator and into a mixing chamber. The saturated air rejoins the remaining dry air and mixes to the desired RH value before flowing into the test chamber. The large chamber will accommodate numerous RH probes or several humidity recorders for efficient, simultaneous calibration. Continuous digital control, using the built-in traceable chilled mirror dew point hygrometer, makes the generator a traceable relative humidity transfer standard. A block diagram showing the operation of the generator is shown in Figure 10.25.

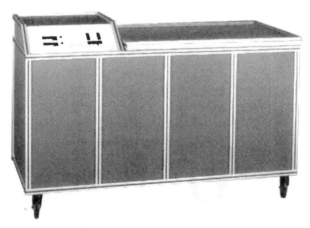

Figure 10.24   Relative humidity generator.
(Courtesy General Eastern)

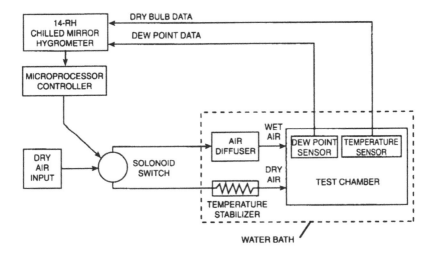

Figure 10.25   Block diagram of divided flow humidity generator.

During the wet air portion of the cycle shown in Figure 10.26, the air is thoroughly saturated by an efficient diffuser. It then passes into the test chamber. During the remainder of the cycle, the dry air enters the chamber, after having first been completely temperature stabilized so that it is at virtually the same temperature as the water in the jacket surrounding the chamber. This stabilization ensures that there is a very small temperature gradient between the incoming air and the chamber itself.

IV. Calibration Using Saturated Salt Solutions[33]

A convenient calibration method consists of the use of saturated salt solutions. At any temperature, the concentration of a saturated solution is fixed and does not have to be determined. By providing excess solute, the solution will remain saturated even in the presence of modest moisture sources and sinks. When the solute is a solid in the pure phase, it is easy to determine that there is saturation. The saturated salt solution, made up

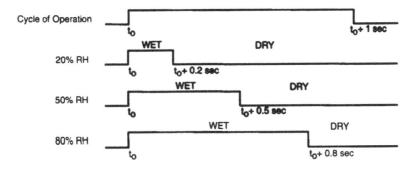

Figure 10.26   Wet/Dry air selection for RH generator.

as a slushy mixture with distilled water and chemically pure salt, is enclosed in a sealed metal or a glass chamber. Wexler and Hasegawa[8] measured the humidity in the atmosphere above eight saturated salt solutions for ambient temperatures 0°C to 50°C (32°F to 122°F), using a dew point hygrometer. Later, Greenspan[33] compiled from the literature data on 28 saturated salt solutions to cover the entire range of relative humidity. Using a data base from 21 separate investigations comprising 1106 individual measurements, fits were made by the method of least squares to regular polynomial equations to obtain the "best" value of relative humidity in air as a function of temperature. These values are summarized in Table 10.4. A graphic representation of the temperature dependence of saturated salts is given in Figure 10.27.

A sealed container in a temperature stable (± 0.25°C or ± 0.45°F) environment can protect the sensors from the influence of nearby people and allow controlled testing of the sensor performance. For extra stability, the test setup should be secured in an enclosure of its own before placing the whole assembly into the temperature-stable environment.

Environmental test chambers are not typically suitable by themselves. The temperature controls cycle the temperature over some dead-band, and the humidity controls on most chambers are primitive at best. A 1°C (1.8°F) variation in temperature can change the %RH by as much as 6%. However, good results can be achieved by first double insulating the humidity sensors in their own environment, thereby de-coupling them from the oscillations of an environmental test chamber.

Saturated salts are the simplest way to generate a known humidity environment that will produce credible results. The technique calls for selecting the proper salt to generate the humidity desired (see Table 10.4), placing the salt in a container, and adding water, but not enough to dissolve the salt. A substantial quantity of undissolved salt should be drifting in the bottom of the container at all times. Only pure salts and distilled water should be used, and the setup should be thoroughly cleaned every time salts are changed. Salt contamination will adversely affect results. Once the test chamber is sealed, the saturated salt solution will create the proper humidity in the air space above the water, but only if given enough time for the water temperature, air temperature, and humidity to all come to equilibrium. There should be no attempt to stir the salt, mix the air, or bubble the air through the solution. Such techniques do not work and cause errors in the test setup and results. Also, the sensor should not be immersed in the liquid. Constant humidity is maintained in the air space above the salt solution.

Other "constant humidity" solutions are available but most should be avoided. Many are highly acidic or alkaline or give off chemical vapors that might harm some sensors. For further information on constant humidity solutions see "Humidity Fixed Points of Binary Saturated Aqueous Solutions," Lewis Greenspan, NBS Journal of Research.[33]

If the salt solution comes into contact with the sensor during testing, the local humidity at the sensor surface will be affected and produce erroneous calibration data. The entire test assembly must be kept in a temperature stable environment for the duration of the test, or the results will be subject to large errors. The temperature stability is much more important than is implied by the temperature coefficient data in Table 10.4. This is because the air has much less thermal mass than the water/salt solution. It can be at a different temperature from that of the liquid. However, humidity values listed in Table 10.4

Table 10.4 Equilibrium relative humidity of saturated salts.[41]

Temperature °C	Lithium Chloride	Potassium Acetate	Magnesium Chloride	Potassium Carbonate	Magnesium Nitrate	Sodium Chloride	Potassium Chloride	Potassium Nitrate	Potassium Sulfate
				% Relative Humidity					
0	11.23 ± 0.54		33.66 ± 0.33	43.13 ± 0.66	60.35 ± 0.55	75.51 ± 0.34	88.61 ± 0.53	96.33 ± 2.9	98.77 ± 1.1
5	11.26 ± 0.47		33.60 ± 0.28	43.13 ± 0.50	58.86 ± 0.43	75.65 ± 0.27	87.67 ± 0.45	96.27 ± 2.1	98.48 ± 0.91
10	11.29 ± 0.41	23.28 ± 0.53	33.47 ± 0.24	43.14 ± 0.39	57.36 ± 0.33	75.67 ± 0.22	86.77 ± 0.39	95.96 ± 1.4	98.18 ± 0.76
15	11.30 ± 0.35	23.40 ± 0.32	33.30 ± 0.21	43.15 ± 0.33	55.87 ± 0.27	75.61 ± 0.18	85.92 ± 0.33	95.41 ± 0.96	97.89 ± 0.63
20	11.31 ± 0.31	23.11 ± 0.25	33.07 ± 0.18	43.16 ± 0.33	54.38 ± 0.23	75.47 ± 0.14	85.11 ± 0.29	94.62 ± 0.66	97.59 ± 0.53
25	11.30 ± 0.27	22.51 ± 0.32	32.78 ± 0.16	43.16 ± 0.39	52.89 ± 0.22	75.29 ± 0.12	84.34 ± 0.26	93.58 ± 0.55	97.30 ± 0.45
30	11.28 ± 0.24	21.61 ± 0.53	32.44 ± 0.14	43.17 ± 0.50	51.40 ± 0.24	75.09 ± 0.11	83.62 ± 0.25	92.31 ± 0.60	97.00 ± 0.40
35	11.24 ± 0.22		32.05 ± 0.13		49.91 ± 0.29	74.87 ± 0.12	82.95 ± 0.25	90.79 ± 0.83	96.71 ± 0.38
40	11.21 ± 0.21		31.60 ± 0.13		48.42 ± 0.37	74.68 ± 0.13	82.32 ± 0.25	89.03 ± 1.2	96.41 ± 0.38
45	11.16 ± 0.21		31.10 ± 0.13		46.93 ± 0.47	74.52 ± 0.16	81.74 ± 0.28	87.03 ± 1.8	96.12 ± 0.40
50	11.10 ± 0.22		30.54 ± 0.14		45.44 ± 0.60	74.43 ± 0.19	81.20 ± 0.31	84.78 ± 2.5	95.82 ± 0.45
55	11.03 ± 0.23		29.92 ± 0.16			74.41 ± 0.24	80.70 ± 0.35		
60	10.95 ± 0.26		29.26 ± 0.18			74.50 ± 0.30	80.25 ± 0.41		
65	10.86 ± 0.29		28.54 ± 0.21			74.71 ± 0.37	79.85 ± 0.48		
70	10.75 ± 0.33		27.77 ± 0.25			75.06 ± 0.45	79.49 ± 0.57		
75	10.64 ± 0.38		26.94 ± 0.29			75.58 ± 0.55	79.17 ± 0.66		
80	10.51 ± 0.44		26.05 ± 0.34			76.29 ± 0.65	78.90 ± 0.77		
85	10.38 ± 0.51		25.11 ± 0.39				78.68 ± 0.89		
90	10.23 ± 0.59		24.12 ± 0.46				78.50 ± 1.0		
95	10.07 ± 0.67		23.07 ± 0.52						
100	9.90 ± 0.77		21.97 ± 0.60						

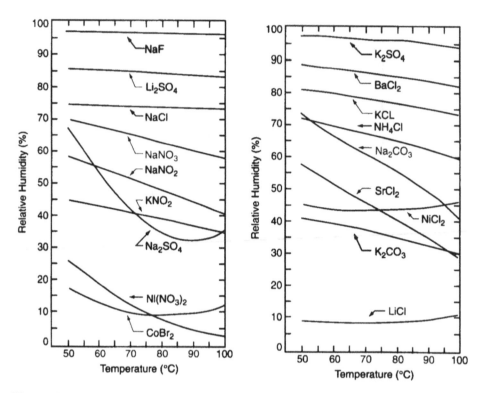

Figure 10.27   Temperature dependence of saturated salts.

are based on the temperature of the liquid and the air in the chamber when in equilibrium. A 1°C (1.8°F) fluctuation at 97.6% RH and 20°C (68°F) will result in an error of 6% RH. The same 1°C (1.8°F) fluctuation at 54.4% RH will cause an error of 3.5% RH. This calibration method can provide a known humidity accurate to ± 2% RH if all the precautions are followed. This method is not recommended for in-process production calibration of humidity sensors. The method is sometimes employed for field calibrations by using a simple saturated salt kit. However, the accuracy of such a field calibration kit is poor.

NPL and some other laboratories encourage the use of saturated salt solutions as a stable transfer medium for one calibrated sensor to then calibrate other sensors, but not as definitive values by themselves. Problems associated with making up the saturated salt solutions often create significant errors which can be undetected if the saturated salt solution itself is not certified by a calibrated sensor.

V. Calibration in the Low PPM Range [45, 46]

Standards laboratories in the USA and overseas can provide accurate and fundamental calibration at water vapor levels down to −70°C (−94°F) frost point, corresponding to 2.55 ppm at atmospheric pressure. As a rule, these laboratories do not offer calibration services at lower frost points. As is shown in Chapter 12 there is an increasing number of applications and requirements for monitoring and measuring water vapor levels in the ppm and ppb range, corresponding to frost points from –70°C to –112°C (–94°F to –170°F), or 2.55 ppm to 1 ppb.

Although aluminum oxide hygrometers are sensitive to humidity at those low levels, they are not uniform and are, therefore, only of any value if they are calibrated at those ppm or ppb levels. It is not recommended, and very inaccurate, to calibrate the sensors at higher frost points and then extrapolate the data to the lower levels because of the lack of uniformity of these sensors, especially at the low end.

Phosphorous pentoxide electrolytic hygrometers employ a fundamental measurement technique (see Chapter 5). But errors are possible due to sample line leaks through connectors, outgassing from the walls of sample lines, back flow, and other problems. Hence it is desirable to employ a calibration system which employs a more fundamental technology.

Users of electrolytic cells are also seeking calibration methods to verify the accuracy of the electrolytic hygrometer. One system which was developed by Panametrics, one of the large manufacturers of aluminum oxide hygrometers, uses flow mixing, i.e., mixing extremely dry gas with small quantities of water. The dry carrier gas must be dryer than the gas to be measured, i.e., below 1 ppb. Such a gas can be produced by passing dry nitrogen through a properly designed cold trap operating at a temperature of, for instance, –140°C (–220°F). Before entering the cold trap, the incoming dry nitrogen gas would typically be at 200 ppb moisture, low enough to avoid blockage of the cold trap with ice particles. The cold trap temperature of –140°C (–220°F) is generated with a commercially available low temperature freezer. The advantage of this approach, compared to the use of liquid nitrogen, is that nitrogen gas can be used without danger of condensation. The moisture content at –140°C (–220°F) is 0.00036 ppb, low enough for the calibrations.

The cold trap consists of 45 meters (148 ft.) of 0.95 cm (3/8 in.) diameter tubing and a filter, all kept at the temperature of –140°C (–220°F). The moisture content of the emergent gas can be calculated using the principles of mass transfer. For an input flow of 15 liters (915 in.^{3}) per minute at 200 ppb moisture, 28 meters (92 ft.) of tubing gives an output of 0.0005 ppb moisture, which corresponds to a frost point of –140°C (220° F).

The first 15 meters (49 ft.) of tubing is a suspended coil used to cool the gas to the trap temperature. The remaining 30 meters (98 ft.) of tubing are used for mass transfer of moisture from the gas to the tubing wall. A warming tube is used to bring the dry gas from the low temperature of the trap to room temperature. Moisture is added to the very dry carrier gas to produce the calibration gas. The entire system must be periodically tested for helium leaks. The system, even when pressurized, is subject to large errors from minute leaks to the environment, since the ambient water vapor pressure is many orders of magnitude higher than for the dry gas and water vapor would flow back into the system. Other errors could be caused by outgassing of metal parts that are exposed to the dry gas. At these low levels, even polished stainless has enough porosity to cause outgassing that can contribute more moisture than is contained in the carrier gas.

Although a calibration system like the one discussed above is one of the best available, it does not provide traceability to any national standards and the accuracy of such calibrations could be challenged by manufacturers and users.

11

WATER VAPOR PRESSURE TABLES

I. General

In this chapter definitions and specifications of water vapor in the atmosphere are presented along with the saturation vapor pressure tables.

These tables convert dew or frost point into the partial water vapor pressure of the gas. The tables are convenient for determining percent relative humidity from dew point and gas temperature measurements, by dividing the saturation vapor pressure at the measured dew point temperature by the saturation pressure related to the measured gas temperature.

Tables 11.1 through 11.6 were abstracted from Reference 5, (Chapter 15) and are reproduced with permission of the Smithsonian Institute. Table 11.7 is based on values given by Keyes in the International Critical Tables.

The vapor pressure tables are presented in Metric and English units, and cover dew points (vapor pressure over water) and frost points (vapor pressure over ice).

Psychrometric charts, conversion charts, and tables for converting from one humidity parameter to another are presented in Chapter 13. Conversions can also be made using the computer disc enclosed in the back cover of this book.

II. Smithsonian Tables
(courtesy Smithsonian Institution Press)

Table 11.1   Definitions and Specifications of Water Vapor in the Atmosphere.
Table 11.2   Saturation Water Vapor Pressure Formulas.
Table 11.3   Saturation Vapor Pressure over Water, Metric Units.
Table 11.4   Saturation Vapor Pressure over Water, English Units.
Table 11.5   Saturation Vapor Pressure over Ice, Metric Units.
Table 11.6   Saturation Vapor Pressure over Ice, English Units.

III. Vapor Pressure of Water Above 100°C

Table 11.7   Vapor Pressures above 100°C.

Table 11.1   Definitions and Specifications.

Definitions and Specifications of Water Vapor in the Atmosphere

The Conference of Directors, International Meteorological organization, Washington, 1947,[1] decided to adopt the following definitions and specifications of the parameters of water vapor in the atmosphere:

(1) **The mixing ratio** r of moist air is the ratio of the mass m_v of water vapor to the mass m_a of dry air with which the water vapor is associated.

$$r = m_v/m_a$$

(2) **The specific humidity,** mass concentration, or moisture content q of moist air is the ratio of the mass m_v of water vapor to the mass $(m_v + m_a)$ of moist air in which the mass of water vapor m_v is contained.

$$q = \frac{m_v}{m_v + m_a}$$

(3) **Vapor concentration** (density of water vapor in a mixture) or **absolute humidity.**— For a mixture of water vapor and dry air the vapor concentration d_v is defined as the ratio of the mass of vapor m_v to the volume V occupied by the mixture.

$$d_v = \frac{m_v}{V}$$

(4) **The vapor pressure** e' **of water vapor in moist air** at total pressure p and with mixing ratio r is defined by:

$$e' = \frac{r}{0.62197 + r}\, p$$

(5) **Saturation.**—Moist air at temperature T and at total pressure p is said to be saturated if its composition is such that it can coexist in neutral equilibrium with a plane surface of pure condensed phase (water or ice) at the same temperature and pressure.

(6) **Saturation mixing ratio.**—The symbol r_w denotes saturation mixing ratio of moist air with respect to a plane surface of pure water. The symbol r_i denotes saturation mixing ratio of moist air with respect to a plane surface of pure ice.

(7) **Saturation vapor pressure in the pure phase.**—The saturation vapor pressure e_w of pure aqueous vapor with respect to water is the pressure of the vapor when in a state of neutral equilibrium with a plane surface of pure water at the same temperature and pressure; similarly for e_i, in respect to ice, e_w and e_i are temperature-dependent functions only; i.e.,

$$e_w = e_w\,(T)$$
$$e_i = e_i\,(T)$$

---

[1]   Resolution 166, International Meteorological Organization, Conference of Directors, Washington, 1947.

(8) **Saturation vapor pressure of moist air.**—The saturation vapor pressure with respect to water e'_w of moist air at pressure p and temperature T is defined by:

$$e'_w = \frac{r_w}{0.62197 + r_w} \, p$$

Similarly, the saturation vapor pressure with respect to ice e'_i of moist air at pressure p and temperature T is defined by:

$$e'_i = \frac{r_i}{0.62197 + r_i} \, p$$

(9) **Relations between saturation vapor pressure of pure phase and of moist air.**— In the meteorological range of pressure and temperature the following relations hold with an error of 0.5 percent or less:

$$e'_w = e_w$$
$$e'_i = e_i$$

(10) **The thermodynamic dew-point temperature** T_d of moist air at temperature T, pressure p, and mixing ratio r, is the temperature to which the air must be cooled in order that it shall be saturated with respect to water at the initial pressure p and mixing ratio r.

(11) **The thermodynamic frost-point temperature** T_f of moist air at temperature T, pressure p, and mixing ratio r, is the temperature to which the air must be cooled in order that it shall be saturated with respect to ice at the initial pressure p and mixing ratio r.

(12) **The dew and frost-point temperatures,** so defined, are related with the mixing ratio and total pressure p by the respective equations:

$$e'_w(T_d) = \frac{r}{0.62197 + r} \, p$$

$$e'_i(T_f) = \frac{r}{0.62197 + r} \, p$$

(13)* **The relative humidity** U (in percent) of moist air is defined by:

$$U = 100 \, \frac{r}{r_w}$$

where r is the mixing ratio of moist air at pressure p, and temperature T and r_w the saturation mixing ratio at the same pressure and temperature.

---

*Note—At the Philadelphia meeting (May 6, 1950) of the International Joint Committee on Psychrometric Data, resolutions were adopted giving definitions of relative humidity different from the above. These may be paraphrased as follows:

(a) In regard to a mixture of air and water vapor under given conditions of barometric pressure and temperature at which saturation of air is *possible*, relative humidity is the ratio of the mol fraction of water vapor in the mixture to the mol fraction of water vapor in a mass of air saturated with water vapor at the given barometric pressure and temperature.

(b) In regard to a mixture of air and water vapor or a sample of pure water vapor mixed with any other substance under given conditions of barometric pressure and temperature at which saturation is *impossible*, relative humidity is the ratio of the partial pressure of water vapor in the mixture or sample to the saturation pressure of pure water at the given temperature.

The Committee adopted (b) subject to verification. The mol fraction of water vapor in any mixture is the number of mols, or molecules, of water vapor in the mixture divided by the total number of mols or molecules of all constituents in the mixture.

$$\text{Mol fraction of water vapor} = \frac{r}{r + (m_w/m)}$$

(14) **Relative humidity at temperatures less than 0°C** is to be evaluated with respect to water. The advantages of this procedure are as follows:

(a) Most hygrometers which are essentially responsive to the relative humidity indicate relative humidity with respect to water at all temperatures.

(b) The majority of clouds at temperatures below 0°C consist of water, or mainly of water.

(c) Relative humidities greater than 100% would in general not be observed. This is of particular importance in synoptic weather messages, since the atmosphere is often supersaturated with respect to ice at temperatures below 0°C.

(d) The majority of existing records of relative humidity at temperatures below 0°C are expressed on a basis of saturation with respect to water.

(15)* **The thermodynamic wet-bulb temperature** T_w of moist air at pressure p, temperature T, and mixing ratio r, is the temperature which this air assumes when water is introduced gradually by infinitesimal amounts at the current temperatures and evaporated into the air by an adiabatic process at constant pressure until saturation is reached.

T_w is determined by the equation:

$$\log \frac{L_v(T_w)}{L_v(T)} = \frac{c_{pv} - c_w}{c_{pv}} \log \frac{c_p + c_{pv} r}{c_p + c_{pv} r_w(T_w)}$$

where:

$L_v(T)$ = heat of vaporization of water at temperature T,
$L_v(T_w)$ = heat of vaporization of water at temperature T_w,
c_w = specific heat of liquid water,
$r_w(T_w)$ = saturation mixing ratio with respect to water at pressure p and
    temperature T_w,
c_p = specific heat of dry air at constant pressure, and
c_{pv} = specific heat of water vapor at constant pressure.

Here c_p and c_{pv} are assumed to be independent of temperature in the interval T_w to T ...

The relationship between T_w as defined and the wet-bulb temperature as indicated by a particular psychrometer is a matter to be determined by carefully controlled experimentation, taking account of the various parameters concerned; e.g., ventilation, size of thermometer bulb, radiation, etc.[2]

(16)* **The thermodynamic equivalent temperature** T_e of moist air at pressure p, temperature T, and mixing ratio r, is the temperature which the air assumes by means of adiabatic condensation at constant pressure of all the water vapor which it contains, the condensed water falling out of the system immediately.

T_e is determined by the equation where:

$L_v(T)$ = heat of vaporization of water at temperature T

$L_v(T_e)$ = heat of vaporization of water at temperature T_e

c_w = specific heat of liquid water

c_p = specific heat of dry air at constant pressure

c_{pv} = specific heat of water vapor at constant pressure.

Here c_p and c_{pv} are assumed to be independent of temperature in the interval T_w to T_e.

---

[2] The Working Subcommittee of the International Joint Committee on Psychrometric Data[3] recommends that thermodynamic wet-bulb temperature be defined as the solution T_w (p, T, r) of the equation:

$$h(p, T, r) - r \cdot h'_w(p, T_w) = h_s(p, T_w) - r_w(p, T_w) \cdot h'_w(p, T_w)$$

where $h(p, T, r)$ = specific enthalpy of moist air; $h'_w(p, T_w)$ = specific enthalpy of pure compressed liquid (or solid) water; and $h_s(p, T_w)$ = specific enthalpy of saturated air at pressure p and temperature T_w. This definition combines the correct energy and weight values for the steady hyphen flow process of injecting pure compressed liquid (or solid) water at pressure p and temperature T_w into a stream of moist air at pressure p, temperature T, and mixing ratio r to bring the air adiabatically to saturation at pressure p and temperature T_w. This is to be regarded as the appropriate idealization of the actual process by which the thin film of water on a wet-bulb thermometer immersed in a stream of moist air maintains, at any rate for a time, a steady value below that of the air itself.

* Definitions (13), (15), and (16) were rescinded and the definitions in note (a) under (13) and in footnote 2 were adopted for relative humidity and for thermodynamic wet-bulb temperature by the WMO Commission for Aerology, first session, Toronto, 1953, and approved by the Executive Committee, WMO, 1953.

Table 11.2 Saturation vapor pressure formulas.

Resolution 164 of the Twelfth Conference of Directors of the International Meteorological Organization (Washington 1947) adopted the *Goff–Gratch[1] formulation* for the saturation vapor pressure in the pure phase over plane surfaces of pure water and pure ice:

$$\log_{10} e_w = -7.90298 \, (T_s/T - 1) + 5.02808 \log_{10} (T_s/T) \tag{1}$$
$$-1.3816 \times 10^{-7} \, (10^{11.344(1-T/T_s)} - 1)$$
$$+8.1328 \times 10^{-8} \, (10^{-8.49140 \, (T_s/T - 1)} - 1) + \log_{10} c_{ws}$$

and

$$\log_{10} e_i = -9.09718 \, (T_o/T - 1) - 3.56654 \log_{10} (T_o/T) \tag{2}$$
$$+ 0.876793 \, (1 - T/T_o) + \log_{10} e_{io}$$

where:

e_w = saturation vapor pressure over a plane surface of pure ordinary liquid water in mb.

e_i = saturation vapor pressure of a plane surface of pure ordinary water ice (mb)

T = absolute (thermodynamic) temperature (°K).

T_s = steam-point temperature (373.16°K).

T_o = ice-point temperature (273.16°K).

e_{ws} = saturation pressure of pure ordinary liquid water at steam-point temperature (1 standard atmosphere = 1013.246 mb).

e_{io} = saturation pressure of pure ordinary water ice at ice-point temperature (0.0060273 standard atmosphere = 6.1071 mb).

The Goff–Gratch formulas are based on integration of the Clausius–Clapeyron equation, considering the deviations from a perfect gas, and on modern experimental data. The stated range of validity of (1) is 0°C to 100°C. Since there is a dearth of experimental data on vapor pressure over supercooled water and the necessary thermodynamic data for an exact integration of the Clausius–Clapeyron equation does not exist, no completely satisfactory formula exists for the vapor pressure over liquid water at temperatures below 0°C. However, direct extrapolation of (1) gives values of e_w in the middle of the range suggested by other investigators and has been adopted for the range 0°C to –50°C, pending further research.

Values for each half degree Centigrade and whole degree Fahrenheit were computed from (1) and (2), and values for each 0.1° were obtained by interpolation (Newton's method), with the exception of the few values in Table 3 for T > 100°C, which were computed from Keyes[2] formula:

$$\log_{10} e_w \text{ (mm of mercury)} = -2892.3693/T$$
$$-2.892736 \log_{10}T - 4.9369728 \times 10^{-3}T + 5.606905 \times 10^{-6} T^2$$
$$-4.645869 \times 10^{-9} T^3 + 3.7874 \times 10^{-12} T^4 + 19.3011421.$$

The small difference between e_w and e_i at 0°C (32°F) arises from the fact that the triple point for water is 0.01°C.

Table 11.3   Saturation vapor pressure over water.

SATURATION VAPOR PRESSURE OVER WATER

Temperature °C.	.0 mb.	.1 mb.	.2 mb.	.3 mb.	.4 mb.	.5 mb.	.6 mb.	.7 mb.	.8 mb.	.9 mb.
					Metric units					
-50	0.06356									
-49	0.07124	0.07044	0.06964	0.06885	0.06807	0.06730	0.06654	0.06578	0.06503	0.06429
-48	0.07975	0.07886	0.07797	0.07710	0.07624	0.07538	0.07453	0.07370	0.07287	0.07205
-47	0.08918	0.08819	0.08722	0.08625	0.08530	0.08435	0.08341	0.08248	0.08156	0.08065
-46	0.09961	0.09852	0.09744	0.09637	0.09531	0.09426	0.09322	0.09220	0.09118	0.09017
-45	0.1111	0.1099	0.1087	0.1075	0.1063	0.1052	0.1041	0.1030	0.1018	0.1007
-44	0.1239	0.1226	0.1213	0.1200	0.1187	0.1174	0.1161	0.1149	0.1136	0.1123
-43	0.1379	0.1364	0.1350	0.1335	0.1321	0.1307	0.1293	0.1279	0.1266	0.1252
-42	0.1534	0.1518	0.1502	0.1486	0.1470	0.1455	0.1440	0.1424	0.1409	0.1394
-41	0.1704	0.1686	0.1669	0.1651	0.1634	0.1617	0.1600	0.1583	0.1567	0.1550
-40	0.1891	0.1872	0.1852	0.1833	0.1815	0.1796	0.1777	0.1759	0.1740	0.1722
-39	0.2097	0.2076	0.2054	0.2033	0.2013	0.1992	0.1971	0.1951	0.1931	0.1911
-38	0.2323	0.2299	0.2276	0.2253	0.2230	0.2207	0.2185	0.2162	0.2140	0.2119
-37	0.2571	0.2545	0.2520	0.2494	0.2469	0.2444	0.2419	0.2395	0.2371	0.2347
-36	0.2842	0.2814	0.2786	0.2758	0.2730	0.2703	0.2676	0.2649	0.2623	0.2597
-35	0.3139	0.3108	0.3077	0.3047	0.3017	0.2987	0.2957	0.2928	0.2899	0.2870
-34	0.3463	0.3429	0.3396	0.3362	0.3330	0.3297	0.3265	0.3233	0.3201	0.3170
-33	0.3818	0.3781	0.3745	0.3708	0.3673	0.3637	0.3602	0.3567	0.3532	0.3497
-32	0.4205	0.4165	0.4125	0.4085	0.4046	0.4007	0.3968	0.3930	0.3893	0.3855
-31	0.4628	0.4584	0.4541	0.4497	0..4454	0.4412	0.4370	0.4328	0.4287	0.4246
-30	0.5088	0.5040	0.4993	0.4946	0.4899	0.4853	0.4807	0.4762	0.4717	0.4672
-29	0.5589	0.5537	0.5485	0.5434	0.5383	0.5333	0.5283	0.5234	0.5185	0.5136
-28	0.6134	0.6077	0.6021	0.5966	0.5911	0.5856	0.5802	0.5748	0.5694	0.5642
-27	0.6727	0.6666	0.6605	0.6544	0.6484	0.6425	0.6366	0.6307	0.6249	0.6191
-26	0.7371	0.7304	0.7238	0.7172	0.7107	0.7042	0.6978	0.6914	0.6851	0.6789
-25	0.8070	0.7997	0.7926	0.7854	0.7783	0.7713	0.7643	0.7574	0.7506	0.7438
-24	0.8827	0.8748	0.8671	0.8593	0.8517	0.8441	0.8366	0.8291	0.8217	0.8143
-23	0.9649	0.9564	0.9479	0.9396	0.9313	0.9230	0.9148	0.9067	0.8986	0.8906
-22	1.0538	1.0446	1.0354	1.0264	1.0173	1.0084	0.9995	0.9908	0.9821	0.9734
-21	1.1500	1.1400	1.1301	1.1203	1.1106	1.1009	1.0913	1.0818	1.0724	1.0631
-20	1.2540	1.2432	1.2325	1.2219	1.2114	1.2010	1.1906	1.1804	1.1702	1.1600
-19	1.3664	1.3548	1.3432	1.3318	1.3204	1.3091	1.2979	1.2868	1.2758	1.2648
-18	1.4877	1.4751	1.4627	1.4503	1.4381	1.4259	1.4138	1.4018	1.3899	1.3781
-17	1.6186	1.6051	1.5916	1.5783	1.5650	1.5519	1.5389	1.5259	1.5131	1.5003
-16	1.7597	1.7451	1.7306	1.7163	1.7020	1.6879	1.6738	1.6599	1.6460	1.6323
-15	1.9118	1.8961	1.8805	1.8650	1.8496	1.8343	1.8191	1.8041	1.7892	1.7744
-14	2.0755	2.0586	2.0418	2.0251	2.0085	1.9921	1.9758	1.9596	1.9435	1.9276
-13	2.2515	2.2333	2.2153	2.1973	2.1795	2.1619	2.1444	2.1270	2.1097	2.0925
-12	2.4409	2.4213	2.4019	2.3826	2.3635	2.3445	2.3256	2.3069	2.2883	2.2698
-11	2.6443	2.6233	2.6024	2.5817	2.5612	2.5408	2.5205	2.5004	2.4804	2.4606
-10	2.8627	2.8402	2.8178	2.7956	2.7735	2.7516	2.7298	2.7082	2.6868	2.6655
-9	3.0971	3.0729	3.0489	3.0250	3.0013	2.9778	2.9544	2.9313	2.9082	2.8854
-8	3.3484	3.3225	3.2967	3.2711	3.2457	3.2205	3.1955	3.1706	3.1459	3.1214
-7	3.6177	3.5899	3.5623	3.5349	3.5077	3.4807	3.4539	3.4272	3.4005	3.3745
-6	3.9061	3.8764	3.8468	3.8175	3.7883	3.7594	3.7307	3.7021	3.6738	3.6456
-5	4.2148	4.1830	4.1514	4.1200	4.0888	4.0579	4.0271	3.9966	3.9662	3.9361
-4	4.5451	4.5111	4.4773	4.4437	4.4103	4.3772	4.3443	4.3116	4.2791	4.2468
-3	4.8981	4.8617	4.8256	4.7897	4.7541	4.7187	4.6835	4.6486	4.6138	4.5794
-2	5.2753	5.2364	5.1979	5.1595	5.1214	5.0836	5.0460	5.0087	4.9716	4.9347
-1	5.6780	5.6365	5.5953	5.5544	5.5138	5.4734	5.4333	5.3934	5.3538	5.3144
-0	6.1078	6.0636	6.0196	5.9759	5.9325	5.8894	5.8466	5.8040	5.7617	5.7197

(continued)

Table 11.3 Continued.

SATURATION VAPOR PRESSURE OVER WATER

Temperature °C.	.0 mb.	.1 mb.	.2 mb.	.3 mb.	.4 mb.	.5 mb.	.6 mb.	.7 mb.	.8 mb.	.9 mb.
					Metric units					
0	6.1078	6.1523	6.1971	6.2422	6.2876	6.3333	6.3793	6.4256	6.4721	6.5190
1	6.5662	6.6137	6.6614	6.7095	6.7579	6.8066	6.8556	6.9049	6.9545	7.0044
2	7.0547	7.1053	7.1562	7.2074	7.2590	7.3109	7.3631	7.4157	7.4685	7.5218
3	7.5753	7.6291	7.6833	7.7379	7.7928	7.8480	7.9036	7.9595	8.0158	8.0724
4	8.1294	8.1868	8.2445	8.3026	8.3610	8.4198	8.4789	8.5384	8.5983	8.6586
5	8.7192	8.7802	8.8416	8.9033	8.9655	9.0280	9.0909	9.1542	9.2179	9.2820
6	9.3465	9.4114	9.4766	9.5423	9.6083	9.6748	9.7416	9.8089	9.8765	9.9446
7	10.013	10.082	10.151	10.221	10.291	10.362	10.433	10.505	10.577	10.649
8	10.722	10.795	10.869	10.943	11.017	11.092	11.168	11.243	11.320	11.397
9	11.474	11.552	11.630	11.708	11.787	11.867	11.947	12.027	12.108	12.190
10	12.272	12.355	12.438	12.521	12.606	12.690	12.775	12.860	12.946	13.032
11	13.119	13.207	13.295	13.383	13.472	13.562	13.652	13.742	13.833	13.925
12	14.017	14.110	14.203	14.297	14.391	14.486	14.581	14.678	14.774	14.871
13	14.969	15.067	15.166	15.266	15.365	15.466	15.567	15.669	15.771	15.874
14	15.977	16.081	16.186	16.291	16.397	16.503	16.610	16.718	13.826	16.935
15	17.044	17.154	17.264	17.376	17.487	17.600	17.713	17.827	17.942	18.057
16	18.173	18.290	18.407	18.524	18.643	18.762	18.882	19.002	19.123	19.245
17	19.367	19.490	19.614	19.739	19.864	19.990	20.117	20.244	20.372	20.501
18	20.630	20.760	20.891	21.023	21.155	21.288	21.422	21.556	21.691	21.827
19	21.964	22.101	22.240	22.379	22.518	22.659	22.800	22.942	23.085	23.229
20	23.373	23.518	23.664	23.811	23.959	24.107	24.256	24.406	24.557	24.709
21	24.861	25.014	25.168	25.323	25.479	25.635	25.792	25.950	26.109	26.269
22	26.430	26.592	26.754	26.918	27.082	27.247	27.413	27.580	27.748	27.916
23	28.086	28.256	28.428	28.600	28.773	28.947	29.122	29.298	29.475	29.652
24	29.831	30.011	30.191	30.373	30.555	30.739	30.923	31.109	31.295	31.483
25	31.671	31.860	32.050	32.242	32.434	32.627	32.821	33.016	33.212	33.410
26	33.608	33.809	34.008	34.209	34.411	34.615	34.820	35.025	35.232	35.440
27	35.649	35.859	36.070	36.282	36.495	36.709	36.924	37.140	37.358	37.576
28	37.796	38.017	38.239	38.462	38.686	38.911	39.137	39.365	39.594	39.824
29	40.055	40.287	40.521	40.755	40.991	41.228	41.466	41.705	41.945	42.187
30	42.430	42.674	42.919	43.166	43.414	43.663	43.913	44.165	44.418	44.672
31	44.927	45.184	45.442	45.701	45.961	46.223	46.486	46.750	47.016	47.283
32	47.551	47.820	48.091	48.364	48.637	48.912	49.188	49.466	49.745	50.025
33	50.307	50.590	50.874	51.160	51.447	51.736	52.026	52.317	52.610	52.904
34	53.200	53.497	53.796	54.096	54.397	54.700	55.004	55.310	55.617	55.926
35	56.236	56.548	56.861	57.176	57.492	57.810	58.129	58.450	58.773	59.097
36	59.422	59.749	60.077	60.407	60.739	61.072	61.407	61.743	62.081	62.421
37	62.762	63.105	63.450	63.796	64.144	64.493	64.844	65.196	65.550	65.906
38	66.264	66.623	66.985	67.347	67.712	68.078	68.446	68.815	69.186	69.559
39	69.934	70.310	70.688	71.068	71.450	71.833	72.218	72.605	72.994	73.385
40	73.777	74.171	74.568	74.966	75.365	75.767	76.170	76.575	76.982	77.391
41	77.802	78.215	78.630	79.046	79.465	79.885	80.307	80.731	81.157	81.585
42	82.015	82.447	82.881	83.316	83.754	84.194	84.636	85.079	85.525	85.973
43	86.423	86.875	87.329	87.785	88.243	88.703	89.165	89.629	90.095	90.564
44	91.034	91.507	91.981	92.458	92.937	93.418	93.901	94.386	94.874	95.363
45	95.855	96.349	96.845	97.343	97.844	98.347	98.852	99.359	99.869	100.38
46	100.89	101.41	101.93	102.45	102.97	103.50	104.03	104.56	105.09	105.62
47	106.16	106.70	107.24	107.78	108.33	108.88	109.43	109.98	110.54	111.10
48	111.66	112.22	112.79	113.36	113.93	114.50	115.07	115.65	116.23	116.81
49	117.40	117.99	118.58	119.17	119.77	120.37	120.97	121.57	122.18	122.79
50	123.40	124.01	124.63	125.25	125.87	126.49	127.12	127.75	128.38	129.01

(continued)

Table 11.3   Continued.

SATURATION VAPOR PRESSURE OVER WATER

Metric units

Temperature °C.	.0 mb.	.1 mb.	.2 mb.	.3 mb.	.4 mb.	.5 mb.	.6 mb.	.7 mb.	.8 mb.	.9 mb.
50	123.40	124.01	124.63	125.25	125.87	126.49	127.12	127.75	128.38	129.01
51	129.65	130.29	130.93	131.58	132.23	132.88	133.53	134.19	134.84	135.51
52	136.17	136.84	137.51	138.18	138.86	139.54	140.22	140.91	141.60	142.29
53	142.98	143.68	144.38	145.08	145.78	146.49	147.20	147.91	148.63	149.35
54	150.07	150.80	151.53	152.26	152.99	153.73	154.47	155.21	155.96	156.71
55	157.46	158.22	158.97	159.74	160.50	161.27	162.04	162.82	163.59	164.38
56	165.16	165.95	166.74	167.53	168.33	169.13	169.93	170.74	171.55	172.36
57	173.18	174.00	174.82	175.65	176.48	177.31	178.15	178.99	179.83	180.68
58	181.53	182.38	183.24	184.10	184.96	185.83	186.70	187.58	188.45	189.34
59	190.22	191.11	192.00	192.89	193.79	194.69	195.60	196.51	197.42	198.34
60	199.26	200.18	201.11	202.05	202.98	203.92	204.86	205.81	206.76	207.71
61	208.67	209.63	210.59	211.56	212.53	213.51	214.49	215.48	216.46	217.45
62	218.45	219.45	220.45	221.46	222.47	223.48	224.50	225.52	226.54	227.58
63	228.61	229.65	230.70	231.74	232.79	233.85	234.91	235.97	237.03	238.11
64	239.18	240.26	241.34	242.43	243.52	244.62	245.72	246.82	247.93	249.04
65	250.16	251.28	252.41	253.54	254.67	255.81	256.95	258.10	259.25	260.40
66	261.56	262.73	263.90	265.07	266.25	267.43	268.61	269.80	271.00	272.20
67	273.40	274.61	275.82	277.04	278.26	279.49	280.72	281.96	283.20	284.45
68	285.70	286.96	288.21	289.48	290.75	292.02	293.30	294.58	295.86	297.15
69	298.45	299.75	301.06	302.37	303.69	305.01	306.34	307.67	309.00	310.34
70	311.69	313.04	314.39	315.75	317.12	318.49	319.87	321.25	322.63	324.02
71	325.42	326.82	328.22	329.63	331.05	332.47	333.89	335.33	336.76	338.20
72	339.65	341.10	342.56	344.03	345.50	346.97	348.45	349.93	351.42	352.91
73	354.41	355.91	357.43	358.94	360.46	361.99	363.52	365.06	366.61	368.15
74	369.71	371.27	372.84	374.41	375.99	377.57	379.16	380.75	382.35	383.95
75	385.56	387.18	388.80	390.43	392.06	393.70	395.34	396.99	398.65	400.31
76	401.98	403.65	405.34	407.02	408.71	410.41	412.11	413.82	415.53	417.25
77	418.98	420.71	422.45	424.20	425.95	427.71	429.47	431.24	433.02	434.80
78	436.59	438.38	440.18	441.99	443.80	445.62	447.45	449.28	451.11	452.96
79	454.81	456.67	458.53	460.40	462.28	464.16	466.05	467.94	469.85	471.76
80	473.67	475.59	477.52	479.45	481.39	483.34	485.29	487.25	489.22	491.19
81	493.17	495.16	497.15	499.16	501.17	503.18	505.20	507.23	509.26	511.30
82	513.35	515.41	517.47	519.54	521.62	523.70	525.79	527.89	529.99	532.10
83	534.22	536.35	538.48	540.62	542.77	544.92	547.08	549.25	551.43	553.61
84	555.80	557.99	560.20	562.41	564.62	566.85	569.08	571.32	573.57	575.83
85	578.09	580.36	582.64	584.93	587.22	589.52	591.83	594.14	596.46	598.79
86	601.13	603.48	605.83	608.19	610.56	612.94	615.32	617.72	620.12	622.52
87	624.94	627.36	629.79	632.23	634.68	637.13	639.59	642.07	644.55	647.03
88	649.53	652.03	654.54	657.06	659.59	662.12	664.66	667.22	669.78	672.34
89	674.92	677.50	680.09	682.69	685.30	687.92	690.55	693.18	695.82	698.47
90	701.13	703.80	706.47	709.16	711.85	714.55	717.26	719.98	722.71	725.45
91	728.19	730.94	733.70	736.47	739.25	742.04	744.84	747.64	750.46	753.28
92	756.11	758.95	761.80	764.66	767.52	770.40	773.29	776.18	779.09	782.00
93	784.92	787.85	790.79	793.74	796.69	799.66	802.63	805.62	808.61	811.62
94	814.63	817.65	820.69	823.73	826.78	829.84	832.91	835.99	839.08	842.17
95	845.28	848.40	851.52	854.66	857.80	860.96	864.12	867.30	870.48	873.68
96	876.88	880.09	883.31	886.55	889.79	893.04	896.30	899.57	902.86	906.15
97	909.45	912.76	916.08	919.42	922.76	926.11	929.47	932.84	936.23	939.62
98	943.02	946.43	949.85	953.28	956.73	960.18	963.65	967.12	970.61	974.10
99	977.61	981.13	984.65	988.19	991.74	995.30	998.87	1002.45	1006.04	1009.64
100	1013.25	1016.87	1020.50	1024.14	1027.80	1031.46	1035.13	1038.82	1042.51	1046.22
101	1049.94	1053.67	1057.41	1061.16	1064.93	1068.70	1072.49	1076.28	1080.09	1083.91
102	1087.74									

Table 11.4 Saturation vapor pressure over water.

SATURATION VAPOR PRESSURE OVER WATER

English units

Temperature °F	Vapor pressure 10⁻³ in. Hg.	Temperature °F	Vapor pressure 10⁻³ in. Hg.	Temperature °F	Vapor pressure 10⁻³ in. Hg.	Temperature °F	Vapor pressure 10⁻³ in. Hg.	Temperature °F	Vapor pressure 10⁻³ in. Hg.
Unit:									
-60.0	1.651	-56.0	2.130	-52.0	2.733	-48	3.488	-44.0	4.424
-59.5	1.705	-55.5	2.198	-51.5	2.818	-47.5	3.594	-43.5	4.556
-59.0	1.761	-55.0	2.268	-51.0	2.906	-47.0	3.703	-43.0	4.692
-58.5	1.818	-54.5	2.340	-50.5	2.997	-46.5	3.815	-42.5	4.831
-58.0	1.877	-54.0	2.414	-50.0	3.089	-46.0	3.930	-42.0	4.973
-57.5	1.938	-53.5	2.491	-49.5	3.183	-45.5	4.049	-41.5	5.121
-57.0	2.000	-53.0	2.569	-49.0	3.281	-45.0	4.170	-41.0	5.271
-56.5	2.064	-52.5	2.650	-48.5	3.384	-44.5	4.294	-40.5	5.425

Temperature °F Unit: in. Hg.

Temperature °F	.0 10⁻³ in. Hg.	.1 10⁻³ in. Hg.	.2 10⁻³ in. Hg.	.3 10⁻³ in. Hg.	.4 10⁻³ in. Hg.	.5 10⁻³ in. Hg.	.6 10⁻³ in. Hg.	.7 10⁻³ in. Hg.	.8 10⁻³ in. Hg.	.9 10⁻³ in. Hg.
-40	5.584									
-39	5.915	5.881	5.847	5.814	5.780	5.747	5.714	5.681	5.649	5.616
-38	6.263	6.227	6.192	6.156	6.121	6.086	6.051	6.017	5.983	5.949
-37	6.630	6.592	6.555	6.517	6.480	6.443	6.406	6.370	6.334	6.298
-36	7.016	6.977	6.937	6.898	6.860	6.821	6.782	6.744	6.706	6.668
-35	7.424	7.382	7.341	7.299	7.258	7.217	7.176	7.136	7.096	7.056
-34	7.849	7.805	7.762	7.719	7.676	7.634	7.592	7.549	7.507	7.466
-33	8.298	8.252	8.206	8.161	8.116	8.071	8.026	7.982	7.937	7.893
-32	8.770	8.721	8.673	8.626	8.578	8.531	8.484	8.437	8.390	8.344
-31	9.270	9.219	9.169	9.118	9.068	9.018	8.968	8.918	8.869	8.819
-30	9.789	9.736	9.683	9.630	9.578	9.526	9.474	9.423	9.372	9.321
-29	10.34	10.28	10.23	10.17	10.12	10.06	10.00	9.951	9.896	9.842
-28	10.91	10.85	10.79	10.73	10.68	10.62	10.56	10.51	10.45	10.40
-27	11.52	11.46	11.40	11.33	11.27	11.21	11.15	11.09	11.03	10.97
-26	12.15	12.08	12.02	11.96	11.90	11.83	11.77	11.70	11.64	11.58
-25	12.82	12.75	12.69	12.62	12.55	12.49	12.42	12.35	12.29	12.22
-24	13.52	13.45	13.38	13.31	13.24	13.17	13.10	13.03	12.96	12.89
-23	14.25	14.17	14.10	14.03	13.95	13.88	13.81	13.73	13.66	13.59
-22	15.02	14.94	14.87	14.79	14.71	14.64	14.56	14.48	14.41	14.33
-21	15.83	15.75	15.66	15.58	15.50	15.42	15.34	15.26	15.18	15.10
-20	16.68	16.59	16.51	16.42	16.34	16.25	16.17	16.08	16.00	15.91
-19	17.56	17.47	17.38	17.29	17.21	17.12	17.03	16.94	16.86	16.77
-18	18.49	18.40	18.30	18.21	18.11	18.02	17.93	17.83	17.74	17.65
-17	19.46	19.36	19.26	19.16	19.07	18.97	18.87	18.78	18.68	18.59
-16	20.48	20.38	20.27	20.17	20.07	19.97	19.87	19.77	19.66	19.56
-15	21.55	21.44	21.33	21.22	21.12	21.01	20.90	20.80	20.69	20.59
-14	22.66	22.55	22.43	22.32	22.21	22.10	21.99	21.88	21.77	21.66
-13	23.83	23.71	23.59	23.47	23.36	23.24	23.12	23.01	22.89	22.78
-12	25.05	24.93	24.80	24.68	24.55	24.43	24.31	24.19	24.07	23.95
-11	26.33	26.20	26.07	25.94	25.81	25.68	25.55	25.43	25.30	25.17
-10	27.66	27.52	27.39	27.26	27.12	26.99	26.86	26.72	26.59	26.46
- 9	29.06	28.92	28.77	28.63	28.49	28.35	28.21	28.07	27.93	27.80
- 8	30.52	30.37	30.22	30.07	29.93	29.78	29.63	29.49	29.35	29.20
- 7	32.04	31.88	31.73	31.58	31.42	31.27	31.12	30.97	30.82	30.67
- 6	33.63	33.47	33.31	33.15	32.99	32.83	32.67	32.51	32.35	32.20
- 5	35.29	35.12	34.95	34.79	34.62	34.46	34.29	34.13	33.96	33.80
- 4	37.03	36.85	36.68	36.50	36.32	36.15	35.98	35.80	35.63	35.46
- 3	38.84	38.66	38.47	38.29	38.11	37.93	37.75	37.57	37.39	37.21
- 2	40.73	40.54	40.35	40.16	39.97	39.78	39.59	39.40	39.21	39.03
- 1	42.71	42.51	42.31	42.11	41.91	41.71	41.51	41.32	41.12	40.92
- 0	44.77	44.56	44.35	44.14	43.94	43.73	43.52	43.32	43.12	42.91

(continued)

Table 11.4 Continued.

SATURATION VAPOR PRESSURE OVER WATER

Tempera-ture °F	.0 in. Hg.	.1 in. Hg.	.2 in. Hg.	.3 in. Hg.	.4 in. Hg.	.5 in. Hg.	.6 in. Hg.	.7 in. Hg.	.8 in. Hg.	.9 in. Hg.
					English units					
0	0.04477	0.04498	0.04519	0.04540	0.04562	0.04583	0.04604	0.04626	0.04647	0.04669
1	0.04691	0.04713	0.04735	0.04757	0.04780	0.04802	0.04824	0.04847	0.04869	0.04892
2	0.04915	0.04938	0.04961	0.04984	0.05008	0.05031	0.05054	0.05078	0.05102	0.05125
3	0.05149	0.05173	0.05197	0.05221	0.05245	0.05269	0.05293	0.05318	0.05343	0.05367
4	0.05392	0.05417	0.05442	0.05467	0.05492	0.05517	0.05543	0.05568	0.05594	0.05620
5	0.05646	0.05672	0.05698	0.05724	0.05750	0.05776	0.05803	0.05829	0.05856	0.05883
6	0.05910	0.05937	0.05964	0.05991	0.06019	0.06046	0.06074	0.06101	0.06129	0.06157
7	0.06185	0.06213	0.06242	0.06270	0.06298	0.06327	0.06355	0.06384	0.06413	0.06442
8	0.06471	0.06500	0.06530	0.06560	0.06589	0.06619	0.06649	0.06679	0.06709	0.06739
9	0.06769	0.06800	0.06830	0.06861	0.06892	0.06923	0.06954	0.06985	0.07017	0.07048
10	0.07080	0.07112	0.07144	0.07176	0.07208	0.07240	0.07272	0.07305	0.07337	0.07370
11	0.07403	0.07436	0.07469	0.07503	0.07536	0.07570	0.07604	0.07638	0.07672	0.07706
12	0.07740	0.07774	0.07809	0.07843	0.07878	0.07913	0.07948	0.07983	0.08018	0.08053
13	0.08089	0.08125	0.08161	0.08197	0.08234	0.08270	0.08307	0.08343	0.08380	0.08417
14	0.08454	0.08491	0.08528	0.08566	0.08603	0.08641	0.08679	0.08717	0.08755	0.08793
15	0.08832	0.08871	0.07810	0.08949	0.08988	0.09027	0.09067	0.09106	0.09146	0.09186
16	0.09226	0.09266	0.09306	0.09347	0.09387	0.09428	0.09469	0.09510	0.09551	0.09592
17	0.09634	0.09676	0.09718	0.09760	0.09802	0.09845	0.09888	0.09931	0.09974	0.10017
18	0.10060	0.10104	0.10147	0.10191	0.10235	0.10279	0.10323	0.10367	0.10411	0.10456
19	0.10501	0.10546	0.10592	0.10637	0.10683	0.10729	0.10775	0.10821	0.10867	0.10913
20	0.10960	0.11007	0.11054	0.11102	0.11149	0.11197	0.11245	0.11292	0.11340	0.11389
21	0.11437	0.11486	0.11535	0.11584	0.11633	0.11683	0.11733	0.11783	0.11833	0.11883
22	0.11933	0.11983	0.12034	0.12085	0.12136	0.12187	0.12238	0.12290	0.12342	0.12394
23	0.12446	0.12499	0.12552	0.12605	0.12658	0.12711	0.12764	0.12818	0.12872	0.12926
24	0.12980	0.13035	0.13090	0.13145	0.13200	0.13255	0.13310	0.13366	0.13422	0.13478
25	0.13534	0.13591	0.13647	0.13704	0.13762	0.13819	0.13877	0.13934	0.13992	0.14051
26	0.14109	0.14168	0.14226	0.14285	0.14345	0.14404	0.14464	0.14524	0.14584	0.14644
27	0.14705	0.14766	0.14827	0.14889	0.14950	0.15012	0.15074	0.15136	0.15198	0.15261
28	0.15324	0.15387	0.15450	0.15514	0.15578	0.15642	0.15706	0.15771	0.15836	0.15901
29	0.15966	0.16032	0.16097	0.16163	0.16230	0.16296	0.16362	0.16429	0.16496	0.16563
30	0.16631	0.16699	0.16767	0.16835	0.16904	0.16973	0.17042	0.17111	0.17181	0.17251
31	0.17321	0.17392	0.17462	0.17533	0.17605	0.17676	0.17747	0.17819	0.17891	0.17963
32	0.18036	0.18109	0.18182	0.18256	0.18330	0.18404	0.18478	0.18553	0.18628	0.18703
33	0.18778	0.18854	0.18929	0.19005	0.19082	0.19158	0.19235	0.19313	0.19390	0.19468
34	0.19546	0.19624	0.19703	0.19782	0.19861	0.19940	0.20020	0.20100	0.20181	0.20261
35	0.20342	0.20423	0.20504	0.20586	0.20668	0.20750	0.20833	0.20916	0.20999	0.21082
36	0.21166	0.21250	0.21334	0.21419	0.21504	0.21589	0.21675	0.21761	0.21847	0.21933
37	0.22020	0.22107	0.22194	0.22282	0.22370	0.22458	0.22547	0.22636	0.22725	0.22814
38	0.22904	0.22994	0.23084	0.23175	0.23266	0.23357	0.23449	0.23541	0.23633	0.23726
39	0.23819	0.23912	0.24006	0.24100	0.24194	0.24289	0.24384	0.24479	0.24575	0.24671
40	0.24767	0.24864	0.24960	0.25058	0.25155	0.25253	0.25352	0.25450	0.25549	0.25648
41	0.25748	0.25848	0.25948	0.26049	0.26150	0.26251	0.26353	0.26455	0.26557	0.26660
42	0.26763	0.26866	0.26970	0.27074	0.27179	0.27284	0.27389	0.27494	0.27600	0.27706
43	0.27813	0.27920	0.28027	0.28135	0.28243	0.28351	0.28460	0.28569	0.28679	0.28789
44	0.28899	0.29010	0.29121	0.29232	0.29344	0.29456	0.29569	0.29682	0.29795	0.29909
45	0.30023	0.30137	0.30252	0.30367	0.30483	0.30599	0.30715	0.30832	0.30949	0.31067
46	0.31185	0.31303	0.31422	0.31541	0.31661	0.31781	0.31901	0.32022	0.32143	0.32265
47	0.32387	0.32509	0.32632	0.32755	0.32879	0.33003	0.33127	0.33252	0.33377	0.33503
48	0.33629	0.33755	0.33882	0.34010	0.34137	0.34266	0.34394	0.34523	0.34653	0.34783
49	0.34913	0.35044	0.35175	0.35306	0.35439	0.35571	0.35704	0.35837	0.35971	0.36105
50	0.36240	0.36375	0.36511	0.36646	0.36783	0.36920	0.37057	0.37195	0.37333	0.37472

(continued)

Table 11.4 Continued.

SATURATION VAPOR PRESSURE OVER WATER

Temperature °F	English units									
	.0 in. Hg.	.1 in. Hg.	.2 in. Hg.	.3 in. Hg.	.4 in. Hg.	.5 in. Hg.	.6 in. Hg.	.7 in. Hg.	.8 in. Hg.	.9 in. Hg.
50	0.36240	0.36375	0.36511	0.36646	0.36783	0.36920	0.37057	0.37195	0.37333	0.37472
51	.37611	.37751	.37891	.38031	.38172	.38314	.38456	.38598	.38741	.38884
52	.39028	.39172	.39317	.39462	.39608	.39754	.39901	.40048	.40195	.40343
53	.40492	.40641	.40790	.40940	.41090	.41241	.41393	.41544	.41697	.41850
54	.42003	.42157	.42311	.42466	.42621	.42777	.42933	.43090	.43248	.43406
55	0.43564	0.43723	0.43882	0.44042	0.44203	0.44364	0.44525	0.44687	0.44849	0.45012
56	.45176	.45340	.45504	.45670	.45835	.46001	.46168	.46335	.46503	.46671
57	.46840	.47009	.47179	.47350	.47521	.47692	.47864	.48037	.48210	.48384
58	.48558	.48733	.48908	.49084	.49260	.49437	.49614	.49792	.49971	.50150
59	.50330	.50510	.50691	.50873	.51055	.51238	.51421	.51605	.51789	.51974
60	0.52160	0.52346	0.52533	0.52720	0.52908	0.53096	0.53285	0.53475	0.53665	0.53856
61	.54047	.54239	.54432	.54625	.54818	0.55013	.55208	.55403	.55600	.55797
62	.55994	.56192	.56391	.56590	.56790	.56990	.57191	.57393	.57595	.57798
63	.58002	.58206	.58411	.58616	.58823	.59029	.59237	.59445	.59654	.59863
64	.60073	.60284	.60495	.60707	.60919	.61133	.61347	.61561	.61777	.61992
65	0.62209	0.62426	0.62644	0.62862	0.63082	0.63302	0.63522	0.63743	0.63954	0.64188
66	.64411	.64635	.64859	.65085	.65311	.65537	.65765	.65993	.66221	.66451
67	.66681	.66912	.67143	.67376	.67608	.67842	.68076	.68312	.68547	.68784
68	.69021	.69259	.69497	.69737	.69977	.70217	.70459	.70701	.70944	.71188
69	.71432	.71677	.71923	.72169	.72416	.72664	.72913	.73163	.73413	.73664
70	0.73916	0.74169	0.74422	0.74676	0.74931	0.75186	0.75443	0.75700	0.75958	0.76217
71	.76476	.76736	.76997	.77259	.77521	.77785	.78049	.78314	.78579	.78846
72	.79113	.79381	.79650	.79919	.80190	.80461	.80733	.81006	.81279	.81554
73	.81829	.82105	.82382	.82659	.82938	.83217	.83497	.83778	.84060	.84343
74	.84626	.84910	.85195	.85481	.85768	.86055	.86344	.86633	.86923	.87214
75	0.87506	0.87799	0.88092	0.88387	0.88682	0.88978	0.89275	0.89573	0.89872	0.90172
76	.90472	.90773	.91075	.91378	.91682	.91987	.92292	.92599	.92906	.93215
77	.93524	.93834	.94145	.94457	.94770	.95084	.95398	.95714	.96030	.96348
78	.96666	.96985	.97305	.97626	.97948	.98271	.98595	.98920	.99246	.99572
79	.99900	1.00228	1.00558	1.00888	1.01220	1.01552	1.01885	1.02220	1.02555	1.02891
80	1.0323	1.0357	1.0391	1.0425	1.0459	1.0493	1.0527	1.0561	1.0596	1.0630
81	1.0665	1.0700	1.0735	1.0769	1.0804	1.0840	1.0875	1.0910	1.0946	1.0981
82	1.1017	1.1053	1.1089	1.1125	1.1161	1.1197	1.1234	1.1270	1.1307	1.1343
83	1.1380	1.1417	1.1453	1.1490	1.1527	1.1564	1.1602	1.1639	1.1677	1.1714
84	1.1752	1.1790	1.1828	1.1866	1.1904	1.1943	1.1981	1.2020	1.2058	1.2097
85	1.2136	1.2175	1.2214	1.2253	1.2292	1.2332	1.2371	1.2411	1.2450	1.2490
86	1.2530	1.2570	1.2610	1.2650	1.2691	1.2731	1.2772	1.2812	1.2853	1.2894
87	1.2935	1.2976	1.3017	1.3059	1.3100	1.3142	1.3183	1.3225	1.3267	1.3309
88	1.3351	1.3393	1.3436	1.3478	1.3521	1.3564	1.3606	1.3649	1.3692	1.3736
89	1.3779	1.3822	1.3866	1.3910	1.3954	1.3998	1.4042	1.4086	1.4130	1.4174
90	1.4219	1.4264	1.4308	1.4353	1.4398	1.4443	1.4489	1.4534	1.4580	1.4625
91	1.4671	1.4717	1.4763	1.4809	1.4856	1.4902	1.4949	1.4995	1.5042	1.5089
92	1.5136	1.5183	1.5230	1.5278	1.5325	1.5373	1.5421	1.5469	1.5517	1.5565
93	1.5613	1.5661	1.5710	1.5759	1.5807	1.5856	1.5905	1.5955	1.6004	1.6053
94	1.6103	1.6153	1.6203	1.6253	1.6303	1.6353	1.6404	1.6454	1.6505	1.6556
95	1.6607	1.6658	1.6709	1.6761	1.6812	1.6864	1.6916	1.6967	1.7019	1.7072
96	1.7124	1.7176	1.7229	1.7282	1.7335	1.7388	1.7441	1.7494	1.7548	1.7601
97	1.7655	1.7709	1.7763	1.7817	1.7871	1.7926	1.7980	1.8035	1.8090	1.8145
98	1.8200	1.8255	1.8311	1.8366	1.8422	1.8478	1.8533	1.8590	1.8646	1.8702
99	1.8759	1.8816	1.8873	1.8930	1.8987	1.9045	1.9102	1.9160	1.9218	1.9276
100	1.9334	1.9392	1.9450	1.9509	1.9568	1.9626	1.9685	1.9745	1.9804	1.9863

(continued)

Table 11.4 Continued.

SATURATION VAPOR PRESSURE OVER WATER

English units

Temperature °F	.0 in. Hg.	.1 in. Hg.	.2 in. Hg.	.3 in. Hg.	.4 in. Hg.	.5 in. Hg.	.6 in. Hg.	.7 in. Hg.	.8 in. Hg.	.9 in. Hg.
100	1.9334	1.9392	1.9450	1.9509	1.9568	1.9626	1.9685	1.9745	1.9804	1.9863
101	1.9923	1.9983	2.0043	2.0103	2.0164	2.0224	2.0285	2.0346	2.0407	2.0486
102	2.0529	2.0590	2.0652	2.0713	2.0775	2.0837	2.0899	2.0961	2.1024	2.1086
103	2.1149	2.1212	2.1275	2.1338	2.1402	2.1465	2.1529	2.1593	2.1657	2.1722
104	2.1786	2.1851	2.1916	2.1981	2.2046	2.2111	2.2176	2.2242	2.2308	2.2374
105	2.2440	2.2506	2.2573	2.2639	2.2706	2.2773	2.2840	2.2907	2.2975	2.3042
106	2.3110	2.3178	2.3246	2.3315	2.3383	2.3452	2.3521	2.3590	2.3659	2.3728
107	2.3798	2.3868	2.3938	2.4008	2.4078	2.4148	2.4219	2.4290	2.4361	2.4432
108	2.4503	2.4574	2.4646	2.4718	2.4790	2.4862	2.4935	2.5007	2.5080	2.5153
109	2.5226	2.5299	2.5373	2.5447	2.5521	2.5595	2.5669	2.5744	2.5818	2.5893
110	2.5968	2.6043	2.6118	2.6194	2.6270	2.6346	2.6422	2.6498	2.6574	2.6651
111	2.6728	2.6805	2.6882	2.6960	2.7037	2.7115	2.7193	2.7271	2.7350	2.7428
112	2.7507	2.7586	2.7665	2.7745	2.7824	2.7904	2.7984	2.8064	2.8145	2.8225
113	2.8306	2.8387	2.8468	2.8550	2.8631	2.8713	2.8795	2.8877	2.8960	2.9042
114	2.9125	2.9208	2.9291	2.9374	2.9458	2.9541	2.9625	2.9709	2.9794	2.9878
115	2.9963	3.0048	3.0133	3.0219	3.0305	3.0390	3.0477	3.0563	3.0649	3.0736
116	3.0823	3.0910	3.0997	3.1085	3.1172	3.1260	3.1348	3.1437	3.1525	3.1614
117	3.1703	3.1792	3.1882	3.1972	3.2062	3.2152	3.2242	3.2333	3.2424	3.2515
118	3.2606	3.2697	3.2789	3.2881	3.2973	3.3065	3.3158	3.3250	3.3343	3.3437
119	3.3530	3.3624	3.3718	3.3812	3.3906	3.4001	3.4096	3.4191	3.4286	3.4381
120	3.4477	3.4573	3.4669	3.4765	3.4862	3.4958	3.5056	3.5153	3.5250	3.5348
121	3.5446	3.5544	3.5643	3.5741	3.5840	3.5940	3.6039	3.6139	3.6239	3.6339
122	3.6439	3.6539	3.6640	3.6741	3.6842	3.6944	3.7046	3.7148	3.7250	3.7352
123	3.7455	3.7558	3.7661	3.7765	3.7869	3.7972	3.8077	3.8181	3.8286	3.8391
124	3.8496	3.8601	3.8707	3.8813	3.8919	3.9025	3.9132	3.9239	3.9346	3.9453
125	3.9561	3.9669	3.9777	3.9885	3.9994	4.0103	4.0212	4.0321	4.0431	4.0541
126	4.0651	4.0762	4.0872	4.0983	4.1095	4.1206	4.1318	4.1430	4.1543	4.1655
127	4.1768	4.1881	4.1994	4.2108	4.2222	4.2336	4.2450	4.2565	4.2680	4.2795
128	4.2910	4.3026	4.3141	4.3257	4.3374	4.3490	4.3607	4.3725	4.3842	4.3960
129	4.4078	4.4196	4.4315	4.4434	4.4553	4.4672	4.4792	4.4912	4.5033	4.5153
130	4.5274	4.5395	4.5517	4.5638	4.5760	4.5882	4.6005	4.6128	4.6251	4.6374
131	4.6498	4.6622	4.6746	4.6871	4.6995	4.7120	4.7246	4.7371	4.7497	4.7624
132	4.7750	4.7877	4.8004	4.8131	4.8258	4.8386	4.8514	4.8643	4.8772	4.8901
133	4.9030	4.9160	4.9290	4.9420	4.9551	4.9681	4.9813	4.9944	5.0076	5.0208
134	5.0340	5.0473	5.0605	5.0738	5.0872	5.1006	5.1140	5.1274	5.1409	5.1544
135	5.1679	5.1815	5.1951	5.2087	5.2223	5.2360	5.2497	5.2635	5.2773	5.2911
136	5.3049	5.3188	5.3327	5.3466	5.3606	5.3746	5.3886	5.4027	5.4167	5.4309
137	5.4450	5.4592	5.4734	5.4876	5.5018	5.5161	5.5305	5.5448	5.5592	5.5736
138	5.5881	5.6026	5.6171	5.6317	5.6463	5.6609	5.6755	5.6902	5.7050	5.7197
139	5.7345	5.7493	5.7642	5.7791	5.7940	5.8090	5.8239	5.8390	5.8540	5.8691
140	5.8842	5.8993	5.9145	5.9297	5.9450	5.9602	5.9755	5.9909	6.0062	6.0217
141	6.0371	6.0526	6.0681	6.0836	6.0992	6.1148	6.1305	6.1461	6.1619	6.1776
142	6.1934	3.2092	6.2251	6.2410	6.0569	6.2729	6.2889	6.3049	6.3210	6.3371
143	6.3532	6.3694	6.3856	6.4018	6.4180	6.4344	6.4507	6.4671	6.4835	6.4999
144	6.5164	6.5329	6.5495	6.5661	6.5827	6.5994	6.6160	6.6328	6.6496	6.6664
145	6.6832	6.7001	6.7170	6.7339	6.7509	6.7679	6.7850	6.8021	6.8192	6.8364
146	6.8536	6.8708	6.8881	6.9054	6.9228	6.9402	6.9576	6.9751	6.9926	7.0101
147	7.0277	7.0453	7.0630	7.0807	7.0984	7.1162	7.1340	7.1518	7.1697	7.1876
148	7.2056	7.2236	7.2416	7.2597	7.2778	7.2959	7.3141	7.3323	7.3506	7.3689
149	7.3872	7.4056	7.4240	7.4424	7.4609	7.4794	7.4980	7.5166	7.5353	7.5540
150	7.5727	7.5915	7.6103	7.6291	7.6480	7.6670	7.6859	7.7049	7.7240	7.7431

(continued)

Table 11.4 Continued.

SATURATION VAPOR PRESSURE OVER WATER

English units

Temperature °F	.0 in. Hg.	.1 in. Hg.	.2 in. Hg.	.3 in. Hg.	.4 in. Hg.	.5 in. Hg.	.6 in. Hg.	.7 in. Hg.	.8 in. Hg.	.9 in. Hg.
150	7.5727	7.5915	7.6103	7.6291	7.6480	7.6670	7.6859	7.7049	7.7240	7.7431
151	7.7622	7.7814	7.8005	7.8198	7.8391	7.8584	7.8777	7.8971	7.9166	7.9361
152	7.9556	7.9752	7.9948	8.0145	8.0342	8.0539	8.0737	8.0935	8.1134	8.1333
153	8.1532	8.1732	8.1932	8.2132	8.2333	8.2535	8.2736	8.2939	8.3141	8.3344
154	8.3548	8.3752	8.3956	8.4161	8.4367	8.4572	8.4778	8.4985	8.5192	8.5399
155	8.5607	8.5815	8.6024	8.6233	8.6442	8.6652	8.6862	8.7073	8.7284	8.7496
156	8.7708	8.7921	8.8133	8.8347	8.8561	8.8775	8.8990	8.9205	8.9420	8.9637
157	8.9853	9.0070	9.0287	9.0505	9.0723	9.0942	9.1161	9.1381	9.1601	9.1821
158	9.2042	9.2263	9.2485	9.2707	9.2930	9.3153	9.3377	9.3601	9.3826	9.4051
159	9.4276	9.4502	9.4728	9.4955	9.5182	9.5410	9.5638	9.5867	9.6096	9.6326
160	9.6556	9.6786	9.7017	9.7249	9.7481	9.7713	9.7946	9.8179	9.8413	9.8647
161	9.8882	9.9117	9.9353	9.9589	9.9826	10.006	10.030	10.054	10.078	10.102
162	10.126	10.150	10.174	10.198	10.222	10.246	10.271	10.295	10.319	10.344
163	10.368	10.392	10.417	10.442	10.466	10.491	10.516	10.540	10.565	10.590
164	10.615	10.640	10.665	10.690	10.715	10.740	10.766	10.791	10.816	10.842
165	10.867	10.892	10.918	10.944	10.969	10.995	11.021	11.046	11.072	11.098
166	11.124	11.150	11.176	11.202	11.228	11.254	11.281	11.307	11.333	11.360
167	11.386	11.412	11.439	11.466	11.492	11.519	11.546	11.572	11.599	11.626
168	11.653	11.680	11.707	11.734	11.761	11.788	11.815	11.843	11.870	11.898
169	11.925	11.953	11.980	12.008	12.035	12.063	12.091	12.119	12.147	12.175
170	12.203	12.231	12.259	12.288	12.316	12.344	12.373	12.401	12.430	12.458
171	12.487	12.515	12.544	12.573	12.601	12.630	12.659	12.688	12.717	12.746
172	12.775	12.804	12.834	12.863	12.892	12.922	12.951	12.981	13.011	13.040
173	13.070	13.100	13.130	13.159	13.189	13.219	13.249	13.279	13.310	13.340
174	13.370	13.400	13.431	13.461	13.492	13.522	13.553	13.584	13.614	13.645
175	13.676	13.707	13.738	13.769	13.800	13.831	13.862	13.893	13.924	13.956
176	13.987	14.019	14.050	14.082	14.113	14.145	14.177	14.209	14.241	14.273
177	14.305	14.337	14.369	14.402	14.434	14.466	14.499	14.531	14.564	14.596
178	14.629	14.662	14.695	14.727	14.760	14.793	14.826	14.859	14.893	14.926
179	14.959	14.992	15.026	15.059	15.093	15.126	15.160	15.194	15.227	15.261
180	15.295	15.329	15.363	15.397	15.431	15.465	15.499	15.534	15.568	15.602
181	15.637	15.672	15.706	15.741	15.776	15.811	15.846	15.881	15.916	15.951
182	15.986	16.021	16.056	16.092	16.127	16.163	16.198	16.234	16.269	16.305
183	16.341	16.377	16.413	16.449	16.485	16.521	16.557	16.594	16.630	16.667
184	16.703	16.739	16.776	16.813	16.849	16.886	16.923	16.960	16.997	17.034
185	17.071	17.108	17.145	17.185	17.220	17.258	17.295	17.333	17.370	17.408
186	17.446	17.484	17.522	17.560	17.598	17.637	17.675	17.713	17.752	17.790
187	17.829	17.868	17.906	17.945	17.984	18.023	18.062	18.101	18.140	18.179
188	18.218	18.257	18.297	18.336	18.376	18.415	18.455	18.494	18.534	18.574
189	18.614	18.654	18.694	18.734	18.774	18.814	18.855	18.895	18.936	18.976
190	19.017	19.058	19.099	19.140	19.181	19.222	19.263	19.304	19.345	19.387
191	19.428	19.469	19.511	19.553	19.594	19.636	19.678	19.720	19.762	19.804
192	19.846	19.888	19.930	19.973	20.015	20.058	20.100	20.143	20.185	20.228
193	20.271	20.314	20.357	20.400	20.443	20.486	20.530	20.573	20.617	20.660
194	20.704	20.748	20.792	20.835	20.879	20.924	20.968	21.012	21.056	21.101
195	21.145	21.190	21.234	21.279	21.324	21.369	21.414	21.459	21.504	21.549
196	21.594	21.639	21.684	21.730	21.775	21.821	21.867	21.912	21.958	22.004
197	22.050	22.096	22.142	22.189	22.235	22.282	22.328	22.375	22.421	22.468
198	22.515	22.562	22.609	22.656	22.703	22.750	22.797	22.844	22.892	22.939
199	22.987	23.035	23.083	23.130	23.178	23.226	23.275	23.323	23.371	23.420
200	23.468	23.516	23.565	23.614	23.663	23.711	23.760	23.809	23.858	23.908

(continued)

Table 11.4 Continued.

SATURATION VAPOR PRESSURE OVER WATER

English units

Temperature °F	.0 in. Hg.	.1 in. Hg.	.2 in. Hg.	.3 in. Hg.	.4 in. Hg.	.5 in. Hg.	.6 in. Hg.	.7 in. Hg.	.8 in. Hg.	.9 in. Hg.
200	23.468	23.516	23.565	23.614	23.663	23.711	23.760	23.809	23.858	23.908
201	23.957	24.006	24.056	24.106	24.155	24.205	24.255	24.305	24.355	24.405
202	24.455	24.505	24.555	24.606	24.656	24.707	24.758	24.808	24.859	24.910
203	24.961	25.012	25.063	25.115	25.166	25.217	25.269	25.321	25.372	25.424
204	25.476	25.528	25.580	25.632	25.685	25.737	25.789	25.842	25.895	25.947
205	26.000	26.053	26.106	26.159	26.212	26.265	26.318	26.371	26.425	26.478
206	26.532	26.586	26.640	26.694	26.748	26.802	26.856	26.910	26.965	27.019
207	27.074	27.129	27.183	27.238	27.293	27.348	27.404	27.459	27.514	27.569
208	27.625	27.681	27.736	27.792	27.848	27.904	27.960	28.016	28.072	28.129
209	28.185	28.241	28.298	28.355	28.411	28.468	28.525	28.582	28.639	28.697
210	28.754	28.811	28.869	28.927	28.985	29.042	29.100	29.158	29.216	29.275
211	29.333	29.391	29.450	29.508	29.567	29.626	29.685	29.744	29.803	29.862
212	29.921									

Table 11.5 Saturation vapor pressure over ice.

SATURATION VAPOR PRESSURE OVER ICE

Metric units

Temperature °C	.0	.1	.2	.3	.4	.5	.6	.7	.8	.9
Unit:	10^{-5} mb.	10^{-5} mb.	10^{-5} mb.	10^{-5} mb.	10^{-5} mb.	10^{-5} mb.	10^{-5} mb.	10^{-5} mb.	10^{-5} mb.	10^{-5} mb.
-100	1.403									
- 99	1.719	1.685	1.651	1.617	1.585	1.553	1.522	1.491	1.461	1.432
- 98	2.101	2.059	2.019	1.979	1.939	1.901	1.863	1.826	1.790	1.754
- 97	2.561	2.511	2.462	2.414	2.366	2.320	2.274	2.230	2.186	2.143
- 96	3.117	3.057	2.997	2.939	2.882	2.826	2.771	2.717	2.664	2.612
- 95	3.784	3.712	3.640	3.571	3.502	3.435	3.369	3.304	3.240	3.178
- 94	4.584	4.497	4.412	4.329	4.246	4.166	4.087	4.009	3.932	3.858
- 93	5.542	5.438	5.336	5.236	5.138	5.041	4.946	4.853	4.762	4.672
- 92	6.685	6.561	6.439	6.320	6.203	6.088	5.975	5.863	5.754	5.647
- 91	8.049	7.902	7.757	7.615	7.475	7.338	7.203	7.070	6.939	6.811
- 90	9.672	9.497	9.324	9.155	8.988	8.825	8.664	8.506	8.351	8.199
- 89	11.60	11.39	11.19	10.98	10.79	10.59	10.40	10.22	10.03	9.850
- 88	13.88	13.63	13.39	13.15	12.92	12.69	12.46	12.24	12.02	11.81
- 87	16.58	16.29	16.00	15.72	15.45	15.18	14.91	14.65	14.39	14.13
- 86	19.77	19.43	19.09	18.76	18.43	18.11	17.79	17.48	17.18	16.88
- 85	23.53	23.13	22.73	22.34	21.96	21.58	21.21	20.84	20.48	20.12
- 84	27.96	27.48	27.02	26.56	26.10	25.66	25.22	24.79	24.36	23.94
- 83	33.16	32.60	32.05	31.51	30.98	30.45	29.93	29.43	28.93	28.44
- 82	39.25	38.60	37.95	37.32	36.69	36.08	35.48	34.88	34.30	33.72
- 81	46.38	45.62	44.86	44.12	43.40	42.68	41.97	41.28	40.59	39.91
- 80	54.72	53.83	52.95	52.08	51.23	50.39	49.56	48.75	47.95	47.16
- 79	64.44	63.40	62.37	61.36	60.37	59.39	58.43	57.48	56.54	55.62
- 78	75.77	74.56	73.36	72.19	71.03	69.89	68.77	67.66	66.57	65.50
- 77	88.94	87.53	86.14	84.78	83.43	82.11	80.80	79.52	78.25	77.00
- 76	104.2	102.6	101.0	99.41	97.85	96.31	94.79	93.29	91.82	90.37
Unit:	10^{-3} mb.	10^{-3} mb.	10^{-3} mb.	10^{-3} mb.-	10^{-3} mb.	10^{-3} mb.	10^{-3} mb.	10^{-3} mb.	10^{-3} mb.	10^{-3} mb.
- 75	1.220	1.201	1.182	1.164	1.146	1.128	1.110	1.093	1.076	1.059
- 74	1.425	1.403	1.382	1.360	1.340	1.319	1.299	1.279	1.259	1.239
- 73	1.662	1.637	1.612	1.587	1.563	1.539	1.515	1.492	1.470	1.447
- 72	1.936	1.907	1.878	1.850	1.822	1.794	1.767	1.740	1.714	1.668
- 71	2.252	2.218	2.185	2.152	2.120	2.088	2.057	2.026	1.995	1.965
- 70	2.615	2.576	2.538	2.501	2.464	2.427	2.391	2.355	2.320	2.286
- 69	3.032	2.988	2.944	2.901	2.858	2.816	2.775	2.734	2.694	2.654
- 68	3.511	3.460	3.410	3.360	3.311	3.263	3.215	3.169	3.122	3.077
- 67	4.060	4.002	3.944	3.887	3.831	3.776	3.721	3.668	3.615	3.562
- 66	4.688	4.621	4.555	4.490	4.426	4.363	4.301	4.239	4.179	4.119
- 65	5.406	5.330	5.255	5.180	5.107	5.035	4.964	4.893	4.824	4.755
- 64	6.225	6.138	6.052	5.968	5.884	5.802	5.721	5.640	5.561	5.483
- 63	7.159	7.060	6.962	6.866	6.771	6.677	6.584	6.493	6.402	6.313
- 62	8.223	8.110	7.999	7.889	7.781	7.674	7.568	7.464	7.361	7.259
- 61	9.432	9.304	9.177	9.053	8.930	8.808	8.688	8.569	8.452	8.337
- 60	10.80	10.66	10.51	10.37	10.24	10.10	9.961	9.826	9.693	9.562
- 59	12.36	12.20	12.03	11.87	11.72	11.56	11.40	11.25	11.10	10.95
- 58	14.13	13.94	13.76	13.58	13.40	13.22	13.04	12.87	12.70	12.53
- 57	16.12	15.91	15.70	15.49	15.29	15.09	14.89	14.70	14.51	14.32
- 56	18.38	18.14	17.91	17.68	17.45	17.22	17.00	16.77	16.55	16.34
- 55	20.92	20.65	20.39	20.12	19.86	19.61	19.36	19.11	18.86	18.62
- 54	23.80	23.50	23.20	22.90	22.61	22.32	22.03	21.75	21.47	21.19
- 53	27.03	26.69	26.35	26.02	25.69	25.37	25.05	24.73	24.42	24.11
- 52	30.67	30.29	29.91	29.53	29.17	28.80	28.44	28.08	27.73	27.38
- 51	34.76	34.33	33.90	33.48	33.06	32.65	32.24	31.84	31.45	31.06
- 50	39.35	38.87	38.39	37.92	37.45	36.99	36.53	36.08	35.64	35.20

(*continued*)

Table 11.5 Continued.

SATURATION VAPOR PRESSURE OVER ICE

Metric units

Temperature °C Unit:	.0 mb.	.1 mb.	.2 mb.	.3 mb.	.4 mb.	.5 mb.	.6 mb.	.7 mb.	.8 mb.	.9 mb
-50	0.03935	0.03887	0.03839	0.03792	0.03745	0.03699	0.03653	0.03608	0.03564	0.03520
-49	0.04449	0.04395	0.04341	0.04289	0.04236	0.04185	0.04134	0.04083	0.04033	0.03984
-48	0.05026	0.04965	0.04905	0.04846	0.04788	0.04730	0.04673	0.04616	0.04560	0.04504
-47	0.05671	0.05603	0.05536	0.05470	0.05405	0.05340	0.05276	0.05212	0.05150	0.05087
-46	0.06393	0.06317	0.06242	0.06168	0.06095	0.06022	0.05950	0.05879	0.05809	0.05740
-45	0.07198	0.07113	0.07030	0.06947	0.06865	0.06784	0.06704	0.06625	0.06547	0.06469
-44	0.08097	0.08003	0.07909	0.07817	0.07725	0.07635	0.07546	0.07457	0.07370	0.07283
-43	0.09098	0.08993	0.08889	0.08786	0.08684	0.08584	0.08484	0.08386	0.08289	0.08192
-42	0.1021	0.1010	0.09981	0.09866	0.09753	0.09641	0.09530	0.09420	0.09312	0.09204
-41	0.1145	0.1132	0.1119	0.1107	0.1094	0.1082	0.1070	0.1057	0.1045	0.1033
-40	0.1283	0.1268	0.1254	0.1240	0.1226	0.1212	0.1198	0.1185	0.1171	0.1158
-39	0.1436	0.1420	0.1404	0.1389	0.1373	0.1358	0.1343	0.1328	0.1313	0.1298
-38	0.1606	0.1588	0.1571	0.1553	0.1536	0.1519	0.1502	0.1485	0.1469	0.1452
-37	0.1794	0.1774	0.1755	0.1736	0.1717	0.1698	0.1679	0.1661	0.1642	0.1624
-36	0.2002	0.1980	0.1959	0.1938	0.1917	0.1896	0.1875	0.1855	0.1834	0.1814
-35	0.2233	0.2209	0.2185	0.2161	0.2138	0.2115	0.2092	0.2069	0.2047	0.2024
-34	0.2488	0.2461	0.2435	0.2409	0.2383	0.2357	0.2332	0.2307	0.2282	0.2257
-33	0.2769	0.2740	0.2711	0.2682	0.2653	0.2625	0.2597	0.2569	0.2542	0.2515
-32	0.3079	0.3047	0.3014	0.2983	0.2951	0.2920	0.2889	0.2859	0.2828	0.2799
-31	0.3421	0.3385	0.3350	0.3315	0.3280	0.3246	0.3212	0.3178	0.3145	0.3112
-30	0.3798	0.3759	0.3720	0.3681	0.3643	0.3605	0.3567	0.3530	0.3494	0.3457
-29	0.4213	0.4170	0.4127	0.4084	0.4042	0.4000	0.3959	0.3918	0.3877	0.3838
-28	0.4669	0.4621	0.4574	0.4527	0.4481	0.4435	0.4390	0.4345	0.4300	0.4256
-27	0.5170	0.5118	0.5066	0.5014	0.4964	0.4913	0.4863	0.4814	0.4765	0.4717
-26	0.5720	0.5663	0.5606	0.5549	0.5493	0.5438	0.5383	0.5329	0.5276	0.5222
-25	0.6323	0.6260	0.6198	0.6136	0.6075	0.6015	0.5955	0.5895	0.5836	0.577
-24	0.6985	0.6916	0.6848	0.6780	0.6713	0.6646	0.6580	0.6515	0.6450	0.6386
-23	0.7709	0.7634	0.7559	0.7485	0.7412	0.7339	0.7267	0.7195	0.7125	0.7055
-22	0.8502	0.8419	0.8338	0.8257	0.8176	0.8097	0.8018	0.7940	0.7862	0.7785
-21	0.9370	0.9280	0.9190	0.9101	0.9013	0.8926	0.8840	0.8754	0.8669	0.8585
-20	1.032	1.022	1.012	1.002	0.9928	0.9833	0.9739	0.9645	0.9553	0.9461
-19	1.135	1.124	1.114	1.103	1.092	1.082	1.072	1.062	1.052	1.042
-18	1.248	1.236	1.225	1.213	1.201	1.190	1.179	1.168	1.157	1.146
-17	1.371	1.358	1.345	1.333	1.320	1.308	1.296	1.284	1.272	1.260
-16	1.506	1.492	1.478	1.464	1.451	1.437	1.424	1.410	1.397	1.384
-15	1.652	1.637	1.622	1.607	1.592	1.577	1.562	1.548	1.534	1.520
-14	1.811	1.795	1.778	1.762	1.746	1.730	1.714	1.698	1.683	1.667
-13	1.984	1.966	1.948	1.930	1.913	1.895	1.878	1.861	1.844	1.827
-12	2.172	2.153	2.133	2.114	2.095	2.076	2.057	2.039	2.020	2.002
-11	2.376	2.355	2.334	2.313	2.292	2.271	2.251	2.231	2.211	2.191
-10	2.597	2.574	2.551	2.529	2.506	2.484	2.462	2.440	2.419	2.397
- 9	2.837	2.812	2.787	2.763	2.739	2.715	2.691	2.667	2.644	2.620
- 8	3.097	3.070	3.043	3.017	2.991	2.965	2.939	2.913	2.888	2.862
- 7	3.379	3.350	3.321	3.292	3.264	3.236	3.208	3.180	3.152	3.124
- 6	3.685	3.653	3.622	3.591	3.560	3.529	3.499	3.468	3.438	3.409
- 5	4.015	3.981	3.947	3.913	3.879	3.846	3.813	3.781	3.748	3.717
- 4	4.372	4.335	4.298	4.262	4.226	4.190	4.154	4.119	4.084	4.049
- 3	4.757	4.717	4.678	4.638	4.600	4.561	4.523	4.485	4.447	4.409
- 2	5.173	5.130	5.087	5.045	5.003	4.961	4.920	4.878	4.838	4.797
- 1	5.623	5.577	5.530	5.485	5.439	5.394	5.349	5.305	5.260	5.217
- 0	6.107	6.057	6.007	5.958	5.909	5.860	5.812	5.764	5.717	5.670

(continued)

Table 11.6 Saturation vapor pressure over ice.

SATURATION VAPOR PRESSURE OVER ICE

English units

Temperature °F	Vapor pressure in Hg.	Temperature °F	Vapor pressure in Hg.	Temperature °F	Vapor pressure in Hg.	Temperature °F	Vapor pressure in Hg.	Temperature °F	Vapor pressure in Hg.
-160.0	1.008×10^{-7}	-136.0	1.536×10^{-6}	-112.0	1.616×10^{-5}	-88.0	1.258×10^{-4}	-64.0	7.652×10^{-4}
-159.5	1.071	-135.5	1.619	-111.5	1.691	-87.5	1.310	-63.5	7.927
-159.0	1.138	-135.0	1.706	-111.0	1.770	-87.0	1.363	-63.0	8.211
-158.5	1.209	-134.5	1.798	-110.5	1.852	-86.5	1.418	-62.5	8.505
-158.0	1.285	-134.0	1.894	-110.0	1.938	-86.0	1.475	-62.0	8.808
-157.5	1.365×10^{-7}	-133.5	1.995×10^{-6}	-109.5	2.027×10^{-5}	-85.5	1.534×10^{-4}	-61.5	9.121×10^{-4}
-157.0	1.450	-133.0	2.101	-109.0	2.120	-85.0	1.596	-61.0	9.444
-156.5	1.539	-132.5	2.212	-108.5	2.217	-84.5	1.660	-60.5	9.778
-156.0	1.634	-132.0	2.328	-108.0	2.319	-84.0	1.727	-60.0	1.012×10^{-3}
-155.5	1.734	-131.5	2.450	-107.5	2.425	-83.5	1.796	-59.5	1.048
-155.0	1.84×10^{-7}	-131.0	2.579×10^{-6}	-107.0	2.535×10^{-5}	-83.0	1.867×10^{-4}	-59.0	1.085×10^{-3}
-154.5	1.953	-130.5	2.714	-106.5	2.650	-82.5	1.941	-58.5	1.123
-154.0	2.072	-130.0	2.856	-106.0	2.770	-82.0	2.018	-58.0	1.162
-153.5	2.196	-129.5	3.004	-105.5	2.895	-81.5	2.098	-57.5	1.202
-153.0	2.329	-129.0	3.160	-105.0	3.025	-81.0	2.181	-57.0	1.244
-152.5	2.470×10^{-7}	-128.5	3.323×10^{-6}	-104.5	3.160×10^{-5}	-80.5	2.266×10^{-4}	-56.5	1.287×10^{-3}
-152.0	2.618	-128.0	3.494	-104.0	3.301	-80.0	2.355	-56.0	1.332
-151.5	2.774	-127.5	3.674	-103.5	3.448	-79.5	2.447	-55.5	1.378
-151.0	2.939	-127.0	3.862	-103.0	3.602	-79.0	2.542	-55.0	1.426
-150.5	3.113	-126.5	4.058	-102.5	3.762	-78.5	2.641	-54.5	1.475
-150.0	3.298×10^{-7}	-126.0	4.265×10^{-6}	-102.0	3.928×10^{-5}	-78.0	2.743×10^{-4}	-54.0	1.525×10^{-3}
-149.5	3.492	-125.5	4.481	-101.5	4.101	-77.5	2.849	-53.5	1.577
-149.0	3.698	-125.0	4.708	-101.0	4.281	-77.0	2.959	-53.0	1.631
-148.5	3.914	-124.5	4.945	-100.5	4.664	-76.5	3.073	-52.5	1.686
-148.0	4.143	-124.0	5.194	-100.0	4.664	-76.0	3.191	-52.0	1.743
-147.5	4.384×10^{-7}	-123.5	5.454×10^{-6}	-99.5	4.867×10^{-5}	-75.5	3.313×10^{-4}	-51.5	1.802×10^{-3}
-147.0	4.639	-123.0	5.726	-99.0	5.079	-75.0	3.439	-51.0	1.863
-146.5	4.907	-122.5	6.011	-985	5.299	-74.5	3.570	-505	1.925
-146.0	5.190	-122.0	6.310	-98.0	5.528	-74.0	3.705	-70.0	1.990
-145.5	5.488	-121.5	6.622	-97.5	5.766	-73.5	3.846	-79.5	2.057
-145.0	5.803×10^{-7}	-121.0	6.949×10^{-6}	-97.0	6.014×10^{-5}	-73.0	3.991×10^{-4}	-49.0	2.126×10^{-3}
-144.5	6.133	-120.5	7.291	-96.5	6.271	-72.5	4.141	-48.5	2.197
-144.0	6.483	-120.0	7.649	-96.0	6.539	-72.0	4.296	-48.0	2.270
-143.5	6.852	-119.5	8.022	-95.5	6.818	-71.5	4.457	-47.5	2.345
-143.0	7.240	-119.0	8.414	-95.0	7.108	-71.0	4.624	-47.0	2.422
-142.5	7.646×10^{-7}	-118.5	8.823×10^{-6}	-94.5	7.409×10^{-5}	-70.5	4.796×10^{-4}	-46.5	2.502×10^{-3}
-142.0	8.076	-118.0	9.251	-94.0	7.722	-70.0	4.974	-46.0	2.585
-141.5	8.416	-117.5	9.698	-93.5	8.047	-69.5	5.158	-45.5	2.670
-141.0	9.005	-117.0	1.017×10^{-5}	-93.0	8.385	-69.0	5.349	-45.0	2.757
-140.5	9.518	-116.5	1.065	-92.5	8.736	-68.5	5.547	-44.5	2.847
-140.0	1.003×10^{-6}	-116.0	1.116×10^{-5}	-92.0	9.102×10^{-5}	-68.0	5.751×10^{-4}	-44.0	2.940×10^{-3}
-139.5	1.058	-115.5	1.170	-91.5	9.482	-67.5	5.962	-43.5	3.036
-139.0	1.117	-115.0	1.226	-91.0	9.876	-67.0	6.180	-43.0	3.134
-138.5	1.178	-114.5	1.284	-90.5	1.028×10^{-4}	-66.5	6.405	-42.5	3.235
-138.0	1.243	-114.0	1.344	-90.0	1.071	-66.0	6.638	-42.0	3.340
-137.5	1.311×10^{-6}	-113.5	1.408×10^{-5}	-89.5	1.115×10^{-4}	-65.5	6.879×10^{-4}	-41.5	3.448×10^{-3}
-137.0	1.382	-113.0	1.474	-89.0	1.161	-65.0	7.128	-41.0	3.559
-136.5	1.457	-112.5	1.544	-88.5	1.208	-64.5	7.386	-40.5	3.673
								-40.0	3.790×10^{-3}

(continued)

Table 11.6 Continued.

SATURATION VAPOR PRESSURE OVER ICE

English units

Temperature °F	.0	.1	.2	.3	.4	.5	.6	.7	.8	.9
Unit:	10^{-3}	10^{-3}	10^{-3}	10^{-3}	10^{-3}	10^{-3}	10^{-3}	10^{-3}	10^{-3}	10^{-3}
	in. Hg.	in. Hg.	in. Hg.	in. Hg.	in. Hg.	in. Hg.	in. Hg.	in. Hg.	in. Hg.	in. Hg.
-39	4.035	4.010	3.985	3.960	3.935	3.911	3.886	3.862	3.838	3.814
-38	4.295	4.268	4.242	4.215	4.189	4.163	4.137	4.111	4.086	4.060
-37	4.570	4.542	4.514	4.486	4.458	4.430	4.403	4.376	4.349	4.322
-36	4.862	4.832	4.802	4.773	4.743	4.714	4.685	4.656	4.627	4.598
-35	5.170	5.138	5.107	5.076	5.045	5.014	4.983	4.952	4.922	4.892
-34	5.497	5.463	5.430	5.397	5.364	5.331	5.299	5.266	5.234	5.202
-33	5.843	5.808	5.772	5.737	5.702	5.668	5.633	5.599	5.565	5.531
-32	6.208	6.171	6.133	6.096	6.059	6.023	5.986	5.950	5.914	5.879
-31	6.595	6.555	6.516	6.477	6.438	6.399	6.360	6.322	6.284	6.246
-30	7.003	6.961	6.919	6.878	6.837	6.796	6.755	6.715	6.675	6.635
-29	7.435	7.391	7.347	7.303	7.259	7.216	7.173	7.130	7.087	7.045
-28	7.891	7.844	7.798	7.751	7.705	7.660	7.614	7.569	7.524	7.479
-27	8.373	8.324	8.274	8.226	8.177	8.129	8.081	8.033	7.985	7.938
-26	8.882	8.830	8.778	8.726	8.675	8.624	8.573	8.523	8.472	8.423
-25	9.420	9.365	9.310	9.256	9.201	9.147	9.094	9.040	8.987	8.934
-24	9.987	9.929	9.871	9.814	9.756	9.700	9.643	9.587	9.531	9.475
-23	10.59	10.52	10.46	10.40	10.34	10.28	10.22	10.16	10.10	10.05
-22	11.22	11.16	11.09	11.03	10.96	10.90	10.84	10.78	10.71	10.65
-21	11.88	11.81	11.74	11.68	11.61	11.54	11.48	11.41	11.35	11.28
-20	12.59	12.52	12.45	12.37	12.30	12.23	12.16	12.09	12.02	11.95
Unit:	in. Hg.	in. Hg.	in. Hg.	in. Hg.	in. Hg.	in. Hg.	in. Hg.	in. Hg.	in. Hg.	in. Hg.
-19	0.01333	0.01325	0.01318	0.01310	0.01303	0.01296	0.01288	0.01281	0.01274	0.01266
-18	0.01410	0.01402	0.01394	0.01386	0.01378	0.01371	0.01363	0.01355	0.01348	0.01340
-17	0.01493	0.01485	0.01476	0.01468	0.01459	0.01451	0.01443	0.01435	0.01426	0.01418
-16	0.01579	0.01570	0.01561	0.01553	0.01544	0.01535	0.01527	0.01518	0.01510	0.01501
-15	0.01670	0.01661	0.01651	0.01642	0.01633	0.01624	0.01615	0.01606	0.01597	0.01588
-14	0.01766	0.01756	0.01746	0.01737	0.01727	0.01717	0.01708	0.01698	0.01689	0.01679
-13	0.01867	0.01857	0.01846	0.01836	0.01826	0.01816	0.01806	0.01796	0.01786	0.01776
-12	0.01974	0.01963	0.01952	0.01941	0.01931	0.01920	0.01909	0.01899	0.01888	0.01877
-11	0.02086	0.02075	0.02063	0.02052	0.02041	0.02029	0.02018	0.02007	0.01996	0.01985
-10	0.02203	0.02191	0.02179	0.02167	0.02155	0.02144	0.02132	0.02120	0.02109	0.02097
- 9	0.02327	0.02314	0.02302	0.02289	0.02277	0.02264	0.02252	0.02240	0.02227	0.02215
- 8	0.02457	0.02444	0.02430	0.02417	0.02404	0.02391	0.02378	0.02365	0.02352	0.02340
- 7	0.02594	0.02580	0.02566	0.02552	0.02538	0.02525	0.02511	0.02497	0.02484	0.02470
- 6	0.02737	0.02722	0.02708	0.02693	0.02679	0.02664	0.02650	0.02636	0.02622	0.02608
- 5	0.02888	0.02873	0.02857	0.02842	0.02827	0.02812	0.02796	0.02781	0.02767	0.02752
- 4	0.03047	0.03031	0.03015	0.02999	0.02983	0.02967	0.02951	0.02935	0.02919	0.02904
- 3	0.03213	0.03196	0.03179	0.03162	0.03146	0.03129	0.03112	0.03096	0.03079	0.03063
- 2	0.03388	0.03370	0.03352	0.03335	0.03317	0.03299	0.03282	0.03265	0.03247	0.03230
- 1	0.03572	0.03553	0.03535	0.03516	0.03497	0.03479	0.03461	0.03442	0.03424	0.03406
- 0	0.03764	0.03744	0.03725	0.03705	0.03686	0.03667	0.03648	0.03629	0.03610	0.03591
0	0.03764	0.03784	0.03804	0.03824	0.03844	0.03864	0.03884	0.03904	0.03925	0.03945
1	0.03966	0.03987	0.04008	0.04029	0.04050	0.04071	0.04092	0.04113	0.04135	0.04156
2	0.04178	0.04200	0.04222	0.04243	0.04265	0.04288	0.04310	0.04332	0.04355	0.04377
3	0.04400	0.04423	0.04446	0.04469	0.04492	0.04515	0.04538	0.04562	0.04585	0.04609
4	0.04633	0.04657	0.04681	0.04705	0.04730	0.04754	0.04779	0.04803	0.04828	0.04853
5	0.04878	0.04903	0.04928	0.04954	0.04979	0.05004	0.05030	0.05056	0.05082	0.05108
6	0.05134	0.05160	0.05187	0.05213	0.05240	0.05266	0.05293	0.05320	0.05347	0.05375
7	0.05402	0.05430	0.05457	0.05485	0.05513	0.05541	0.05569	0.05597	0.05626	0.05654
8	0.05683	0.05712	0.05741	0.05770	0.05799	0.05828	0.05858	0.05887	0.05917	0.05947
9	0.05977	0.06007	0.06038	0.06068	0.06099	0.06130	0.06161	0.06192	0.06223	0.06255
10	0.06286	0.06317	0.06349	0.06381	0.06413	0.06445	0.06477	0.06510	0.06542	0.06575

(continued)

Table 11.6 Continued.

SATURATION VAPOR PRESSURE OVER ICE

English units

Temperature °F	.0 in. Hg.	.1 in. Hg.	.2 in. Hg.	.3 in. Hg.	.4 in. Hg.	.5 in. Hg.	.6 in. Hg.	.7 in. Hg.	.8 in. Hg.	.9 in. Hg.
10	0.06286	0.06317	0.06349	0.06381	0.06413	0.06445	0.06477	0.06510	0.06542	0.06575
11	0.06608	0.06641	0.06674	0.06708	0.06741	0.06775	0.06809	0.06843	0.06877	0.06911
12	0.06946	0.06981	0.07016	0.07051	0.07086	0.07121	0.07157	0.07192	0.07228	0.07264
13	0.07300	0.07336	0.07372	0.07409	0.07445	0.07482	0.07519	0.07556	0.07594	0.07631
14	0.07669	0.07707	0.07745	0.07783	0.07822	0.07860	0.07899	0.07938	0.07977	0.08016
15	0.08056	0.08096	0.08136	0.08176	0.08216	0.08256	0.08297	0.08338	0.08379	0.08420
16	0.08461	0.08502	0.08544	0.08586	0.08628	0.08670	0.08713	0.08755	0.08798	0.08841
17	0.08884	0.08927	0.08971	0.09014	0.09058	0.09102	0.09147	0.09191	0.09236	0.09281
18	0.09326	0.09371	0.09417	0.09463	0.09509	0.09555	0.09601	0.09648	0.09695	0.09742
19	0.09789	0.09839	0.09884	0.09932	0.09980	0.1003	0.1008	0.1013	0.1018	0.1022
20	0.1027	0.1032	0.1037	0.1042	0.1047	0.1052	0.1057	0.1063	0.1068	0.1073
21	0.1078	0.1083	0.1088	0.1093	0.1098	0.1104	0.1109	0.1114	0.1119	0.1125
22	0.1130	0.1136	0.1141	0.1147	0.1152	0.1158	0.1163	0.1169	0.1175	0.1180
23	0.1186	0.1192	0.1197	0.1203	0.1208	0.1214	0.1220	0.1226	0.1231	0.1237
24	0.1243	0.1249	0.1255	0.1261	0.1267	0.1273	0.1279	0.1285	0.1291	0.1297
25	0.1303	0.1309	0.1315	0.1322	0.1328	0.1334	0.1341	0.1347	0.1353	0.1360
26	0.1366	0.1372	0.1379	0.1385	0.1392	0.1398	0.1405	0.1411	0.1418	0.1424
27	0.1431	0.1438	0.1445	0.1451	0.1458	0.1465	0.1472	0.1479	0.1486	0.1493
28	0.1500	0.1507	0.1514	0.1521	0.1528	0.1535	0.1542	0.1549	0.1557	0.1564
29	0.1571	0.1578	0.1585	0.1593	0.1600	0.1608	0.1615	0.1622	0.1630	0.1637
30	0.1645	0.1653	0.1660	0.1668	0.1676	0.1684	0.1692	0.1699	0.1707	0.1715
31	0.1723	0.1731	0.1739	0.1747	0.1755	0.1763	0.1771	0.1779	0.1787	0.1795
32	0.1803									

Table 11.7 Vapor pressure of water above 100°C.

Based on values given by Keyes in the International Critical Tables.

Temp. °C	Pressure mm	Pounds per sq. in.	Temp. °F	Temp. °C	Pressure mm	Pounds per sq. in.	Temp. °F
100	760.00	14.696	212.0	145	3116.76	60.268	293.0
101	787.51	15.228	213.8	146	3203.40	61.944	294.8
102	815.86	15.776	215.6	147	3292.32	63.663	296.6
103	845.12	16.342	217.4	148	3382.76	65.412	298.4
104	875.06	16.921	219.2	149	3476.24	67.220	300.2
105	906.07	17.521	221.0	150	3570.48	69.042	302.0
106	937.92	18.136	222.8	151	3667.00	70.908	303.8
107	970.60	18.768	224.6	152	3766.56	72.833	305.6
108	1004.42	19.422	226.4	153	3866.88	74.773	307.4
109	1038.92	20.089	228.2	154	3970.24	76.772	309.2
110	1074.56	20.779	230.0	155	4075.88	78.815	311.0
111	1111.20	21.487	231.8	156	4183.80	80.901	312.8
112	1148.74	22.213	233.6	157	4293.24	83.018	314.6
113	1187.42	22.961	235.4	158	4404.96	85.178	316.4
114	1227.25	23.731	237.2	159	4519.72	87.397	318.2
115	1267.98	24.519	239.0	160	4636.00	89.646	320.0
116	1309.94	25.330	240.8	161	4755.32	91.953	321.8
117	1352.95	26.162	242.6	162	4876.92	94.304	323.6
118	1397.18	27.017	244.4	163	5000.04	96.685	325.4
119	1442.63	27.896	246.2	164	5126.96	99.139	327.2
120	1489.14	28.795	248.0	165	5256.16	101.638	329.0
121	1536.80	29.717	249.8	166	5386.88	104.165	330.8
122	1586.04	30.669	251.6	167	5521.40	106.766	332.6
123	1636.36	31.642	253.4	168	5658.20	109.412	334.4
124	1687.81	32.637	255.2	169	5789.04	112.116	336.2
125	1740.93	33.664	257.0	170	5940.92	114.879	338.0
126	1795.12	34.712	258.8	171	6085.32	117.671	339.8
127	1850.83	35.789	260.6	172	6233.52	120.537	341.6
128	1907.83	36.891	262.4	173	6383.24	123.432	343.4
129	1966.35	38.023	264.2	174	6538.28	126.430	345.2
130	2026.16	39.180	266.0	175	6694.08	129.442	347.0
131	2087.42	40.364	267.8	176	6852.92	132.514	348.8
132	2150.42	41.582	269.6	177	7015.56	135.659	350.6
133	2214.64	42.824	271.4	178	7180.48	138.848	352.4
134	2280.76	44.103	273.2	179	7349.20	142.110	354.2
135	2347.26	45.389	275.0	180	7520.20	145.417	356.0
136	2416.34	46.724	276.8	181	7694.24	148.782	357.8
137	2488.16	48.113	278.6	182	7872.08	152.221	359.6
138	2560.67	49.515	280.4	183	8052.96	155.719	361.4
139	2634.84	50.950	282.2	184	8236.88	159.275	363.2
140	2710.92	52.421	284.0	185	8423.84	162.800	365.0
141	2788.44	53.920	285.8	186	8616.12	166.609	366.8
142	2867.48	55.448	287.6	187	8809.92	170.356	368.6
143	2948.80	57.020	289.4	188	9007.52	174.177	370.4
144	3031.64	58.622	291.2	189	9208.16	178.057	372.2

(continued)

Table 11.7 continued.

Temp. °C	Pressure mm	Pressure Pounds per sq. in.	Temp. °F	Temp. °C	Pressure mm	Pressure Pounds per sq. in.	Temp. °F
190	9413.36	182.025	374.0	235	22967.96	444.128	455.0
191	9620.08	186.022	375.8	236	23382.92	452.152	456.8
192	9831.36	190.107	377.6	237	23802.44	460.264	458.6
193	10047.20	194.281	379.4	238	24229.56	468.523	460.4
194	10265.32	198.499	381.2	239	24661.24	476.871	462.2
195	10488.76	202.819	383.0	240	25100.52	485.365	464.0
196	10715.24	207.199	348.8	241	25543.60	493.933	465.8
197	10944.76	211.637	386.6	242	25994.28	502.647	467.6
198	11179.60	216.178	388.4	243	26449.52	511.450	469.4
199	11417.48	220.778	390.2	244	26912.36	520.400	471.2
200	11659.16	225.451	392.0	245	27381.28	529.467	473.0
201	11905.40	230.213	393.8	246	27855.52	538.638	475.8
202	12155.44	235.048	395.6	247	28335.84	547.926	476.6
203	12408.52	239.942	397.4	248	28823.76	557.360	478.4
204	12666.16	244.924	399.2	249	29317.00	566.898	480.2
205	12929.12	250.008	401.0	250	29817.84	576.583	482.0
206	13197.40	255.196	402.8	251	30324.00	586.370	483.8
207	13467.96	260.428	404.6	252	30837.76	596.305	485.6
208	13742.32	265.733	406.4	253	31356.84	606.342	487.4
209	14022.76	271.156	408.02	254	31885.04	616.556	489.2
210	14305.48	276.623	410.0	255	32417.80	626.858	491.0
211	14595.04	282.222	411.8	256	32957.40	637.292	492.8
212	14888.40	287.895	413.6	257	33505.36	647.888	494.6
213	15184.80	293.626	415.4	258	34059.40	658.601	496.4
214	15488.04	299.490	417.2	259	34618.76	669.417	498.2
215	15792.80	305.383	419.0	260	35188.00	680.425	500.0
216	16104.40	311.408	420.8	261	35761.80	691.520	501.8
217	16420.56	317.522	422.6	262	36343.20	702.763	503.6
218	16742.04	323.738	424.4	263	36932.20	714.152	505.4
219	17067.32	330.028	426.2	264	37529.56	725.703	507.2
220	17395.64	336.377	428.0	265	38133.00	737.372	509.0
221	17731.56	342.872	429.8	266	38742.52	749.158	510.8
222	18072.80	349.471	431.6	267	39361.92	761.135	512.6
223	18417.84	356.143	433.4	268	39986.64	773.215	514.4
224	18766.68	362.888	435.2	269	40619.72	785.457	516.2
225	19123.12	369.781	437.0	270	41261.16	797.861	518.0
226	19482.60	376.732	438.8	271	41910.20	810.411	519.8
227	19848.92	383.815	440.6	272	42566.08	823.094	521.6
228	20219.80	390.987	442.4	273	43229.56	835.923	523.4
229	20596.76	398.276	444.2	274	43902.16	848.929	525.2
230	20978.28	405.654	446.0	275	44580.84	862.053	527.0
231	21365.12	413.134	447.8	276	45269.40	875.367	528.8
232	21757.28	420.717	449.6	277	45964.04	888.799	530.6
233	22154.00	428.388	451.4	278	46669.32	902.437	532.4
234	22558.32	436.207	453.2	279	47382.20	916.222	534.2

(continued)

Table 11.7 continued.

Temp. °C	Pressure mm	Pounds per sq. in.	Temp. °F	Temp. °C	Pressure mm	Pounds per sq. in.	Temp. °F
280	48104.20	930.183	536.0	330	96512.40	1866.245	626.0
281	48833.80	944.291	537.8	331	97758.80	1890.346	627.8
282	49570.24	958.532	539.6	332	99020.40	1914.742	629.6
283	50316.56	972.963	541.4	333	100297.20	1939.431	631.4
284	51072.76	987.586	543.2	334	101581.60	1964.267	633.2
285	51838.08	1002.385	545.0	335	102881.20	1989.398	635.0
286	52611.76	1017.345	545.0	336	104196.00	2014.822	636.8
287	53395.32	1032.497	548.6	337	105526.00	2040.540	638.6
288	54187.24	1047.810	550.4	338	106871.20	2066.552	640.4
289	54989.04	1063.314	552.2	339	108224.00	2092.710	642.2
290	55799.20	1078.980	554.0	340	109592.00	2119.163	644.0
291	56612.40	1094.705	555.8	341	110967.60	2145.763	645.8
292	57448.40	1110.871	557.6	342	112358.40	2172.657	647.6
293	58284.40	1127.036	559.4	343	113749.20	2199.550	649.4
294	59135.60	1143.496	561.2	344	115178.00	2227.179	651.2
295	59994.40	1160.102	563.0	345	116614.40	2254.954	653.0
296	60860.80	1176.856	564.8	346	118073.60	2283.171	654.8
297	61742.40	1193.903	566.6	347	119532.80	2311.387	656.6
298	62624.00	1210.950	568.4	348	121014.80	2340.044	658.4
299	63528.40	1228.439	570.2	349	122504.40	2368.848	660.2
300	64432.80	1245.927	572.0	350	124001.60	2397.799	662.0
301	65352.40	1263.709	573.8	351	125521.60	2427.191	663.8
302	66279.60	1281.638	575.6	352	127049.20	2456.730	665.6
303	67214.40	1299.714	577.4	353	128599.60	2486.710	667.4
304	68156.80	1317.937	579.2	354	130157.60	2516.837	669.2
305	69114.40	1336.454	581.0	355	131730.80	2547.258	671.0
306	70072.00	1354.971	582.8	356	133326.80	2578.119	672.8
307	71052.40	1373.929	584.6	357	134945.60	2609.422	674.6
308	72048.00	1393.181	586.4	358	136579.60	2641.018	676.4
309	73028.40	1412.139	588.2	359	138228.80	2672.908	678.2
310	74024.00	1431.390	590.0	360	139893.20	2705.093	680.0
311	75042.40	1451.083	591.8	361	141572.80	2737.571	681.8
312	76076.00	1471.070	593.6	362	143275.20	2770.490	683.6
313	77117.20	1491.203	595.4	363	144992.80	2803.703	685.4
314	78166.00	1511.484	597.2	364	146733.20	2837.357	687.2
315	79230.00	1532.058	599.0	365	148519.20	2871.892	689.0
316	80294.00	1552.632	600.8	366	150320.40	2906.722	690.8
317	81373.20	1573.501	602.6	367	152129.20	2941.698	692.6
318	82467.60	1594.663	604.4	368	153960.80	2977.116	694.4
319	83569.60	1615.972	606.2	369	155815.20	3012.974	696.2
320	84686.80	1637.575	608.0	370	157692.40	3049.273	698.0
321	85819.20	1659.472	609.8	371	159584.80	3085.866	699.8
322	86959.20	1681.516	611.6	372	161507.60	3123.047	701.6
323	88114.40	1703.854	613.4	373	163468.40	3160.963	703.4
324	89277.20	1726.339	615.2	374	165467.20	3199.613	705.2
325	90447.60	1748.971	617.0				
326	91633.20	1771.897	618.8				
327	92826.40	1794.969	620.6				
328	94042.40	1818.483	622.4				
329	95273.60	1842.291	624.2				

12

APPLICATIONS

There are numerous requirements and applications for measuring humidity. Through the last 40 years, the awareness and need for such measurements has been rapidly increasing. Humidity measurements are now considered very important in terms of cost savings and product quality, and are often required for safety reasons by agencies such as the Environmental Protection Agency (EPA), Federal Aviation Administration (FAA), Food and Drug Administration (FDA), National Fire Protection Agency (NFPA), the Nuclear Protection Agency (NPA), and similar organizations in Europe and Asia.

When confronted with such requirements and regulations, the most important decision is to select the most suitable sensor and measurement method. Humidity can be measured in many ways, and excellent humidity instruments can be found in the marketplace. However, no instrument is good for all applications and if an otherwise excellent instrument is used in the wrong application, unsatisfactory results will be obtained.

It is highly advisable to carefully review each application and discuss it with the manufacturer or a humidity expert. Manufacturers of a large variety of sensors are likely to give a more objective appraisal than companies making only one or two types of instruments, and who may be inclined to promote their own product, even if it is not the best solution for the application at hand.

It is not intended to review all conceivable applications in this book. Instead, this chapter will list a number of the more common applications and discuss the types of sensors used. Some important applications will be discussed in detail; others will be briefly outlined, indicating the need and purpose of such measurements. For the listed applications the most commonly used sensor types are indicated. This in no way precludes any other sensors. Applications of each type vary widely and sometimes sensors which are not mentioned have also been successfully used. Though all conceivable applications can not be listed here, it is hoped that the summary will be useful to anyone who is or will be involved with humidity measurements, by finding some example in this chapter which is somewhat similar and will give an indication of how to solve the problem.

I. Heat Treating

Heat treating is an operation that is used to change the mechanical properties of a metal, such as its strength, hardness, or ductability. Heat treatments involve heating and cooling of a metal or an alloy with regard to temperature, atmosphere, and time to achieve the desired qualities.

Atmosphere control in heat treat furnaces is often accomplished by means of infrared analyzers capable of measuring and controlling CO, CO_2, and methane. Dew point instrumentation is used for control of water vapor, which sometimes can also provide an indirect measurement of other parameters.

A. Humidity Instruments Used For Heat Treating

1. Dew Cup

This simple and manual instrument, described in Chapter 3, is inexpensive and easy to operate, but could lead to large errors due to operator inexperience or interpretation. Inaccuracies are prevalent, especially at low frost points. The dew cup is still used by heat treaters as a low cost device, but is often not acceptable, especially when measurements are to be certified for regulatory agencies.

2. Fog Chamber

The fog chamber is described in Chapter 8. Like the dew cup, it is a manual and simple-to-operate device, but not very accurate. The fog chamber is popular and widely used by heat treaters but is not suitable in applications where continuous monitoring is needed. It usually does not meet requirements of regulatory agencies. Both dew cup and fog chamber measurements give one-time data only and are not suited for continuous monitoring or process control functions.

3. Lithium Chloride Sensor

Though it is reliable, simple, and inexpensive, the lithium chloride sensor has the limitation of a slow response time. In addition, although the sensor is rugged and operates in contaminated atmospheres, it is susceptible to contamination by ammonia, which makes it unsuitable for control of most carbonitriding atmospheres. Cleaning of the sensor, though not very difficult, is time consuming. The measurement range is limited to 12% RH on the low end. This sensor is at present rarely used in heat treat applications.

4. Chilled Mirror Hygrometer

Although expensive and somewhat difficult to install, the chilled mirror hygrometer is well suited for automatic, continuous dew point control of furnace atmospheres. It offers high accuracy, excellent repeatability, and is less expensive and easier to service than infrared analyzers. Compared to dew cups and fog chamber instruments, the condensation hygrometer provides much higher accuracy and repeatability and can easily be used in automatic control systems. The instrument can provide a continuous, 24-hour-a-day, record of the dew point in the furnace atmosphere and therefore can allow the heat treater to do a much higher quality job. However, the cost of such an instrument is much higher, and for measurements of very low dew points, sensor cooling is often required, which is an inconvenience and represents additional cost. Furthermore, the chilled mirror sensor requires periodic cleaning. This represents additional operating costs and the danger of mirror damage if cleaning is not properly conducted. Although hardly used at this time, the cycling chilled mirror (CCM) sensor (see Chapter 3) offers the potential of a much longer mirror cleaning maintenance cycle and could therefore prove to be more suited for heat treat applications where significant contamination is present and where high accuracy and continuous monitoring are required.

5. Aluminum Oxide Hygrometer

The aluminum oxide hygrometer is often used in heat treat applications involving low dew/frost points below −40°C (−40°F), such as exothermic gas ovens. Although accuracies and repeatability for aluminum oxide hygrometers are considerably less than for chilled mirror systems, they are more convenient and less expensive to use at frost points below −40°C (−40°F). If the gas is relatively clean, acceptable results can be obtained using some of the more advanced aluminum oxide probes.

6. Piezoelectric Hygrometer

The Piezoelectric sensor has been used in heat treat applications where accurate measurements at very low frost point or ppm levels are required. They are more accurate than aluminum oxide hygrometers, but at considerably higher expense. Piezoelectric sensors are at this time not widely used by heat treaters.

7. Infrared Analyzer

The infrared analyzer, often used in heat treat applications, provides good accuracy and the capability of directly measuring carbon oxides and other gases. However, these instruments are more expensive than chilled mirror hygrometers, and require considerable maintenance and calibration. They also require maintenance and recalibration at modest to frequent intervals.

8. Electrolytic Hygrometer

As discussed in Reference 2, Chapter 15, electrolytic cells have also found some applications in heat treating. The cells are only useful if the gas to be monitored is clean and filtered. These cells are primarily attractive when very low ppm levels are to be measured. Overall, heat treat applications of electrolytic hygrometers have been limited.

B. Heat Treatment of Steel

1. General

Inert blanketing gases are used during the heat treatment of steel to prevent oxidation. The moisture content of the blanketing gas is indicative of the oxygen concentration. The higher the moisture content, the higher the oxygen concentration. Thus, high moisture content indicates that deleterious oxidation will occur unless remedial action is taken. Moisture is measured in blanketing gases that are used in the heat treatment of steel. The sensor is inserted directly in the gas line just before it enters the heat treatment oven. The blanketing gas consists mainly of hydrogen and nitrogen gases at moderate pressures.

Using this technique, the moisture content, and hence the oxygen content, are monitored accurately and continuously. When moisture content exceeds critical concentrations, remedial action is taken to prevent the loss of large amounts of valuable high-alloyed steel.

In the past, gas samples were taken and analyzed in the laboratory by adsorption on a desiccant and measuring the weight gain. Presently such applications can be successfully handled using aluminum oxide or chilled mirror hygrometers. The chilled mirror instrument is more accurate and is traceable to national standards laboratories, though they are more expensive and cumbersome to use at very low frost point levels. The typical dew/frost point range to be covered can be as low as −70°C (−94°F).

2. Exothermic Gas

In aluminum heat treating or annealing, too much moisture in the furnace atmosphere causes oxidation of some of the alloying constituents. In such applications, it is common practice to measure every gas line going into and coming out of each furnace. Exit gas can be measured by using a cooling coil to reduce the gas temperature to less than 70°C (158°F). The amount of moisture is extremely critical anytime metal is exposed to processing, including heat treating, polishing, carburizing, welding of titanium, stainless steel, and other alloys. Even razor blade and surgical tool manufacturers benefit from moisture measurement instrumentation.

Moisture is measured in an exothermic base gas, which is an inert gas that is generated when natural gas is burned with a controlled, limited amount of air. (Base gas composition is mostly nitrogen with 0.2% to 0.5% combustibles and 0% to 0.2% maximum oxygen.)

In an exothermic gas generator, natural gas is combusted to form 87% N_2, 10% CO_2, and 1.5% H_2. Exothermic atmospheres are used for bright annealing of ferrous and nonferrous metals.

The gas is often fed into an oven used for annealing aluminum. Because it is important that the moisture content be very low, the gas is put through a desiccant bed dryer before being fed into the furnace. The moisture sensor is placed directly in the gas feed line. The typical dew point range to be covered is −25°C to −55°C (−13°F to −67°F). Contaminants include a high level of hydrocarbons.

An electrolytic hygrometer can be used, but it has some disadvantages, including the need for:

- A pump to draw a sufficient flow of gas

- A separate hygrometer for each measurement point

- A long dry-down time for the P_2O_5 cell

- Added maintenance due to contaminants, resulting in relatively high costs

Chilled mirror hygrometers have been used with some success, but are inconvenient and costly, when the dew/frost points are below −40°C (−40°F).

Aluminum oxide sensors have also been frequently used. Chilled mirror hygrometers offer the best and most accurate measurements. The choice usually depends on the moisture range to be covered.

In the event of high moisture levels, the feed gas can be re-circulated through the desiccant bed dryers before entering the furnace. Continuous monitoring of moisture reduces downtime and maintenance costs, resulting in products with a better finish.

3. Methods of Heat Treatment

a. Carburizing and Carbonitriding

Carburizing and carbonitriding are two processes that can be used to produce a hard surface layer of ferrous alloys. Both processes introduce a hardening agent or agents into the surface of the alloy, thereby modifying the composition of the surface layer. The final properties of the surface layer and interior are achieved by appropriate heat treatment involving carburizing or carbonitriding.

b. Carburizing

Carburizing is a process to increase the carbon content of the skin of a low-carbon plain or alloy steel. When the steel is heated to its austenitizing temperature, it absorbs carbon in the presence of CO_2 gas. The amount of carbon absorbed depends upon the temperature, time of exposure, carbon potential of the carburizing medium (gas), and composition of the original steel. The carbon potential of the carburizing atmosphere can be determined by measuring either the carbon dioxide (CO_2) content or the water (H_2) content of the gas. Carbon dioxide and carbon potential are commonly measured with an infrared analyzer. Carbon potential as well as oxygen may be determined directly by a hot-wire analyzer. Oxygen analyzers are also commonly used to determine carbon potential due to the inverse relationship between oxygen and water. However, measuring the dew point of the atmosphere is one of the most accurate ways to determine carbon potential. Many curves have been generated relating carbon dioxide concentration to dew point.

In the carburizing process, an austenitized ferrous metal is brought into contact with an environment of sufficient carbon potential to cause absorption of carbon at the surface and, by diffusion, to create a carbon concentration gradient between the surface and the interior of the metal. As indicated by the definition, two factors may control carburizing. Either the carbon absorption reaction at the surface or the diffusion of carbon in the metal will determine the rate of carburizing. Carburizing is done at elevated temperatures, generally in the range of 68°C to 80°C (155°F to 176°F), although temperatures as low as 63°C (145°F) and as high as 1093°C (2000°F) have been used. Above 80°C (176°F), furnace equipment has increasingly limited life, and this factor is chiefly responsible for limiting carburizing to temperatures to below about 80°C (176°F), except for special applications.

c. Mechanism of Carburizing

In the carburizing process, free carbon is absorbed into the surface layer of a piece of metal (usually steel) that has a relatively low carbon content. The free carbon is derived from a gaseous or liquid substance in intimate contact with the metal surface by means of chemical reactions that do not directly involve the metal but that may be catalyzed by the presence of the metal. Absorption of carbon into the surface layer sets up a concentration gradient, and carbon atoms move by diffusion away from the surface. Carburizing is performed almost exclusively on steel, and is performed chiefly to produce a hard, wear-resistant case overlying a tough core. Final properties are developed by appropriate heat treatment following absorption and diffusion of carbon. Dew point hygrometers are used with a suitable sampling system and with the sensor operating at ambient temperature, unless the dew point to be measured is above ambient temperature. In that case, the sensor must be heated to a temperature above the dew point, which can be easily accomplished by mounting the sensor in thermally-controlled housing. Dew points in carburizing applications are usually in the 0 to 2°C (32 to 36°F) range.

d. Contaminants

The following contaminants are generally present and must be dealt with:

- Particulates, such as carbon

- Evaporated oil from parts which were not properly degreased

- Glycols, sulphur, and various alcohols in an enriching gas

These contaminants, if present in substantial amounts, could cause a chilled mirror hygrometer to read high due to the modification of mirror surface vapor pressure.

e. Filters

The presence of the contaminants shown above indicates that virtually all carburizing heat treat applications should be furnished with a particulate filter, and sometimes with a hydrocarbon filter as well.

f. Carbonitriding

Nitriding is used to harden steel and make it more fatigue- and corrosion-resistant. It also allows the steel to stay hard at temperatures up to the nitriding temperature of 510°C to 566°C (950°F to 1050°F). Nitriding is applied to high reliability gears, bushings for conveyor rollers to handle abrasive alkaline materials, anti-friction bearings, and gun parts. The atmosphere used for nitriding is commonly dissociated ammonia, which permeates the surface with nitrogen.

Carbonitriding is a process in which austenitized ferrous metal is brought into contact with an atmosphere containing both carbon and nitrogen, which elements are simultaneously absorbed and diffused into the metal to produce the desired end product. Typically, carbonitriding is carried out at a lower temperature and for a shorter time than carburizing to obtain a hard, wear-resistant metal that is shallower than is usual in production carburizing.

g. Mechanism of Carbonitriding

In carbonitriding, ammonia is added to the gas-carburizing atmosphere and dissociates to produce hydrogen and mono-atomic nitrogen, the latter being absorbed into the work surface along with carbon from the carburizing gas. The addition of nitrogen has important effects. It inhibits the diffusion of carbon, creating a very hard surface that is easily polished and highly wear resistant, and causes the formation of nitrides, the particle hardness of which leads to even more wear resistance than is attributable to maximum matrix hardness alone. As in carburizing, properties of the carbonitrided case are usually achieved under non-equilibrium conditions. The concentrations of ammonia and enriching gas in the atmosphere in part determine the nitrogen and carbon contents at the metal surface. Time and temperature determine the final surface concentrations of nitrogen and carbon, and determine total depth. Again, suitable sampling systems are required and dew point sensors are usually operated at room ambient temperature. Typical dew points in carbonitriding applications are in the range of about –50°C to –75°C (–58°F to –103°F). A further complicating factor is the usual presence of trace amounts of uncracked ammonia. When combined with water condensed on the mirror of a condensation hygrometer, it causes the formation of ammonia hydroxide, which results in drifts in the dew point measurement or instrument control problems. It is therefore important that before an expensive three-stage optical condensation hygrometer is selected, careful analysis is made of the ammonia content of the gas and possible means of reducing its concentration and effects on the dew point sensor. In most cases, ammonia contents of less than 10 ppm would cause no unacceptable sensor problems. However, because of the low dew/frost points involved, most carbonitriding applications are handled with aluminum oxide hygrometers rather than chilled mirror systems.

h. Contaminants and Other Maintenance Factors

Carbonitriding involves the following maintenance considerations:

- Carbonitriding atmospheres are generally much cleaner than carburizing atmospheres and only small amounts of particulates are present.

- The presence of anhydrous ammonia could cause mirror surface damage in a condensation type sensor if a conventional gold-coated mirror is used. Therefore, a rhodium or platinum mirror should be specified.

- If more than 10–20 ppm of uncracked ammonia is present, the application must be carefully reviewed, i.e., it may not be feasible to use a chilled mirror hygrometer. Aluminum oxide hygrometers have been used more successfully in such applications, though lower accuracy and repeatability must be accepted, and the sensors should be periodically recalibrated.

- Sometimes a hydrogen atmosphere is used which causes about 11°C (20°F) loss of depression in a chilled mirror sensor, due to the high heat capacity of H_2. Hence, more extensive sensor cooling would be required to get down to dew points below –50°C (–58°F) and dew points below –62°C (–80°F) may not be feasible. In such cases aluminum oxide sensors are usually preferred.

C. Other Metal Operations

1. Galvanizing

Galvanizing is a process where long, thin sheets of pickled steel are heat treated on a conveyor and then dipped into a molten zinc bath to coat the surface. Typically five regions of the conveyor perform the heat treatment of steel to its desired characteristics. These are the back, radiant tube, control cool, jet cool, and snout.

2. Nickel Alumide Coating

Nickel alumide coating is used on aircraft engine turbine parts. The parts are sprayed with a powder and then placed in a furnace for curing and aging. Bottled argon and hydrogen are used in this process. Dew points of the gases used in the batch furnace can be as low as −90°C (130°F) for argon and −76°C (−105°F) for hydrogen. The typical dew point range is −90°C (−130°F) to −40°C (−40°F). There are very few contaminants.

3. Pickling

This is the chemical removal of surface oxides and scale from metals by acid solutions. The metals are generally dipped in sulfuric or hydrochloric acids with water. Prior to pickling, the parts must be thoroughly cleaned of dirt or oil. After pickling, the parts must be completely neutralized by an alkaline and then a clear rinse. This will help prevent subsequent rusting.

4. Sintering

In this process, metal parts made from compressed metal powders are heated to produce the desired characteristics of the part. Reasons for using powder metallurgy include the ability to:

- Fabricate refractory or reactive metals

- Homogeneously combine dissimilar materials

- Produce metal of controlled porosity or permeability

- Produce large numbers of certain small parts at lower cost than by conventional techniques

The sintering atmosphere must be carefully controlled to prevent oxidation or combustion. The furnace must also be able to remove any absorbed or other gases that may be liberated from parts during sintering. Hydrogen, dissociated ammonia, or cracked hydrocarbon atmospheres are commonly used. The dew point range is −50°C to −40°C (−58°F to −40°F).

D. Summary

The use of continuously monitoring dew point hygrometers in heat treat applications is difficult and challenging due to sensor contamination, high temperatures and the wide dew point measurement ranges required. Therefore, the installation must be carefully designed to minimize these errors. The use of a chilled mirror optical dew point hygrometer in conjunction with the proper filters and a good sampling system will provide reliable, continuous long-term moisture measurement, though the cost is high, and frequent sensor maintenance is often required. Aluminum oxide sensors provide a good alternative, especially at frost points below −40°C (−40°F). They are easier to use and maintain, but are less accurate. In some applications, the electrolytic cell and Piezoelectric hygrometer have been successfully used. Where no high accuracy is needed, and periodic measurements are adequate, the fog chamber instrument is widely used.

II. Semiconductors[5]

A. General

Sophisticated microcircuits (chips) involve extremely high packing densities of individual elements and are subject to damage due to contamination and moisture. The probability of failures resulting from a single defect increases as the distance between components on the chip decreases. Since large investments are required in capital equipment, processing times, and materials, the cost of failures has increased sharply. Hence it has become extremely important to control the atmospheres in which such sophisticated chips are manufactured.

Among the contaminants that are most likely to cause semi-conductor failures are metallic ions, other charged ions, oxygen, and water. Reliable humidity trace moisture instrumentation is often critically needed to assure a high level of quality and reliability of the product and process. Levels of moisture that can cause damage are often very low (in the 1 ppm range and below). To measure and monitor such low levels of moisture, highly specialized instrumentation is required.

B. Moisture Penetration

There are numerous ways moisture can enter into the manufacturing process. Many procedures in the manufacture of a chip involve the use of gases. Some of them may be reactive, such as hydrogen, silane, and arsine. Others like nitrogen and argon are inert. Most suppliers of "dry" gases take extraordinary steps to remove water from their products, but not all them have the expertise and instrumentation to properly verify the absence of water down to a few parts per million or parts per billion and the micro-chip manufacturer has a need to employ its own techniques to assure quality.

Moisture contamination is always present during the production, packaging, transport, and use of gases. Valves, fittings, and piping at each manufacturing process are sources of leaks where water vapor from the atmosphere can enter the process. This is especially true where equipment is operated under partial or high vacuum.

C. Moisture Contamination

Regardless of where or how trace amounts of water enter the process, they cause reduced product yields and non-uniform device characteristics. The following are examples of potential problems:

1. In epitaxial reactors, water entrained in the carrier or dopant gases can cause point defects through oxidation at the hot silicon surface.

2. In organometallic vapor deposition of certain semiconductors, particularly aluminum gallium arsenide, the presence of even a minute quantity of water can cause oxidation of the aluminum, resulting in a large number of defects in the film. Most of such reactors use upstream catalytic gas purifiers made from platinum-palladium.

3. Oxidation processes are sensitive to moisture content in the vapor stream and very small changes in water content can dramatically affect the oxidation kinetics. This is especially important when growing thin gate layers in metal oxide semiconductor field-effect transistor (MOSFET) devices, since the device performance is strongly influenced by defect density in the thin oxide.

4. During deposition of metal, the presence of moisture increases the danger of high-resistance conductive pathways.

5. Moisture in the gas stream of a chemical-vapor-deposition (CVD) reactor leads to the formation of particulate oxides, which settle on the surface of the water and cause defects in the oxide film. By sequestering dopant atoms as oxides, water molecules can also cause unwanted changes in resistivity profiles during epitaxy and diffusion.

6. In the packaging of hermetically-sealed devices, moisture can cause corrosion of leads and metal semiconductor contacts.

D. Damage Caused By Moisture

The amount of moisture that can reduce yields or reliability depends to some extent on the process in use and on the nature of the device under production. For example, water is commonly introduced, or generated in situ during the growth of relatively thick passivation-oxide films on silicon. However, if a wet-dry-wet process is incorporated, it is important to establish the baseline water concentration during the deposition step. In the case of epitaxial growth, a defect density that may be acceptable for a large-area device can result in the failure of a chip that contains on the order of 10^5 discrete elements. It is not feasible to determine exactly what level of moisture is damaging to a particular process. However, generally manufacturers and researchers agree that it is necessary to monitor gas streams in at least the ppm range, and in some instances at ppb levels. In addition, committees setting standards for the industry have for some time been reducing the criteria for acceptable levels of moisture. Table 12.1 presents a summary of proposed standards for commonly used gases. There have been several documented cases in which entire semiconductor process lines have failed due to moisture penetration. There appear to be definite relationships among the moisture content of gases, moisture-bound metallics and particulates, and semiconductor production yields. Studies conducted with a number of semiconductor manufacturers over the past 10 years have shown that yield declines as the level of moisture and moisture-bound metallics rises. It is also evident that the moisture content of gases is one of the keys to solving production problems in the semiconductor industry. Monitoring moisture-bound trace metallics appears to be more important than tightening limits on such gaseous impurities as oxygen in nitrogen lines.

Table 12.1 Proposed SEMI standards for moisture content of gases.[5]

Gas	Form	Max. Acceptable H_2O Level (ppm)
Argon	Bulk liquid	1
Arsine	Cylinder	4
Hydrogen chloride	Cylinder	10
Hydrogen	Bulk liquid	3
Hydrogen	Cylinder	3
Nitrogen	Bulk liquid	1
Nitrogen	Cylinder	1
Phosphine	Cylinder	2
Silane	Cylinder	3
Ammonia	Cylinder	5
Nitrous oxide	Cylinder	3
Oxygen	Cylinder	2
Oxygen	Bulk liquid	2
Dichlorosilane	Cylinder	—
Helium	Cylinder	1
Carbon tetrafluoride	Cylinder	5
Sulfur hexafluoride	Cylinder	8
Hexafluoroethane	Cylinder	7
Tungsten hexafluoride	Cylinder	—
Boron trifluoride	Cylinder	—

E. Instrumentation

Standard moisture-detection methods are not always sensitive enough for the task. The following commercially available instruments have been upgraded to meet requirements of the semiconductor industry.

1. Piezoelectric Moisture Monitor

The Piezoelectric moisture monitor, which is discussed in more detail in Chapter 5, offers good accuracy and repeatable measurements in the low ppm range, but is more expensive and somewhat more difficult to operate than the electrolytic and aluminum oxide types. The analyzer is microprocessor controlled and capable of measuring trace quantities of moisture in the low ppm range and has a sensitivity of about 0.02 ppm. Process gas suppliers, as well as semiconductor process and quality control engineers, often rely on automatic analyzers to ensure reliability of their products.

Piezoelectric automatic moisture analyzers represent an advance over previous instruments. The single most important difference lies in the design of the detection system. Conventional equilibrium detectors, e.g., dew point instruments and electrolytic or aluminum oxide probes, are inherently slow to measure small quantities of water. The Piezoelectric instrument uses readily measured changes in very small masses and does not depend on equilibrium to yield results. The heart of such a device is a crystal, calibrated to resonate at 9 MHz, which is coated with a polymer that absorbs moisture (see Chapter 5).

In use, the crystal is first exposed to the wet gas for approximately 30 seconds and is then exposed to a pre-dried reference gas for the same amount of time. Because the crystal is heavier when wet, its operating frequency changes. Resonant frequency is very easy to measure with great accuracy at sub-ppm moisture levels. This mode of operation always maximizes the partial pressure difference of water (by exposing it to the dried gas), and as a result, it is possible to arrive at accurate readings within minutes rather than hours, as would be required in chilled mirror hygrometers. Frequency tuned moisture detectors offer more than fast response and sensitivity. Among the other features that contribute to the effectiveness of this type of device are the following:

- The temperature and pressure of samples can both be controlled internally by the analyzer. Measurements are therefore not affected by ambient changes.

- Internal calibration, using a sample of the gas under test and the internally generated moisture, allows the accuracy of the instrument to be verified in less than 10 minutes.

2. Electrolytic Hygrometer

This instrument can measure levels below 1 ppm and in the low ppb range. Calibration in this range is very difficult (see Chapters 5 and 10). The phosphorous pentoxide cell often provides better accuracy than the aluminum oxide sensor and is a fundamental measurement method. However, measurements tend to be rather slow and the cell is sensitive to contaminants. Accuracy is strongly affected by changes in sample flow rate and leaks in the sampling system. The electrolytic cell, despite these limitations, is often used in semi-conductor applications.

3. Aluminum Oxide Hygrometer

This type of instrument is capable of measuring well below 1 ppm of moisture if properly operated and calibrated. However, accuracies at these levels are modest at best and repeatability is limited. This method is more convenient than most others and is often used, but it is not ideal.

4. Chilled Mirror Hygrometer

This instrument provides by far the most accurate and repeatable measurements, but it has a slow response in the range in which it is to be used. It is also expensive and difficult to operate at ppm levels. For dew point levels below −70°C (−94°F) (less than 10 ppm) extensive sensor cooling is required which adds to the cost and inconvenience. It is not useful much below 1 ppm. The chilled mirror hygrometer has therefore not found many applications in semiconductor manufacturing applications, except as a calibration transfer standard instrument to calibrate other trace moisture instruments at moisture levels in the ppm range. The cryogenic hygrometer discussed in Chapter 3 could be very useful but is presently priced well above the other instrument types and has therefore, as of the mid-1990s, rarely been used in semiconductor manufacturing applications.

III. Water Activity Measurements

A. Definition

Water activity (Aw), is defined as the free moisture available in a material as opposed to the chemically-bound moisture. It is directly related to equilibrium relative humidity (%ERH). Quite simply, %ERH is expressed in terms of 0–100%, and water activity in terms of 0–1. While water activity represents a very useful assessment of the free moisture of a material or substance for a wide variety of quality purposes, it does not necessarily reflect the total moisture content percentage which is an entirely different measurement requiring the use of other principles. The total water content percentage equals the sum of bound water and free water. In simple terms, water activity is the equilibrium relative humidity created by a sample of material in a sealed air space, and expressed on a scale of 0–1 for 0–100% ERH.

B. Measuring Water Activity

Since more and more quality restrictions are being placed upon finished products, pro-cessing, and preservation of materials, applications for water activity measurement have increased rapidly and are becoming more diverse and more firmly established as part of documented quality procedures. Wherever there is a need to monitor or control free water in a material, an application to measure water activity exists. The free water is considered to be the water that is free to enter and leave the material by absorption/desorption. Generally, the main areas of need for water activity testing all have different reasons for making these measurements.

1. Food Processing

The physical-chemical binding of water, as measured by water activity (Aw), can be defined as:

$$Aw = \frac{p}{p_o} = \frac{\%ERH}{100} \tag{12.1}$$

where:

p = vapor pressure of water exerted by the food

p_o = vapor pressure of pure water

$\%ERH$ = Equilibrium Relative Humidity of the food

In essence, the Aw of a food determines whether it will gain or lose moisture in a given environment. For example, if a cracker and a slice of meat were independently transferred into an atmosphere of 75% RH, we would find that over time the meat will lose water and the cracker will gain water. They will come not to equal water content, but rather to equal Aw. The basis of protection by packaging is to slow this process down as much as possible within economic as well as food safety and quality limits. To do this, one must know both the sorption isotherms (moisture content versus Aw) of the food product and the transport properties of the packaging film.

Food products need to meet ongoing improved standards in quality and micro-biological safety. The rise in demands for products with little preservatives and additives has to be met without the sacrifice of micro-biological stability, shelf life, and texture/taste quality. Among other measurements, such as pH testing, the Aw value needs to be controlled since it is this free water that is available for micro-organisms to grow, multiply, and cause spoilage.

Micro-organism growth can be controlled by suppressing water activity with different product formulations and by using additives which may be natural or artificial. By controlling Aw, product shelf life and safety can be improved.

While product safety is the main need for Aw testing in food samples, the overall quality in terms of taste and texture also needs to be considered. Remembering that free moisture is available to move within a product until an equilibrium is reached, it is evident that this can also spoil quality. For example, in a layer cake where the sponge, cream and jam layers are all in contact with one another, the free moisture (Aw) can equilibrate throughout the cake. This can result in spoilage of the sponge by the moisture from the cream or jam. If the Aw values of each component are reasonably well balanced, this can be avoided. A variety of different products can be tested for Aw using different sampling techniques, and most samples are required to be tested at precisely-controlled temperatures to avoid errors.

Figure 12.1   Chilled mirror hygrometer for water activity measurements. *(Courtesy Decagon Devices, Inc.)*

Additionally, many food products in shops now have to be stored below 5°C (41°F) which suggests that any *Aw* testing with an emphasis on shelf life should be carried out at the specified storage temperature. Only small samples can be tested for *Aw* at controlled temperatures. For practical reasons, some typical samples requiring temperature controlled *Aw* sample testing are:

- Cooked Meats

- Cheeses

- Yogurts

- Fillings

Instrumentation used for water activity measurements includes the chilled mirror hygrometer, such as the one shown in Figure 12.1, and various relative humidity instruments, an example of which is shown in Figure 12.2. All of these instruments are specially designed for water activity measurements and include special containers for the specimens that are to be tested. The chilled mirror hygrometer provides the fastest and most accurate measurements, but is more expensive than the RH types which are widely used and also offer good performance.

Figure 12.2   %RH instrument for measuring water activity. *(Courtesy Rotronic, Inc.)*

2. Paper Industry

a. Introduction

Machinery for printing paper or cardboard, coating paper with aluminum, and other types of unique applications, are quite sensitive to the properties of the product to be trans-formed and to any variation of some physical phenomena. One parameter that has long been recognized as influencing the properties of paper and cardboard is moisture. Equilibrium relative humidity and the relative humidity of the storage and work areas are responsible for changes occurring in the moisture content of the product. Various studies have demonstrated the importance of equilibrium relative humidity (%ERH) control.

b. Influence of Equilibrium Relative Humidity on Paper

Differences between the %ERH of paper and the room relative humidity (%RH), result in changes in the moisture content of paper. This must be avoided to prevent problems during the paper conversion or printing process. It is difficult to eliminate any difference between the ambient %RH and the %ERH, however, controls or limits can be established. A %ERH of 50% is ideal for paper since any %RH changes, say in the range of 40%–60% will have little effect on the moisture content.

From time to time, printers experience difficulties due to static electricity, such as paper sheets sticking together. This usually happens when the air and paper are too dry. When the paper and air are in the 40%–45% range this problem seldom occurs.

c. Dimensional Changes

Paper fibers absorb or desorb water depending on the ambient relative humidity. This causes swelling or shrinking of the fibers, which affects the diameter of these fibers more than their length. In a sheet of paper most fibers run parallel to the running direction of the paper machine. Accordingly, dimensional changes which are the results of moisture variations, are more important along the axis that is perpendicular to the running direction of the paper machine, than along the axis parallel to it. At approximately 50% ERH a humidity change of 10% ERH results in a change of typically 0.1%–0.2% in the length of the paper. Such a humidity difference gives a dimensional variation of 1 to 2 mm (39.4 to 78.8 mil) on a 1 × 1 meter (3.28 × 3.28 ft.) paper and could therefore cause poor and inaccurate printing. Paper running through an offset press usually gains water since it is moistened in the process. The change in the moisture content depends not only on the %ERH of the paper (40%–60%), but also on the ambient %RH.

d. Deformations of Paper Due to Humidity

Paper in stacks or rolls shows deformation if too much moisture is exchanged with the surrounding air through the edges of the stack or roll. This is due to the uneven distribution of this moisture as it is exchanged with the ambient air during storage or transport. Water-vapor-tight packaging protects the paper and it should not be removed without first checking %ERH of paper and %RH in the ambient environment. Differences up to +5% RH will not cause problems, while a difference of 8%–10% RH could be critical.

e. Temperature Induced Deformation of Paper

Temperature exerts a minimal influence on paper; however, any large temperature difference between the paper and the ambient air will have almost the same results as a humidity difference. This is due to the fact that %RH in the air layer in the immediate surroundings of the paper stack or roll is modified by the paper temperature. Assuming an ambient air of approximately 50% RH, a temperature difference of ± 1°C (± 1.8°F) will result in a humidity variation of +3% RH.

Thus, it is evident that when temperature differences approach 3°C to 4°C (5.4 to 7.2°F), problems can occur.

f. Curling of Paper Sheets

Paper fibers do not all run exactly in the same direction across the thickness of a sheet of paper. Large moisture variations could result in unequal dimensional changes on both sides of the paper sheet, resulting in curling. This is a more prominent problem with coated stock, since both sides of the paper are not similar. When working with such paper, humidity variations should be kept to less than 10% RH.

3. Pharmaceuticals

Some applications for Aw testing in pharmaceuticals relate to microbiological considerations, but many concern processing to achieve correct moisture levels which facilitates filling of capsules, tablet forming, and packaging. For example, gelatin capsules are often Aw maintained to allow them to be sufficiently supple for filling without cracking or breaking in the process. There are many effervescent powder manufacturers who control total moisture within specified limits, by measuring Aw as a quality check because it is in some cases quicker and more practical.

4. Chemical Industry

Aw testing is a vital part of research and development with soap and hygiene products. Due to the different chemical components of these products, free water can migrate from within a product, and component substances can be carried to the surface using water as a vehicle. This process is known as "efflovescence" and can be a common problem with soap bars.

C. Instrumentation for Water Activity Measurement

1. Relative Humidity Sensors

Several manufacturers have developed instruments specifically for water activity measurements, employing convenient sample compartments for the specimens to be tested. These use secondary type sensors, usually of the capacitive polymer type or some other resistance or capacitance measurement principle. An instrument of this type is shown in Figure 12.2.

High accuracies are generally required, hence the sensors must be carefully calibrated, preferably against a traceable laboratory standard. It is of the utmost importance
that the measurements are made at a stable, fully equilibrated temperature, since any RH
measurement is highly temperature dependent. With great care, accuracies in the order of
± 1% RH are attainable. In order to accomplish this, the instrument must be frequently
recalibrated.

2. Chilled Mirror Hygrometer

To obtain the highest precision with direct traceability to national standards, chilled
mirror hygrometers have been developed for this purpose. Such an instrument measures
dew point, and a very precise temperature measurement must also be made in order to
accurately calculate the %RH, i.e., water activity, for the specimen. The calculation presents
no problem and is customarily made using a programmed microprocessor. Maintaining a
stable equilibrium temperature has been a significant problem. Chilled mirror sensors
have some so-called "self heating" which causes small changes in the temperature in the
sample compartment which may cause significant measurement errors. Much progress
has been made in the last ten years, and accurate fundamental water activity instruments
to measure water activity are now available, though at higher cost than the RH types.
The accuracies offered are only slightly higher than for the RH instruments, but the main
advantage is that these measurements are fundamental and traceable. A chilled mirror
hygrometer specially designed for water activity measurements in food is shown in
Figure 12.1.

IV. Natural Gas

A. General

The presence of excessive amounts of water in natural gas presents many problems.
For this reason, the maximum amount of water that the gas can contain is always specified
and there is a great need for reliable instrumentation that can measure the trace water
content in natural gas.

Natural gas is almost always saturated with water when it leaves the ground so that
slight changes in gas temperature or pressure can cause this water to condense or form
hydrates. Liquid water can be very corrosive to steel gas pipelines. Hydrates, which
resemble wet snow, must be avoided since they plug pipe lines and valves, interrupting the
steady flow of gas. Because of these problems, the allowable limit on water content is generally set at 80–96 grams/10^3m^3 (5 to 6 pounds of water per million standard cubic feet).

B. Measurement Technology

The measurement of water content in natural gas can be made "on-line" or in the laboratory. Laboratory measurements offer the following advantages:

- Monitoring in one central test location

- Potential for using one instrument, such as a gas chromatograph, for several analysis

Disadvantages are:

- Handling of the sample gas may affect the measurements and cause time delays because of the need to take samples to a laboratory at regular intervals

On-line measurements offer the advantage of:

- Real-time readings

- Reduced sample handling effects

- Dedicated analysis

- Possibility of process control

Disadvantages of on-line monitoring are:

- Such measurements are often required in remote locations

- The requirement for well designed sampling systems

Contamination and the presence of hydrates in natural gas have created many problems with instrumentation used to measure water content or dew point in natural gas. As of this date, there are no instruments or sensors on the market that may be considered ideal for this application. Chilled mirror hygrometers have been tried, but with little success because of contamination effects. They are therefore rarely used in natural gas applications. Aluminum oxide sensors have been used with some degree of success when used with a well designed sampling system and if regular calibration and readjustment of sensors can be provided. The method is acceptable in most cases, but is not considered an ideal solution.

Electrolytic phosphorous pentoxide (P_2O_5) hygrometers have also been used with some success, though they require frequent maintenance and "rejuvenation" of the electrolytic cell with P_2O_5. A large number of P_2O_5 cells are in use for natural gas measurements, but they are not considered ideal for the application either. Piezoelectric quartz crystal technology is well known and has been used for many years in communications and timekeeping equipment. This technology can be useful in natural gas applications, but is expensive. An instrument that may be considered ideal for the natural gas market is not currently available.

V. Medical Applications

There are many medical applications for measuring humidity, some of which will be discussed briefly below.

A. Perspiration Measurements

One way the skin of the human body can provide information on the state of health is through temperature and perspiration characteristics. Perspiration and moisture levels at the skin surface have been monitored by various medical and psychological institutions using water activity measurement probes which are used to analyze the behavior at the skin's surface, for example, during menopause of a female patient. The interpretation of these tests is not available, but it is nevertheless an interesting new application for Aw (water activity) testing, reflecting the many diverse uses of this technology.

Perspiration and humidity measurements have also played a major role in research laboratories where clothing is tested for use by military personnel under various climactic conditions, such as in the tropics and in the Antarctic, and when used under different conditions of physical stress and exercise.

Attempts to measure physical parameters of the human body often lead researchers to highly imaginative adaptations of standard industrial instruments. One example is the need for an instrument to measure perspiration rates during various states of stress and exertion of the human body. The following characteristics are required:

- Sensitivity to changes in moisture

- Fast response

- Independence of body temperature variations

- Portability

- High accuracy

Based on these requirements, a chilled mirror dew point hygrometer has been successfully used for this purpose.

Using this instrument, a 3.8 cm (1 1/2 inch) diameter by 3.8 cm (1 1/2 inch) long hollow plastic cap is taped to the body. Dry carrier gas is then introduced into the side of the cap and drawn off from the top. The gas flow causes evaporation from the skin which shows up as a rise in dew point. Flexible tubing is used to connect the cap to the dew point measuring system.

A complete test series includes two sensing caps located on different locations of the body. The subject is placed under different types of physical exertion, from rest to extreme activity. By correlating the dew point, initial sample gas moisture, and flow rate, the actual perspiration is determined. The continuous gas flow allows an indication of increasing and decreasing perspiration.

To establish a baseline for moisture increases, a dry carrier gas is first introduced to the system directly from the gas supply. A large step change in humidity is encountered during the switch over from the dry gas to the test gas. Good sensitivity is required of the hygrometer to enable it to respond to this change as well as to track the minor fluctuations in perspiration caused by the changing exertion rate of the subject.

The use of a dew point sensor to measure high evaporation rate perspiration provides an economical solution to this potentially difficult problem. Dew point and RH instruments are usually only considered in applications for measuring and controlling the environment. But there are many areas in manufacturing and research where humidity technology can be adapted to solve unique problems.

B. Incubators

Accurate measurement and control of temperature and humidity in incubators for prematurely born babies is very important. Bulk Polymer RH sensors, both capacitive and resistance types, are frequently used for such applications. The sensors are also used in hatching chambers for chickens and various other animals, as well as for research purposes. In some special cases where humidity control is critical, chilled mirror sensors have been selected.

C. Artificial Hearts

Although the implantation of artificial hearts has had limited success to date, considerable research is being conducted to advance this technology. Careful humidity and temperature control during fabrication of such hearts have been shown to be very important. Both chilled mirror and bulk polymer sensors have been used for this research.

D. Medical Gases

Health care providers need a reliable, functional medical gas system which can perform when required and during emergencies. Hence good design, quality components, and a prescribed and sensible maintenance plan are mandatory. As a minimum, a typical medical gas system provides oxygen, nitrogen, nitrous oxides, compressed air, and vacuum lines to various patient rooms, operating rooms, nurseries, and intensive care facilities.

Each medical gas supply system consists of source equipment, such as nitrogen tanks, or gas generating systems, alarms, piping, and end-use connections. Various local, state and federal codes are implemented to assure the quality and safety of all systems. Because of the effects on patient health and safety, it is critical that medical gas systems are free of contaminants and operating at carefully controlled moisture levels. Typical maximum allowable contaminant levels, including humidity, are shown in Table 12.2.

A typical operating room with gas delivery system is shown in Figure 12.3. One reason for testing the dryness of the compressed air supply is the danger of moisture deposits in the air lines. This condition can cause bacteria to form, which, in turn, can lead to patients experiencing bronchial discomfort or diseases. Strict regulations have been submitted by the National Fire Protection Agency (NFPA). The same agency has also recently released regulations for carbon monoxide. It has been found that in some instances traces of CO have entered the compressed air lines due to engine emissions or other sources at

Table 12.2   Contaminant levels.

Dew point	3.9°C (39°F) @ 445 kPa (50 lbs./psi$_g$)
Carbon monoxide	less than 10 ppm
Carbon dioxide	± 500 ppm
Gaseous hydrocarbons	less than 25 ppm (as methane)
Halogenated hydrocarbons	less than 2 ppm

the inlet of air compressors. Such traces of CO are dangerous to human life and could be fatal, even when present in very small amounts.

One important use of medical compressed air is to assist respiratory functions of patients. This calls for air free of moisture and undesirable vapors. Such compressed air systems are usually operated at 890 kPa to 1068 kPa (100 psi$_g$ to 120 psi$_g$) and must provide dew points controlled in the range of 0°C to 25°C (−18°F to 77°F).

In previous years, little attention was given to dew point control in medical gas applications. When its importance became widely recognized, various secondary measurement systems were employed to measure and control dew point. Results were mixed,

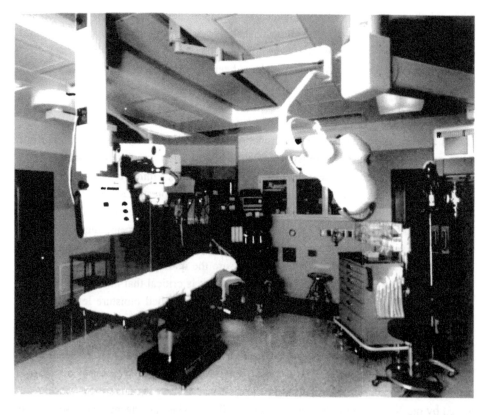

Figure 12.3   Operating room with gas delivery system.
(Courtesy SSOE, Inc.)

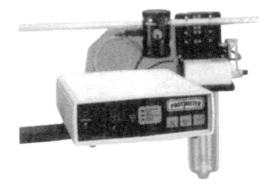

Figure 12.4   CCM medical gas hygrometer, including CO analyzer.
(Courtesy Protimeter, Inc.)

mainly due to the lack of adequate maintenance of the sensors and the frequent need for calibration, which was not often performed at the prescribed intervals, or not at all.

At today's levels of sophistication, frequent medical malpractice litigation, and tight regulations, there is a strong demand for fundamental dew point measurement instrumentation capable of operating within calibration limits for long periods of time, with little or no maintenance. The conventional chilled mirror hygrometer is not ideal for this application because of its need for frequent mirror cleaning. The cycling chilled mirror (CCM) hygrometer with its lower maintenance requirements is now widely accepted for medical gas applications. Conventional chilled mirror hygrometers utilizing special techniques for long-term operation without mirror maintenance, such as a new automatic mirror cleaning method using compressed air, have also been developed. This technique uses some of the available compressed air to periodically blow contaminants off the mirror surface.

Figure 12.5   Conventional medical gas hygrometer with self cleaning feature.
(Courtesy General Eastern Instruments)

A typical CCM hygrometer used for medical gas applications is shown in Figure 12.4. A conventional condensation hygrometer with a compressed air self-cleaning feature is shown in Figure 12.5.

E. ETO Sterilizers

Many medical instruments, including one-time use items for urological and blood work, are sterilized after packaging by exposure to ethylene oxide (ETO) gas. These medical devices are usually packaged in a sealed plastic or polymer envelope and boxed in cardboard cartons for shipment. Palletized cartons are then placed inside a humidity-controlled warehouse room. Humidity controlled in the range of 60%–70% RH helps precondition the packaging, and enhances penetration of the ETO gas.

The pallets are transferred to sterilizer chambers, which are sealed and evacuated to about 500–700 mm Hg (19.7"–27.5" Hg) vacuum. This vacuum removes the gases inside the packaging to help draw the ETO inside. During the vacuum cycle, steam is injected into the chamber to continually moisten the environment and add some heat. The moisture must be monitored throughout this phase to insure sufficient ETO penetration. If the packaging dries out, it could take up to four times as long to sterilize the same amount of material during the ETO gas phase.

After several hours at a moist vacuum, the chamber is filled with bacteria-killing ETO gas. This may be 100% ETO or a mixture with 88% Freon gas or, in a few cases, 88% carbon dioxide. Because ETO is a highly toxic and explosive gas, which was at one time used for making bombs during World War II, a gas like Freon is often added to render it non-explosive. However, new restrictions on the release of chlorofluorocarbons have caused some facilities to return to 100% ETO, which is easier to scrub out of the exhaust than Freon. This is a less expensive option, even considering the additional cost for explosion-proof or intrinsically safe equipment.

During the gas cycle, the humidity level is maintained at 50% RH to 70% RH. A normal gas cycle will last from two to six hours. If the moisture is too low, the gas cycle could be as long as 6 to 24 hours, which severely limits the throughput of product through the sterilizer.

Moisture measurements in ETO have been very difficult to make in the past because the ETO may combine with water vapor to form ethylene glycol, which is very corrosive. This could strongly affect the life and calibration of the sensor. In the past, such chambers have often been monitored with sulfonated polystyrene sensors (Pope cells). While not ideal, they appeared to give better results than most other types of sensors. Some of the more recently developed resistive RH sensors have given better performance. Chilled mirror sensors have also been attempted, but with little or no success.

Because of environmental concerns, the ETO sterilizers are loosing popularity, and more and more are succeeded by hydrogen peroxide systems (H_2O_2).

VI. Museums

Preserving centuries-old art is becoming more and more important. In recent years, there has been an increasing awareness of the need to maintain environmental conditions in which museum artifacts are stored and displayed. Pieces of art have, in the past, often been subjected to candle soot, salt deposits, moisture, and other contaminants. Today's conditions present additional threats, which include automobile emissions and chemical pollutants. However, nothing poses a greater threat to such art than condensation resulting from rapid humidity and temperature changes. Primarily it is relative humidity and temperature that are the major concerns as changes in these conditions affect the stability of delicate and perishable objects. Organic material, such as wood, leather, and canvas, is most susceptible to damage resulting from poor relative humidity and temperature conditions. Paintings can crack as a result of low relative humidity, and leather and fabrics can develop mold growth at relative humidity levels above 60% RH.

Since it is the equilibrium relative humidity of the objects themselves which are to be considered, it seems logical to directly monitor their water activity. Unfortunately, this is often impractical since sampling of artifacts is sometimes impossible and probe attachment or insertion is unacceptable. Coupled with this, water generally moves very slowly in and out of large objects due to their relative masses.

While control of the ambient relative humidity to control an object's ERH% is by far the most common solution, there have been several conservation projects which have required direct water activity of the artifact to be measured. One example is at the Victoria & Albert Museum in London. A project was set up to assess discoloration of marble statues, which is believed to possibly result from migration of salts (efflovescence) by free water movement within the marble. The movement of free water would occur as a result of changes in the relative humidity of the surrounding air affecting the marble. In practice, it is difficult to see changes in longer than a 24-hour periods due to the large masses of marble being examined. For this reason, a long-term monitoring system was put into operation.

Another example is Michelangelo's Sistine Chapel frescoes in the Vatican. Completed in 1512, the paintings were affected by smoky oil lamps, interior temperature fluctuations, and condensation from the breath of worshippers. The frescoes have recently been cleaned and restored. To forestall new damage that could inadvertently be caused by the millions of visitors to the Vatican every year, a climate control system was installed.

Many months of temperature and air flow research were conducted with the aid of computer-controlled modeling. It was found that although lamp soot was no longer a problem, temperature variations and moisture were. Studies made in the chapel vault showed that a geyser of warm, moist air was being forced up toward the ceiling every time the doors were opened. Because the surface temperature of the frescoes is lower than that of the interior air, the dwelling air movements often deposit moisture and particulates as they spread across the ceiling, cool, and then descend. In addition, condensed water can also attract airborne pollutants including chemicals. The climate control system therefore had to provide a clean, stable interior atmosphere by greatly reducing or eliminating airborne substances, minimizing temperature differences across the surfaces of the paintings, and preventing respiratory condensation. These requirements resulted in the need for monitoring of surface and air temperature, and dew points, by thoroughly cleaning the fresh air admitted into the chapel and controlling the air flow. The system that was ultimately installed included 13 chilled mirror hygrometers, 19 surface temperature sensors, and 21 air temperature sensors.

Although percent relative humidity (%RH) is to be monitored, chilled mirror dew point measurements were chosen because of the accuracies and measurement stabilities that can be obtained using these systems. Dew point measurements are then combined with air temperature measurements and converted into %RH. Studies have shown that 55% RH ± 5% and air temperatures of 20°C (68°F) in the winter and 25°C (77°F) in the summer are optimum.

Many other museums, such as the Museum of Fine Arts and Isabella Gardner Museum in Boston, Metropolitan Museum in New York, and the Getty Museum in California which exhibit centuries-old art pieces, paintings, sculptures, books, papers, and other rare objects, have installed climate control systems, consisting of relative humidity and temperature sensors, to preserve valuable art.

In special applications where very precious art is involved, the more expensive chilled mirror sensors are used. In many other cases the modern, reliable, and accurate bulk polymer transmitters (resistive as well as capacitive types) are used. A good calibration system is needed to periodically check the secondary type sensors and to recalibrate them as required. Portable calibration systems for this purpose are presently on the market and can easily be used with fundamental chilled mirror instrumentation to assure accuracy and traceability.

VII. Dryers

A. General

A considerable amount of energy is expended in removing moisture from final and in-process products in a wide range of industries. On average, drying accounts for about 12% of total energy consumed in industrial areas such as food, agriculture, chemicals, textiles, paper, ceramics, building materials, and timber. In general, the operation of dryers, which come in a wide variety of shapes and sizes, is poorly understood. This frequently results in sub-optimal performance.

Humidity of the air within a convective dryer can have a significant effect on its operating performance and the characteristics of the dried product. In general, thermal efficiency of most industrial convective dryers is poor; a figure of 30% is not unusual. Most heat supplied to the dryer is lost in the exhaust gas. If the flow rate of the latter can be reduced, and the humidity of the exhaust air correspondingly increased, there will be a reduction in the heat requirement of the dryer, even if the exhaust temperature has to be raised to avoid possible condensation in the gas cleaning equipment. In practice, there are practical constraints on the degree by which air flow can be reduced. For example, in fluidized bed dryers, a sufficient air velocity is needed to maintain fluidization. Depending on the nature of the drying solids, the elevated humidity could also limit the final moisture content of solids. The use of exhaust air recirculation, which also results in an elevated humidity within the dryer, suffers from similar limitations.

Many industrial dryers have relatively crude control systems. In some cases, the temperature of inlet air is maintained at a preset value, regardless of changes in throughput and inlet moisture content. Other dryers are supplied with a thermocouple, which senses the temperature of the exhaust air. The control system responds, for example, to a decrease in throughput by reducing the heat supply to the dryer and hence the temperature of the inlet air. This approach is wasteful in terms of energy since the thermal efficiency is correspondingly reduced. From an energy viewpoint, it is better to keep inlet air temperature at its design value and respond to the reduction in throughput by adjusting the exhaust damper setting to reduce air supply. This can be accomplished by monitoring outlet air humidity and using this as the control parameter, which is maintained constant. The feasibility of installing such a scheme on an industrial dryer depends on the availability of an inexpensive but reliable hygrometer capable of operating at exhaust air conditions in sometimes dusty environments. Despite potentially attractive financial benefits, instrument limitations and a general lack of awareness of the principles involved, have thus far limited the application of these techniques.

Changes to operating conditions that affect humidity should not be undertaken without due consideration of the effect this may have on characteristics of the final product. In particular, humidity can affect drying kinetics and equilibrium moisture content. With certain products, particularly food substances, humidity may also affect physical characteristics that are deemed to determine product quality.

The instantaneous drying rate (kg or lbs H_2O/kg or lbs dry solids) of a solid can be determined from the local gradient of a plot of its moisture content × (kg H_2O/kg dry solid) against time (t). In general, the drying rate is dependent upon complex interactions between a number of different parameters. These may include temperature, humidity, and velocity of the air passing over the drying material, as well as physical characteristics of the solids. In general, it has been determined that the variation of the drying rate with time or moisture content falls into three distinct periods. Initially, there is an inductive period, in which drying rates adjust to prevailing conditions. This is followed by the constant-rate period, in which drying is controlled by conditions external to the solid. Here, the drying rate changes only slightly with time and moisture content. Finally, during the falling-rate period, which is observed below the so-called critical moisture content, internal drying mechanisms predominate and the drying rate progressively decreases with time. With certain materials, more than one falling-rate period can be distinguished.

In the constant-rate drying period, either the rate of heat transfer to the solid or the rate of mass transfer away from the solid can determine the drying rate. Measuring the temperature of the solid can give a clear indication as to which of these is the limiting process. A temperature equal to the wet bulb temperature of the surrounding air is characteristic of heat transfer control. On the other hand, if the temperature attains the dry bulb temperature of the air, mass transfer control is indicated. Under these latter conditions, the drying rate can be estimated from the basic psychromatic theory and is proportional to the difference between humidities at the solid-gas interface and in the bulk of the air.

Humidity of the drying air can also limit the final moisture content attained by solids. Hygroscopic materials exhibit an equilibrium moisture content, which is dependent on both temperature and humidity. For a given set of prevailing conditions, it is not possible to dry the solids below this value. Papadakis et al.[22] described correlations that enable the equilibrium moisture content to be estimated from a limited number of laboratory measurements.

In summary, a better control of humidity within an industrial dryer can lead to worthwhile fuel savings resulting from improvements in thermal efficiency. To date, control schemes based on measuring the humidity of exhaust air have not been implemented on a large scale, principally because of the lack of suitable hygrometers.

Hygrometers commonly used for such drying applications include chilled mirror instruments (good but expensive) and aluminum oxide hygrometers (easier to use but less accurate).

B. Nitrogen Polyester Chip Dryer

The presence of moisture in polyester chips, which are used for yarn manufacturing, causes fibers to knit together improperly. This results in defects in the woven fabric. To prevent this from happening, moisture is measured in the nitrogen process gas of the polyester-chip dryer, often using an aluminum oxide hygrometer or sometimes a chilled mirror system.

A sensor is inserted into the atmosphere above the polyester chips. After equilibrium has been established between the vapor pressure of the gas and that of the solid, a dew point reading is taken. An empirical calibration curve results from the correlation of several different dew point readings with the laboratory determinations of the moisture content for the solid. This empirical calibration curve is then used to monitor the dryness of the chips on an absolute basis.

Permanent installation of a suitable hygrometer has resulted in substantial manufacturing savings in cost and labor, increases in the quality of the final product, and reduction of scrap produced in the process. In such applications the water vapor range is generally 5 ppm_v to 400 ppm_v, corresponding to a dew point range of $-30°C$ to $-65°C$ ($-22°F$ to $-85°F$) at ambient temperature, and at an operating pressure of about 101.4 kPa (14.7 psi).

C. Nylon Chip Dryer

The presence of moisture in nylon and polyester polymer chips causes severe difficulties during the remelting and fiber spinning stages. These problems are mainly associated with a loss of strength in the final product. To solve these problems, moisture is measured in the nylon chip dryer.

Nylon and polyester polymers are usually made in rod form and then cut into chips that measure approximately 3×3 mm (1/8 in. $\times$ 1/8 in.). Before the polymer can be spun into fiber, it must be thoroughly dried to approximately 0.01% moisture content. The drying process is accomplished either by forcing hot/dry N_2 through a bed of chips or by putting the chips in a large vessel, creating a vacuum, and moderately heating the vessel. Moisture content is critical because the polymer will be remelted and spun at a high temperature of approximately 300°C (572°F) and high pressure, sometimes as high as 55 MPa (8000 psi). In the past, the moisture content of chips was measured by using laboratory techniques. Samples were drawn during the drying cycle and sent to the laboratory for testing. Presently, suitable "in-process" humidity sensors are available and have proven to be very beneficial in performing measurements on a continuous basis.

Aluminum oxide and phosphorous pentoxide sensors are capable of measuring extremely low amounts of water vapor and can be used to determine inferentially the amount of moisture in the chips by correlating laboratory data with data obtained from a sensor installed to read the humidity above the chips. Electrolytic-type hygrometers have in many cases shown some limitations. Aluminum oxide sensors are often better suited for this application. The advantage of continuous moisture monitoring is that the possibility of product loss is considerably reduced. There are several other places in the polymer process where moisture measurements can be performed. When certain stages of polymerization are exposed to oxygen and water, the result could be degradation of the polymer. All manufacturing processes that extrude plastic or polymer materials can benefit from the installation of a moisture sensor to perform continuous monitoring.

The water content to be measured is typically 0.01% to 0.10% by weight in the solid polymer at an operating temperature of 70°C (158°F) and pressure of 1.37 kPa to 108 kPa (0.2 psi$_a$ to 15.7 psi$_a$).

D. Drying Plastic Resin Pellets

Moisture is measured in the air used to dry a hopper of plastic resin pellets in injection-molding and extrusion processes. The presence of excess moisture in the air, which is used to dry plastic pellets, has several adverse effects upon the plastic products that are produced.

Excess moisture:

- Causes clear plastic products to have a cloudy appearance

- Decreases the strength of the product

- Results in a poor finish on the product surface

The product defects listed above reduce product yield while escalating production costs. For a single injection-molding machine, an aluminum oxide single-channel hygrometer or chilled mirror sensor can be used. However, because most injection-molding facilities have several machines, it could be more cost-effective to use a multi-channel hygrometer, i.e., an aluminum oxide instrument (chilled mirror hygrometers are not practical in multi-channel configurations). In addition to the hygrometer electronics and sensor, a sampling system is also recommended. A cooling coil may be used to reduce the sample temperature because the air leaving the dryer is at a very high temperature. A filter is needed to remove any desiccant carry-over from the dryer or particulate matter from the plastic resin. The complete sampling system consists of:

1. Inlet ball valve

2. Cooling coil

3. Micron size sintered stainless steel filter

4. Sample cell

5. Back diffusion coil

The air used to dry the plastic resin pellets is contained in a closed loop consisting of a dual-bed regenerative desiccant-type dryer including a pre-cooler, a post-heater, and a hopper containing the plastic resin. The dryer produces hot/dry air that is blown through a duct to the hopper. The hot/dry air removes moisture from the plastic resin in the hopper and circulates it back to the dryer. In the dryer, the air is cooled, the moisture is absorbed by the desiccant, and then the air is reheated. A hygrometer sampling system is typically installed at the inlet to the hopper. The sampling system cools and filters the sample air as it flows to the moisture sensor, and then vents it out to the atmosphere.

In the past, desiccant dryer beds were switched with solenoid valves controlled by a timer. Switching the beds in this manner could cause one of two problems:

1. Allow wet air to flow to the hopper if the bed becomes saturated before the timer set-point is reached.

2. Allow beds to be switched before this is necessary, thus causing them to be regenerated prematurely.

Using an aluminum oxide or chilled mirror hygrometer, which is permanently installed in the sample system at the hopper inlet, any dryer failure is immediately detected. If air at the hopper inlet is not sufficiently dry, the hygrometer alarm contacts will switch the dryer beds, thus preventing wet air from entering the hopper. This configuration of the hygrometer maximizes dryer efficiency, and minimizes drying time by feeding only dry air to the hopper inlet.

The typical frost point range is $-10°C$ to $-50°C$ ($14°F$ to $-58°F$) at an operating pressure of 101.4 kPa (14.7 psi$_a$) and an operating temperature of $95°C$ to $250°C$ ($203°F$ to $482°F$).

E. Dual Tower Regenerative Desiccant Dryer

Compressed-air lines, located downstream of a dual-tower regenerative desiccant dryer, require moisture measurement to ensure proper drying of the stream. A hygrometer is used to switch drying towers as required and to sound an alarm when the moisture level exceeds a maximum permissible level. Excessive moisture in the air stream causes freezing of the outside gas lines and damage to the equipment or the process for which the air is being used. Excessive moisture can also be an indication of improper functioning of the dryer.

A moisture sensor is placed directly in the air stream or in a by-pass immediately downstream of the dryer. The low alarm set point is set at the maximum desired moisture level. When this level is reached, the alarm is used to automatically switch from one drying tower to the other. The high set point alarm is set to the maximum moisture level that is safe. If this level is reached, i.e., in the case of a malfunction in the dryer, an alarm sounds and the appropriate personnel are alerted.

In the past, the dew point was either not measured or was measured manually with a dew cup. The drying towers were switched on a preset time cycle, regardless of whether or not this was necessary. Users are becoming more concerned about the reliability of their compressed-air drying systems. Aluminum oxide hygrometers are used for most of these applications. They offer two advantages. First, they do not cause any damage to the system, and there is no loss of production resulting from high moisture. Secondly, these hygrometers provide the most cost-efficient use of the dryer by lowering the consumption of utilities and increasing the life of the dryer and desiccant. The dew point temperature is in the range of –60°C to 0°C (–76°F to 32°F) at an operating pressure of 650 kPa to 1.2 MPa (80 psi_g to 150 psi_g).

F. Paper Dryers

During the manufacturing process of paper goods, it is important to extract a certain amount of moisture from the product to insure good and consistent quality. The paper dryer usually consists of a series of rollers which pass the paper product across hot air. The heated paper loses moisture in the form of water vapor. By monitoring the humidity in the exhaust or in the dryer, the speed of the rollers can be controlled and optimized. This will result in:

- Optimum drying time

- Prevention of over-or-underdrying

- Improved product quality

- Minimum energy cost.

One problem is that large amounts of particulate contaminants are often encountered and suitable filters must be used to prevent sensor clogging or measurement errors. Chilled mirror and aluminum oxide sensors have often been successfully used for this application. Also see Section III.2 on water activity measurements for the paper industry.

G. Dry Snack Food Processing

Steam from the snack food drying process moves via a damper control to the post frying chamber. In order to insure consistent product quality, it is necessary to control the steam flow and thus the drying process. The amount of steam that enters the drying chamber directly affects the drying rate and other factors in the drying process. Good control over the drying process assures consistent levels of product crispness and taste.

Temperatures in the drying chamber are usually in the range of 95°C to 175°C (203°F to 347°F). A suitable sampling system must be used in order to make reliable measurements with a dew point hygrometer. High temperature capacitive polymer RH sensors have also been used for this application, without a sampling system.

VIII. Gases

A. Carbon Dioxide Gas

The presence of moisture in carbon dioxide gas can cause freeze-ups in process lines at high pressures. Moisture sensors can be used at various locations in a carbon dioxide process line. One sensor may be placed in a by-pass line which is vented to the atmosphere. Other sensors are usually operated directly in the gas line at pressures of about 2.2 MPa (300 psi_g). Results at both pressures are consistent and reliable. Nevertheless, operation at direct line pressure is generally more advantageous in terms of sensitivity and overall system response time. Using an appropriate moisture detecting system, maintenance costs and downtime can be virtually eliminated. In the case of high moisture levels, carbon dioxide gas can be recirculated through molecular sieve dryers before line freeze-up occurs.

Typically, water vapor detection is required in the 1 ppm_v to 10 ppm_v range, which corresponds to a dew point range of $-60°C$ to $-75°C$ ($-76°F$ to $-103°F$) at 1 atm. Operating pressures are in the range of 205 kPa to 2.2 MPa (15 psi_g to 300 psi_g). Aluminum oxide hygrometers are most commonly used for this application.

B. Gases Containing Hydrogen Sulfide

The presence of moisture in gas streams containing trace amounts of hydrogen sulfide produces corrosive by-products. These by-products can destroy expensive process pipes and fittings. In addition, a high moisture content can render the process unusable or inefficient because of direct poisoning of the catalyst.

Generally, a number of sensors are used in various operating and regenerating process streams. The sensors are installed directly in the gas line. Since gas temperatures are in excess of 60°C (140°F), a cooler must be used in series with the sensor to decrease the sensor temperature to less than 60°C (140°F).

The aluminum oxide sensor appears to be well suited for this application. Phosphorous pentoxide hygrometers have also been used, but electrolytic cells are easily damaged, requiring frequent servicing and additional costs. When permanently installed, a suitable sensor, such as an aluminum oxide probe, has a wide dynamic range that allows for monitoring of both operation and regeneration processes. Multiple sensors with a single readout allow continuous monitoring of a number of process streams at modest cost. High maintenance costs are greatly reduced by effective moisture monitoring.

The typical range to be monitored is 100 ppm_v to 2,000 ppm_v, corresponding to a dew point range of 0°C to $-45°C$ (32°F to $-49°F$) at an operating temperature of 50°C (122°F) and pressure of 3.5 MPa (500 psi_g).

Due to its sensitivity to trace amounts of hydrogen sulfide, a chilled mirror hygrometer is not recommended for this application.

C. Steam Leaks in Process Gas Lines

Moisture measurements can be used to determine whether or not steam leaks occur within steam-jacketed air pipes. The presence of moisture indicates the occurrence of leaks in the system gas lines that are heated by steam. The presence of steam in process lines causes corrosion of the metal and/or subsequent contamination of the product.

A common procedure is to mount a humidity sensor in a sample cell. A sampling line, which includes a cooling coil, connects the sample-cell inlet directly to the process stream. The cooling coil is necessary to reduce the temperature of the test gas to less than the maximum temperature the sensor can be operated at, i.e., +90°C (194°F) for a chilled mirror sensor, +60°C (140°F) for an aluminum oxide probe. A steam leak can be immediately detected by an increase in the dew point, and an alarm system is used to provide an audible warning when a steam leak occurs.

Chilled mirror and aluminum oxide hygrometers have been used for this application. This has prevented maintenance problems arising from corrosion of pipes by detecting a steam leak in the early stages of development. In this way, costly downtime and repairs can be kept to a minimum.

The required water vapor range is about 1000 ppm_v to 20,000 ppm_v, corresponding to a dew point range of +20°C to −20°C (68°F to −4°F).

D. Argon in Glove Boxes

The presence of moisture in the argon blanket gas for a glove box causes serious contamination problems. Moisture causes product contamination, unwanted side reactions, and difficulties in quality control.

The problem can be solved by placing sensors in the glove box at ambient conditions to measure the moisture content of the argon gas at both the entrance and exit. A large increase between these readings demonstrates the existence of moisture build-up and indicates that appropriate quality control action should be taken.

Monitoring of moisture present in the glove box can eliminate the loss of expensive materials resulting from moisture build-up in production samples.

A water vapor range of 0.001 ppm_v to 10 ppm_v or dew point range of −50°C to −110°C (−58°F to −167°F) is needed, which can usually be accomplished satisfactorily with an aluminum oxide or electrolytic hygrometer.

E. Controlled Atmosphere Glove Boxes

A controlled atmosphere glove box, sometimes called a dry box, is an enclosure that maintains a leak-free environment so that experiments can be carried out under controlled conditions, such as positive or negative pressure, parts-per-million levels of oxygen and moisture, or use of inert gases.

Controlled atmosphere glove boxes are widely used for work involving air-sensitive inorganic, organic, organometallic, and biochemical materials. Lithium batteries, catalyses hemoglobin research, or other procedures utilizing oxygen- or moisture-sensitive materials can thus be properly manipulated. To establish and maintain these low levels, a drying train capable of removing moisture and oxygen from the gas circulating through the glove box is required. Moisture levels in the low ppm range are to be monitored. Aluminum oxide sensors are often used for this application. The electrolytic cell can also be used.

F. Blanketing Gas in Radionuclide Glove Boxes

Radionuclides are generally processed in glove boxes with an inert blanketing gas. Presence of moisture in this blanketing gas can cause a number of problems. For example:

- High dew points are normally indicative of leaks in the system. Such conditions lead to contamination of the external working area.

- Absorbed water on certain radionuclide sources attenuates the radioactive emission of such sources.

- High moisture content in a blanketing gas affects the calibration of nuclear equipment as a result of attenuation.

It is therefore important to measure moisture in the glove boxes which are used to prepare sources or conduct experiments with radionuclides. The sensor is permanently mounted within the glove box either using a bulkhead connector or by wall-mounting the sensor with a threaded opening. The dew point of the gas must be monitored continuously. An aluminum oxide hygrometer is versatile and can operate in either a dynamic or a static system, under pressure, or in a vacuum. This makes it an invaluable tool for this type of application. For example, a hygrometer can be used to continuously determine the moisture content of the air at a pressure from 0.01 mm Hg Pa to 101.4 kPa (0.01 mm Hg to 1 atm) during the calibration of a nuclear instrument. The ultimate purpose of this nuclear instrument is to determine air density in the atmosphere as a function of altitude.

Electrolytic hygrometers with sample pump systems have been utilized, though with limited success. These systems do not have the versatility of the aluminum oxide hygrometer. In addition, since a sampling system is required, the possibility of external radioactive contamination is an ever-present concern.

The accurate determination of moisture content in static systems can easily be accomplished. The need for sampling systems is eliminated and permanent mounting of the hygrometer inside the glove box prevents the possibility of external contamination.

A dew/frost point range of −30°C to −76°C (−22°F to −105°F) is typical, at an ambient temperature of 24°C (75°F) and operating pressure of 101.4 kPa (14.7 psi).

G. Hydrogen-Rich Hydrocarbon Streams

The presence of moisture at high levels in a hydrocarbon stream causes freeze-ups in a process cold box. A sensor is placed into a dryer ahead of the cold box. Controlling moisture in the dryer at a dew point below −30°C (−22°F) ensures that freeze-ups will not occur in the cold box. Continuous monitoring of the moisture content in the hydrocarbon stream with a suitable hygrometer, usually an aluminum oxide probe, eliminates the downtime and high labor costs associated with process line freeze-ups. Generally, in such applications, the dew point range is –30°C to –110°C (–22°F to –167°F).

H. Ethylene Gas for Polyethylene Production

High moisture content in ethylene feed gases reduces catalyst activity, thereby decreasing the overall yield of polyethylene reactors. For this reason, it is desirable to measure moisture in an ethylene gas stream which is used in the production of high-density polyethylene.

This application involves an integrated polyethylene unit which produces high density polyethylene by the conventional method of de-methanizing natural gas and thermal cracking of the heavier components. The resulting ethylene feed gas is then passed through an activated aluminum drying column. Moisture sensors are used at various locations in the gas lines to determine the moisture content of the ethylene.

Until recently, no suitable hygrometers were available for this purpose. Recent developments, especially in the field of aluminum oxide sensors, have resulted in satisfactory moisture monitoring. The effect of moisture on the catalyst activity in the feed stream is ascertained by continuous monitoring. The effect of various quantities of water in liquids such as hexane, which are added as solvents to the feed stream in small quantities to form copolymers, can now be discovered.

For this application a dew/frost point range below –50°C (–58°F) must be covered at an operating temperature of about 38°C (100°F) and pressure of 3.2 MPa (450 psi_g).

I. Recycle Gas in Catalytic Reforming Processes

To obtain optimum catalyst performance, it is necessary to maintain the moisture content in the recycle gas stream at a specified level. Excessive moisture poisons the catalyst by removing halogen, thus altering the balance between its acid and metal functions. A moisture level higher or lower than that specified for a catalyst impairs certain reactions, rendering the process less economical by reducing the percent yield of desired end products.

Moisture is therefore measured in recycle gas streams in catalytic-reforming processes which are used in the production of gasoline components and other aromatic hydrocarbons. Typical humidity levels to be measured are 5 to 50 ppm_v at operating temperatures of 27°C to 48°C (81°F to 119 °F). Aluminum oxide sensors are commonly used for this application.

J. Cylinder Gases

Cylinder gases are generally supplied as clean and uncontaminated products. The presence of water vapor is most undesirable and is a quality defect. Both the cylinder gas supplier and the user are interested in measuring the trace moisture content of the gas to determine its quality. Guarantees made, and levels to be measured, are normally in the low ppm and, in some cases, ppb range. A measurement of excess moisture is not necessarily caused by an inferior product; it can also be the result of problems with leaks or the gas handling system. Particulate or chemical contamination is usually not a problem since such gases are very clean.

Electrolytic and aluminum oxide hygrometers are frequently used for cylinder gas measurements. In some cases, the Piezoelectric hygrometer may be used. If high accuracy and traceability are crucial, the chilled mirror hygrometer is selected, though the latter is expensive an difficult to use at the low ppm levels. Cylinder gas manufacturers obviously prefer fast-response instruments to cut test times and costs.

IX. Meteorological Applications

A. Weather Forecasting

Temperature and humidity measurements are of vital importance to weather bureaus for computer analysis and weather forecasting. Ground-based as well as aerial sensors are used. Aerial sensors are usually mounted on aircraft or small balloons, the so-called "sondes" which telemeter data back to earth. For ground-based systems, sensors are mounted in suitable weather proof housings or aspirators and installed on towers. Many systems are required to obtain the necessary data needed for accurate weather forecasting. In earlier years, aspirated dry/wet bulb psychrometers were in common use. Often encountered problems were freeze-up of the water supply in winter months, contamination, and the need for frequent maintenance. In the 1930s and 1940s, many psychrometric systems were replaced with lithium chloride "dew cells," which require less maintenance. However, dew cells (saturated salt sensors) have a slow response and, when contaminated, produce inaccurate measurements. In installations where high accuracy and traceability are required, the chilled mirror hygrometer has often been chosen, although its cost is considerably higher. Maintenance requirements are also a draw back. In the 1980s, significant advances were made in the technology of bulk polymer RH sensors, resulting in sensing elements suitable for outdoor use. Because of its broader operating temperature range, the capacitive bulk polymer sensor has been more commonly used than resistive types. The need for periodic calibration remains a point of concern. The CCM chilled mirror hygrometer offers promise as a low maintenance chilled mirror sensor, though at a higher cost. Such a system has not yet been field tested and is unproven at this point in time.

Infrared hygrometers have been successfully used in areas with high levels of chemical and particulate contaminants. Their cost is even higher than for chilled mirror hygrometers, and therefore applications of infrared hygrometers have been limited.

B. Airports

Humidity information is used to compute the "lift" encountered by airplanes taking off, or landing on an air field. Improper information can cause unstable landings, damage to landing gear, or accidents.

Another problem of great concern and which has caused fatal accidents in the past, is "wind shear." This condition is very difficult to predict. Warning signals are only given a very short time before wind shear actually occurs. Scientific know-how of this phenomena is incomplete at the present time. However, enough is known to be able to predict wind shear conditions with reasonable certainty several minutes before they occur. Humidity measurements play an important role in this. Research measurements are mostly obtained using chilled mirror hygrometers. Lyman–Alpha and infrared hygrometers have also been used in related research projects.

C. Plant Site Locations

Meteorological measurements, often made using chilled mirror hygrometers, help determine the best site for a nuclear power plant or fossil fuel plant emitting toxic fumes. Such meteorological information determines the most common wind directions which allow the plant to be located where the least amount of danger exists to population areas with regard to air pollution, hazardous gases, or radiation from a possible nuclear radiation leak or accident.

D. Aircraft Noise Pollution

As discussed in the section on automotive applications, nitric oxide emissions from automobile exhausts are dependent on relative humidity of the intake air. Likewise, noise produced by aircraft is dependent on the relative humidity of the atmosphere in the immediate vicinity of the aircraft. EPA specifications are based on a certain relative humidity, usually 50%. When tests are conducted under environmental conditions when the relative humidity is higher or lower, and this is almost always the case, relative humidity must be accurately measured to allow precise corrections to be made for the field measurements. Since this data must be approved by the EPA and FAA, such measurements are always specified to be traceable to NIST, or in overseas situations, traceable to applicable foreign standards.

E. Upper Atmosphere Measurements

1. Cloud Studies

Several research programs are funded by defense agencies with the objective to analyze the effects of humidity and other parameters on data transmissions in the upper atmosphere and through clouds. High-flying aircraft have been used for this purpose, utilizing specially-designed chilled mirror hygrometers. Lyman–Alpha hygrometers and, in some instances, infrared systems have also been used for this purpose. Measurements are generally conducted at very low pressures and very low ambient temperatures. Typical dew/frost points to be measured are as low as –80°C (–112°F). See also Chapter 6 on Lyman–Alpha hygrometers.

2. Ozone Depletion Studies

Extensive research, spanning more than two decades, has resulted in almost indisputable evidence that there is a gradual thinning of the ozone layer in the stratosphere. This causes damaging ultra-violet radiation which is otherwise absorbed by the ozone layer, to penetrate to the earth and thereby cause changes in its environment. This is known to lead to medical dangers. It has been shown that even a small amount of far ultraviolet radiation, can cause skin cancer when an individual is exposed to the sun for a significant period of time. The level of dangerous ultraviolet radiation varies from location to location on earth and is greater in places where there is a so called "ozone hole."

Measurements, including water vapor detection, are made at regular intervals, from many sites all over the world, to obtain statistical information over long time periods (decades). Specially-designed cryogenic chilled mirror hygrometers are used and flown on high altitude balloons. Dew/frost point measurement below –80°C (–112°F) are required. A commonly used system incorporates a freon or dry ice-cooled mirror with heater (see Chapter 3). The frost point is then measured by heating the mirror, using a small heater imbedded in the mirror, and clearing the mirror surface of any ice. The exact temperature at which the ice layer on the mirror disappears is by definition the frost point. The approach used, is the same as for the cryogenic hygrometer discussed in Chapter 3.

X. Other Applications

A. Industrial

1. Instrument Air

Large amounts of instrument air are used in the power generation industry to operate valves, dampers, and pneumatic instrumentation. Instrument air is also used in may other industries for a multitude of applications. Contaminants and moisture will cause wear on moving parts, leave deposits that can cause problems operating various systems, deteriorate "O" rings and seals, and cause freeze-ups when exposed to winter time temperatures. To eliminate excessive moisture in such instrument air lines, power plants often install "refrigerative" dryers. These employ two towers (containers) filled with desiccant material that are alternated between being in operation or in refrigeration. They are regenerated by heating up the tower and purging the dry air. The regeneration cycle is timer controlled and not performance driven. When operating properly, the dew point of the output is about –35°C (–31°F). The instrument is set to alarm and notify the operators when the dew point approaches an unsafe condition. It is also customary to monitor changes in dew point to set the timer cycles on the dryer. When the dew point is too high after regeneration, the cycle needs to be lengthened or the desiccant must be replaced.

Instrument air systems are also widely used in chemical plants, food and drug processing, pulp and paper mills, petrochemical processing, and many other industries.

Frozen air lines can reduce production or shut down operations for many hours. The presence of moisture in the air used as a blanketing gas for sensitive instruments causes corrosion problems, rapid change of calibration, and loss of instrument sensitivity. Humidity sensors are usually installed at the source of the instrument air equipment. Alarm systems are always added to alert operators when the dew point deviates from a predetermined level. Close moisture control can greatly reduce the need for major repairs of delicate in-process instrumentation and is thus very cost-effective.

Generally, the water vapor pressure to be monitored for these applications is below 1000 ppm_v, corresponding to dew points below $-20°C$ ($-4°F$). Both chilled mirror and aluminum oxide hygrometers are often used for instrument air applications.

2. Evacuation and Gas Refill in Oil-Filled Transformers

It is desirable to continuously monitor moisture during vacuum evacuation and gas refill of oil-filled transformers. The moisture content of blanketing gases of oil-filled transformers directly affects the dielectric properties of the oil as a result of water vapor dissolution.

During this process, transformers are evacuated and filled with oil. After evacuation nitrogen is bled in, and the pressure is brought to atmospheric level. A single sensor is used to monitor the actual dew point during both the evacuation and nitrogen repressurization. The sensor is mounted in a standard sample cell connected directly to the evacuation and refill line.

Transformer units which are sealed before they are thoroughly dried often require rework, and the cost of materials and testing for these large electrical units can be considerable. Accurate, continuous measurement of moisture content allows the manufacturer to eliminate these problems.

In this application, the water vapor range is approximately 1 ppm_v to 100 ppm_v, corresponding to a dew/frost point range of $-40°C$ to $-76°C$ ($-40°F$ to $-105°F$) over a pressure range of 1 micron to 1 atm. Aluminum oxide probes are often preferred for this application.

3. Pharmaceuticals

As discussed in the section on water activity measurements, there is a great need for humidity measurements in pharmaceutical companies, specifically for controlling the environment in chambers where tablets, medicines and other pharmaceutical products are manufactured and dried. Although many applications could be taken care of with polymer type %RH sensors, chilled mirror hygrometers are sometimes preferred because of their fundamental nature, accuracy, and traceability, which is important to satisfy regulatory agencies.

4. Dried Foods

Foods contain moisture in three states: free moisture, hydrogen-bonding, and ionic-bonding. Free moisture contributes to the bulk of the food and affects its packaging and shipping. More importantly, free moisture allows micro-organisms to proliferate, thus causing spoilage of food. The goal is to eliminate free moisture, while retaining ionic-bonded moisture and most of the hydrogen-bonded moisture. If the bonded moisture is removed, undesirable changes in flavor and color occur which may render the product unsaleable (see also Section III on water activity).

The procedure consists of an empirical calibration of the dew point, versus the moisture content of dried food with a known weight percentage of water. Once this empirical calibration is conducted, routine determinations of the absolute moisture content of the dried food substance is possible. Generally a batch-type analysis is performed using a sealed vessel containing a moisture sensor and a test sample utilizing the permanent installation of a hygrometer, controlling the range of moisture present in the dried foods. Shorter delivery times, lower overhead from a packaging and storage viewpoint, and less food spoilage have resulted from accurate moisture measurements and control, while high quality of taste and appearance has been maintained.

The water vapor range to be controlled is 0.001 to 1000 ppmv, corresponding to a dew/frost point range of –20°C to –110°C (–4°F to –167°F) at ambient temperature and an operating pressure of 103.4 kPa (15 psi). Aluminum oxide sensors are often selected for this purpose. Electrolytic and Piezoelectric instruments are possible alternatives.

5. Radar Wave Guides

Moisture measurements are required in service air systems aboard naval ships, including air systems for radar wave guides. The subject air is used to pressurize the ship's radar wave guide. Moisture content is critical because the dielectric constant has to be kept to a minimum. The moisture content of air throughout the rest of the ship is also imperative. When very high pressures and flows are used (such as missile launches and aircraft servicing), excessively high dew points can cause freeze-up of pipelines and valves.

In general, humidity sensors are used to monitor the dew point of the compressed air that has been dried by molecular sieve desiccant bed dryers. This application requires that the hygrometer meet certain naval specifications. Hygrometers used for this purpose in the US must satisfy all of the requirements dictated by the U.S. Navy shock and vibration specification (MIL. SPEC. 901C).

In some cases, electrolytic hygrometers have been used, but have been found to be less satisfactory due to the necessity of frequent servicing and close regulation of flow rates and temperature. Continuous moisture monitoring using aluminum oxide or chilled mirror sensors can be used to measure the moisture content, indicating an excessive water vapor level. This allows preventive maintenance to be conducted before critical water concentrations are reached. The possibility of serious malfunctions of radar wave guides and other equipment due to excessive moisture can be virtually eliminated.

The dew/frost point range to be monitored is normally –29°C to –51°C (–20°F to –60°F) at ambient temperature and at a pressure of 377 kPa to 471 kPa (40 psi_g to 50 psi_g).

6. Plate Glass Production

Plate glass is produced by pouring molten glass onto molten tin in a furnace. A blanketing gas of 20% hydrogen and 80% nitrogen is used to prevent oxidation. During the process, carbon in the tin and silicon from heating elements combine to form silicon carbide, which then collects on furnace walls. Elemental tin also collects on furnace walls. When the dew point exceeds approximately –28°C (–18°F), the tin/silicon carbide scale has been found to cause imperfections by breaking loose and falling into the glass. For this reason, moisture is measured in the glass-making blanketing atmospheres.

A sampling system, consisting of a filter and cooler, is employed to cool the gas to less than 60°C (140°F) and to prevent the deposition of scale on the sensor. Infrared analyzers have been used in the past. Unfortunately, vapors from the process frequently contaminate the sample cell, causing erroneous dew point measurements and requiring frequent service. By using an aluminum oxide instrument instead of an infrared analyzer, more accurate and continuous results can often be obtained, while maintenance costs are reduced. Since operators are alerted to conditions which could cause imperfections, product quality and process yield are improved.

The dew/frost point range to be covered is –20°C to –76°C (–4°F to –105°F) at an operating temperature of 30°C (86°F) and operating pressure of about 115 kPa (2 psi$_g$).

7. Hydrogen-Cooled Electric Generators

Hydrogen is often used to cool large stationary generators because of its high heat capacity and low viscosity. Hydrogen has a high heat capacity and therefore removes excess heat efficiently. Hydrogen also has a low viscosity, allowing higher capacity operation of the generators while maintaining efficient cooling. The hydrogen must be kept dry to maintain both its heat capacity and viscosity. Ambient moisture is a contaminant which reduces the heat capacity and increases the viscosity of the cooling hydrogen. In addition, excess moisture increases the danger of arcing in the high voltage, high current generators which could seriously damage the generators and cause ignition of the explosive hydrogen.

Cooling hydrogen is continuously circulated through the generator and through a molecular sieve dryer. To be certain that the hydrogen in the generator is dry, a probe or sample cell assembly is installed in the return line from the generator to the dryer. Detection of moisture content above a predetermined acceptable level will trip the alarm relay and annunciator.

Aluminum oxide systems with compatible moisture sensors are commonly used for this application. A special alarm relay, capable of carrying 125 VDC at 0.1 amp, is often available and may be required in some installations for interface with alarm annunciators.

Lithium chloride hygrometers and in-line psychrometers have been used in the past. These systems require a high degree of maintenance and have proven to be less reliable. Chilled mirror hygrometers have also been used in this application.

The dew/frost point range to be covered is +10°C to –50°C (50°F to –58°F) at operating temperatures of 30°C to 40°C (86°F to 104°F) and at operating pressures of 105 kPa to 620 kPa (0.5 to 75 psi$_g$).

8. Packaging of Orthopedic Casting Material

Moisture is measured in the nitrogen which is used as a blanketing gas during the packaging of orthopedic casting material within a glove box.

The presence of moisture in the nitrogen blanketing gas has numerous adverse effects on this product. These include:

- Causing material to set or harden during packaging

- Decreasing the shelf life of the product after it has been packaged

The nitrogen blanket gas is recirculated from the glove box to a dual-tower re-generative molecular sieve dryer and is then fed back into the glove box. To compensate for any losses, the glove box is purged at a slow rate with dry nitrogen.

Aluminum oxide moisture sensors are often used, and are installed through the wall of the glove box near the nitrogen entrance and exit ports. The sensors monitor moisture content of the nitrogen entering and exiting to detect any build-up of moisture within the glove box. A few applications have also been handled using capacitive polymer sensors.

With the humidity sensor installed permanently in the glove box, any build-up of moisture can be caught at an early stage, thus avoiding the above-mentioned problems. Alarm contacts allow the operator the freedom to perform other functions simultaneously.

Of interest is a water vapor range of 1 ppm_v to 100 ppm_v or dew/frost point range of $-40°C$ to $-75°C$ ($-40°F$ to $-103°F$) at an operating pressure of 101 kPa (14.7 psi_a) and at nominal ambient temperature.

9. Compressed Air in Ozone Generators

Handling and treatment of water for drinking, industrial processes, or other uses, as well as the treatment of wastewater, require much concern and often large expenditures. The purification of fresh water for human consumption requires particular attention to the removal of various naturally occurring bacteria and organic substances.

Ozone (O_3) is one of the most powerful oxidizers available. It produces O_2 when it has done its job. It kills all dangerous bacteria and decomposes many organic substances that are otherwise difficult to remove from water. Ozone works 3,000 times faster than chlorine and leaves no residue. Treatment of water is one of the most effective methods for preparation of clear, fresh water. This is especially critical for surface water, i.e., water obtained from lakes and rivers, which is more likely to be contaminated with unwanted microorganisms and chemicals. Other methods of treatment, such as chlorine, require much larger exposure times to remove toxins and bacteria. The handling of corrosive chlo-rine gas is also avoided and chlorine levels in drinking water are minimized by the use of ozone. A concern with chlorine treatment is the production of carcinogenic chlorocarbons, especially trihalomethanes, that are restricted in drinking water by EPA regulations.

Ozone treatment was first used by Werner von Siemens in 1857. Ozone is also very simple to prepare in a continuous way by electrolysis of Oxygen (O_2) in a chamber. Air or pure O_2, is passed through a cell containing two electrodes. A high-voltage (about 12,000 V) discharge between the electrodes produces ozone from the oxygen present.

The supply gas used for ozone production, either air or oxygen, has a recommended maximum dew point of $-50°C$ ($-58°F$). Ozone generation is improved markedly by low-ering the dew point to $-70°C$ ($-94°F$). Moisture measurement is desired in water treatment plants incorporating ozonizers because, at higher dew point temperatures, ozone generation is adversely affected in several ways. First, the efficiency of the ozonizer is reduced; i.e.

less ozone is produced from the same electrical discharge. Secondly, because ozone is such a powerful oxidizer, it will also oxidize nitrogen (N_2) in the air to form nitric oxide (NO_2). Moisture reacts with nitric oxide to produce nitrous (HNO_2) and nitric (HNO_3) acids that are highly corrosive to piping and other equipment in the ozonation system. Thirdly, increased moisture levels increase the possibility of arcing by the electrodes, causing severe damage to the ozone generator. As an example, raising the dew point from −50°C (−58°F) to −40°C (−40°F) can mean as much as a 15% decrease in ozone production efficiency, as well as a substantial increase in nitric acid that significantly shortens the lifetime of the electrodes.

To eliminate problems with moisture, the air is dried prior to entering the ozone generator. To monitor moisture effectively in this application, a suitable trace moisture detector is used, usually an aluminum oxide instrument with the probe mounted directly in the pipeline between the dryer or O_2 source, and the generator, as is shown in Figure 12.6. In many cases, it is desirable to measure at both the ozone generator inlet to detect leakage in the piping system and provide optimum point-of-use process control, and at the generator outlet. The dew/frost point range of concern is −50°C to −65°C (−58°F to −85°F) at ambient temperature and a pressure of 239 kPa to 790 kPa (20 psi_g to 100 psi_g). Aluminum oxide hygrometers are most often used for this application.

10. Steel Warehouses

After the production of steel is completed, it is usually stored in a warehouse before being used in the next fabrication cycle. Many of these warehouses are expansive buildings with large doors for access by trucks, railroad trains, and vessels used to transport the steel.

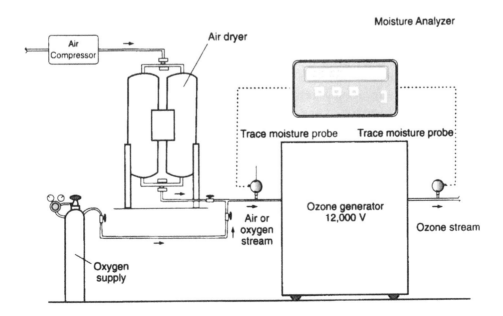

Figure 12.6   Moisture measurement in ozone purification of water.

While in storage, the steel could be damaged by corrosion, which can easily happen if condensation occurs.

Most of the time the dew point of the outside air is lower than the surface temperature of the steel, so condensation will not occur. However, condensation happens when a rainstorm, the formation of fog, or an increase in outside air temperature, raises the dew point above the steel temperature. The steel should be protected against this condition, which can be accomplished by closing doors, turning on a dehumidifier, or heating the steel. To conserve energy, the warehouse would only be dehumidified, or the steel heated, if such weather conditions exist.

A chilled mirror dew point hygrometer with a temperature sensor can be effectively used for this purpose. The temperature sensor must be installed in a small hole drilled into a large representative sample block of steel and is used to obtain the temperature of the steel in storage.

The dew point of the surrounding air is then measured and compared with the steel temperature. Signal outputs of these temperatures are fed to a differential comparator which sounds an alarm whenever the dew point temperature approaches within a few degrees of the steel temperature. The dehumidifiers or heaters can then be turned on automatically to prevent damage and to operate in the most energy efficient way.

11. Injection Molding

Small plastic parts and consumer products are often injection molded. The raw plastic is first heated and melted, and is then forced under pressure into steel dies which shape the molten plastic into various commercial products. The raw plastic is purchased from the manufacturer in the form of small pellets. These are usually delivered in large containers, which are stored until needed.

In some cases, the plastic pellets are too wet, either as received from the manufacturer or from condensation during storage. When this occurs, the plastic will not flow properly in the mold, which is a serious problem since many parts will then have to be rejected. Therefore, the pellets are first directed into a hot air dryer just before molding, at a temperature of 80°C to 93°C (175°F to 200°F).

Although a large improvement in product uniformity can be obtained from drying the pellets, the operator has no way of knowing when they are dry enough to mold. Most dryers are automatically run for a fixed period of time. Some pellets that are very wet are not dried sufficiently, and these parts must be rejected to prevent improper molding. Conversely, much money can be wasted by drying material that is already sufficiently dry.

This problem can be solved by using a chilled mirror, or aluminum oxide hygrometer to measure the dew point at the dryer output duct. A high temperature sampling system must also be used to reduce the air sample temperature to bring it into the measurement range of sensor. A sampling system allows the temperature of the air to be decreased by drawing the air through a certain length of tubing prior to measuring. Since the ambient temperature outside the tubing is much lower, the sample temperature equilibrates toward ambient. In effect, it is a "heat exchanger." The dew point in the tubing remains constant, and may be measured at the other end. Since the pellets are often dusty and dirty, a filter must be installed ahead of the sensor (see Figure 12.7).

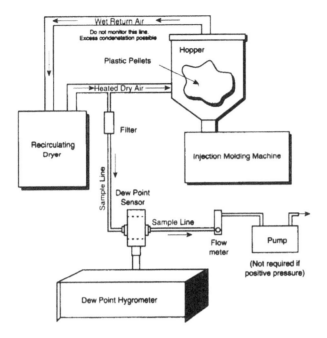

Figure 12.7   Measuring moisture in pellet dryer.

Experiments have shown that a dew point between –18°C and –23°C (0°F and –10°F) at the dryer output is optimum for the equipment and types of plastics being used. The drying time varies depending upon ambient air conditions and the dryness of the plastic material in stock. When the dryer output is reduced to the desired level, the operator can load his molding machines.

The addition of a dew point instrument allows the manufacturer to save money two ways: The energy cost can be greatly reduced since material is no longer dried excessively, and the yield is improved since it is not necessary to reject material which does not flow properly. A hygrometer often pays for itself in a very short time. Several types of sensors have been used for this purpose including aluminum oxide, chilled mirror and polymer types.

12. Pigments for Paints and Plastics

The presence of moisture in vacuum-tumbler dried pigments, which are dispersed in paints and plastics, causes undesirable changes in the physical and chemical properties of the product. The problem can be alleviated by measuring moisture in the pigments used for such paints and plastics. The correlation between dew point measurements and pigments of known moisture content provides an empirical calibration in percent moisture. The calibration is performed under vacuum conditions of the tumbler dryer.

Continuous monitoring of the moisture present in the system using a suitable hygrometer, such as a chilled mirror type or aluminum oxide instrument, has eliminated the loss of production time due to slow laboratory turnaround on pigment analysis. The dew/frost point range to be monitored is usually –20°C to –40°C (–4°F to –40°F).

13. Circuit Breakers Using Sulfur Hexafluoride

Sulfur Hexafluoride (SF_6) is often used as an insulating gas in large electrical circuit breakers. Excess moisture in this gas could cause arcing, which could trip the circuit breaker and cut off power in a distribution network.

A moisture sensor may be installed in the circuit breaker or in the SF_6 fill line and used to monitor the SF_6 dew point. The dew point measurement may then be converted into ppm_v and compared with the gas manufacturer's specifications.

Electrolytic and aluminum oxide sensors have been used for this purpose, the latter probably being more widely used at the present time.

The typical range to be covered is 1 to 1000 ppm_v in the SF_6 gas, corresponding to a frost point range of –76°C to –20°C (–105°F to –4°F) at pressures up to 1.8 MPa (250 psi_g) and at ambient temperature.

14. Battery Manufacturing

Many manufactured products are affected by ambient humidity. Some are processed in controlled chambers, others such as batteries must be manufactured in humidity controlled rooms. In most cases, this can be handled by a properly designed air conditioning system, but in some cases, such as for the manufacture of lithium batteries, the entire manufacturing area must be controlled at a very low humidity level, sometimes well below –20°C (–4°F). Substantial product losses could be suffered if this condition is not maintained. A complicating factor is that doors are occasionally opened and closed to allow personnel to enter or leave the room. The very dry conditions could also lead to discomfort on the part of the employees.

Dry rooms incorporating large air dryers can be constructed to control humidity and temperature during the assembly process. Once the assembly process is completed and sealed, humidity control is no longer needed and the products may be removed safely from the controlled atmosphere.

In the past, dry room humidity measurements were often made using a "dew cup." However these are not very accurate, primarily because of operator inconsistencies and difficulties observing frost on the cooled surface. Chilled mirror hygrometers, though much more expensive, have been used successfully in areas where humidity control is critical and product losses could be very substantial.

15. Manufacture of China and Dinnerware

During the final decorative firing process in the manufacture of china and dinnerware, moisture is added to the kiln, thereby releasing boron from the surface layers of the glaze as boric acid. This results in a very durable acid resistant glaze. Most manufacturers use different glazes for different products which requires different ideal moisture levels in the kilns in order to produce the highest quality surface.

To make such measurements, the hot gas is drawn from the interior of the gas fired kiln using a vacuum pump and is then cooled to a temperature below 100°C (212°F), but above the gas dew point temperature. A suitable sampling system must be used for this purpose. The dew point sensor is installed in a manifold of the heated sampling system and is used to monitor the moisture of the sample gas continuously. The dew point information can be used to automatically control the kiln moisture level during the decorative firing process, resulting in higher product quality and lower cost.

Kiln temperatures can be as high as 500°C (about 950°F). Humidity levels are high (approaching 100% in some cases). Such measurements are difficult to make and require a well designed heated sampling system to make certain that all components of the sampling system and the sensor itself that come in contact with the sample air are at a temperature above the prevailing dew point, but below the maximum operating temperature of the system components and the sensor.

Modern high temperature RH sensors, capable of operating above 100°C (212°F) and at high RH levels, are most convenient for this application. If a higher accuracy and fundamental measurement is desired, the chilled mirror hygrometer may be the best choice, but the cost including sampling system would be much higher and the sensor cannot be operated above 95°C (203°F).

16. Gas Turbines

Gas turbines are usually connected to electrical generators and used in power plants and cogeneration installations. Most turbines are operated using natural gas. Normally steam or ammonia is injected into the intake air stream in order to control the flame temperature, which in turn determines the emission rate of oxides and nitrogen (NOx). In order to properly set the correct rate of steam or ammonia injection, and to prevent icing, which can cause catastrophic failures, the dew point of the upstream intake air must be closely monitored. EPA regulations limit the emissions of gas turbines which are dependent on the relative humidity of the intake air. RH measurements can be used to make corrections in the emissions detected to establish compliance with EPA regulations. Such measurements must be accurate, traceable to humidity standards, and be acceptable to the regulating agencies. Various types of dew point instruments have been used in the past, but with limited success, mostly because of the difficulty of frequent maintenance and/or need for calibration. The cycling chilled mirror (CCM) hygrometer has in recent years been proven to be an excellent solution and is presently used in many installations. The use of certain polymer type RH sensors have also been attempted.

17. Clean Rooms

Many products manufactured and processed in a clean room environment are moisture sensitive. For this reason, clean room specifications often include control levels of typically 35% relative humidity for year-round operation. The proper %RH level must be maintained within ± 2 %RH at temperatures below 20°C (68°F). A relative humidity of 35% at 20°C (68°F) corresponds to a dew point of about 4°C (40°F). The effects of higher humidity levels in close tolerance environments can be detrimental to product quality and production schedules.

In semiconductor manufacturing, when the humidity level fluctuates in a wafer fabrication area, many problems can occur. Bake-out times typically increase, and the entire process usually becomes harder to control. Humidity levels above 35% RH make the components vulnerable to corrosion.

In pharmaceutical manufacturing facilities, high humidity causes fine powders to adsorb moisture, clogging the powder feed to the tablet press. Variations in humidity cause difficult adjustments in bed temperature and spraying rates, resulting in heat damage and moisture intrusion. Humidity in air ductwork creates moist areas for bacterial colonies to grow and cause process contamination.

Two common approaches to humidity control are air conditioning and desiccants. Air conditioning lowers the temperature of a surface exposed to the clean room air stream below the dew point of that air stream. Excess water vapor condenses and the resulting air is dehumidified. The air must then be reheated to the proper control temperature and routed to the clean room. Standard refrigeration equipment can produce dew points of 5°C (+40°F) on a reliable basis.

In a desiccant system, the process air stream passes through a desiccant medium. The desiccant adsorbs moisture directly from the air stream, and the resulting dehumidified air is routed to the clean room. Measurement of %RH is not very difficult and many approaches are possible. The most commonly used are RH elements like the ones used in air conditioning applications. In some cases where accuracy and stability is of utmost importance, chilled mirror hygrometers have been installed to measure dew point and temperature from which %RH is derived using a microprocessor calculation.

B. Automotive

1. Automobile Exhaust Emissions

Nitric oxide emissions from automobile exhaust systems known as "NOx" are closely regulated by the Environmental Protection Agency (EPA) in the USA and by similar institutions in many overseas countries. Aside from difficulties in measuring and controlling these emissions in trace amounts, they are also affected by, and dependent on, the relative humidity of the engine intake air. EPA specifications are based on a certain relative humidity and corrections are allowed for measurements made at higher or lower relative humidity levels. To validate such corrections, accurate humidity measurements must be made and, more importantly, these measurements must be certified and traceable to national standards (such as NIST in the USA). For this reason, these measurements are customarily made using chilled mirror hygrometers. A typical test set up is shown in Figure 12.8. The humidity range to be covered is usually within 20% RH to 80% RH and measurements are not difficult to make. Most complexities are related to accuracy, repeatability, and traceability.

Prior to 1983, the wet/dry bulb psychrometer was used as the humidity reference. However, in April 1983 the U.S. Environmental Agency adopted the chilled mirror dew point hygrometer as its official standard and all correction factors were thereafter calculated with data from this type of instrument.

Automotive manufacturers use a number of dynamometer test cells where motor vehicles are tested under various environmental conditions.

The vehicle under test is subjected to different conditions of temperature, humidity, engine speed, engine load, and fuel air mixtures. Each of these parameters is fed into a data-logging computer, which calculates performance data and NOx emissions.

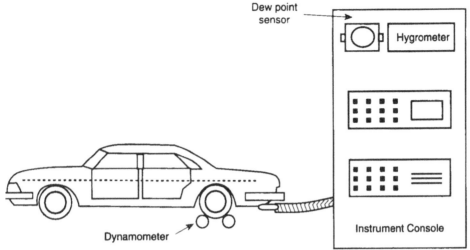

Dew point
sensor

Hygrometer

Instrument Console

Dynamometer

Figure 12.8   Hygrometer for automobile emissions testing.

2. Engine Testing

Combustion engines require the proper mixture of ambient air and fuel. The efficiency and performance of such engines are affected by the moisture content of the air. Carburetor settings could be adjusted based on humidity measurements and thereby optimized for more fuel-efficient operation or better performance. Humidity measurements are therefore quite important for engine testing and research. The most commonly used instrument for research in this field is the chilled mirror hygrometer.

3. Electronic Components and Failure Analysis

The presence of moisture under various climactic conditions is believed to be a contributing factor in the failure of certain electrical automobile accessories. Moisture is measured in conditioned air that is used for failure analysis of electronic components.

A humidity instrument is mounted inside the automobile to measure changes in environmental conditions. A test series is run to determine the effect of various levels of moisture on the response of electrical and electronic equipment in the automobile.

Sling psychrometers were used in the past. This method could not provide the continuous, reliable data required for a valid experiment. Today's systems usually employ an aluminum oxide or chilled mirror hygrometer (conventional or CCM). The need for better quality control instrumentation has been clearly demonstrated and requires proper control of moisture content in the failure analysis performed on electrical and electronic automobile equipment in the field.

4. Paint Spray Booths

When automobile bodies are painted, conditions of humidity are critical. If water vapor levels in the air are too high, moisture can become trapped between the metal and the paint. When the painted shell is subsequently baked (or cured), the moisture migrates outward and if the paint is still elastic, bubbles appear on the surface which is most undesirable.

The solution to this problem is to maintain the ambient air at the appropriate humidity level prior to injecting it into the spray booths, or to control the humidity in the paint area using a humidity sensor situated in the return air duct. Most manufacturers prefer to measure humidity in the return air ducts. In the return air, the sensor is not only subjected to dust particles of the paints but also to the volatile vapors that are generated. For this reason, sensors must often meet intrinsic safety requirements.

Reliable humidity measurements in paint booths are very difficult to make because of the many contaminants. Few RH sensors survive for long periods of time. Chilled mirror sensors have also been tried, but with little success. The use of CCM type chilled mirror sensors has the potential of providing an improvement, though such sensors are unlikely to be immune to the heavy contamination levels in spray booths.

C. Laboratory Standards

Chilled mirror hygrometers are often used as transfer standards. They are widely used in calibration laboratories for calibrating lower-level instruments and to support research when precise and traceable measurements are desired.

The conventional chilled mirror hygrometer is a fundamental measurement device. Once calibrated, this instrument retains its calibration over long periods of time, provided it is properly maintained and kept in good operating condition through regular mirror cleaning. To make sure that traceability is maintained, it is customary to have the instrument at least annually recalibrated and compared with one of the national standards (NIST, NPL, NRLM, PTB, or other standards laboratories). The instrument is then given a certificate and is used to provide measurements that can be certified and accepted by regulatory agencies.

Though psychrometers and phosphorous pentoxide instruments have also been used for this purpose, the chilled mirror hygrometer is considered to be superior and is presently the most commonly used instrument for laboratory calibration. Certified dew/ frost point measurements can be made in the range of +95°C to −70°C (203°F to −94°F) with accuracies of ± 0.1°C to ± 0.2°C (± 0.18°F to ± .36°F). At lower frost points, calibration using a chilled mirror hygrometer becomes difficult and time consuming. Other methods such as the electrolytic hygrometer, may also be considered as discussed in Chapter 10. The more recently developed cryogenic hygrometer, though quite expensive, is very accurate, faster, and offers another good alternative for low frost point calibration (see Chapter 3). The CCM hygrometer has certain limitations and is in most instances not the ideal choice.

D. Nuclear Reactors

1. Helium Cooled Reactors

Helium is often used as a reactor coolant for nuclear installations because of its excellent thermal properties and chemical inertness. If the helium gas is contaminated with moisture, it causes degradation of the graphite moderator rods and changes the process heat balance. It is therefore important to measure moisture in the helium gas which is used for nuclear reactors.

The helium is measured in either the dryer line or the pressurization line. Moisture content is monitored continuously since leakage in the heat exchanger system can occur at any time. If the gas system is clean, no sampling arrangement is required. However, if particulate matter is present, a suitable filter should be placed in the line upstream of the sensor.

A complete reactor shutdown can result from heavy moisture content in the helium coolant. In addition, graphite moderator-rod replacement at inconvenient intervals causes considerable loss of time and materials, as well as loss of reactor efficiency. Continuous monitoring of the level of moisture in the gas by a trace moisture hygrometer helps prevent these upset conditions.

Phosphorous pentoxide moisture monitors have been used to determine moisture content in the helium line. However, the possibility of cell flooding and particle contamination raise some doubts about the reliability of such a system. Aluminum and silicon oxide sensors have been found to be more satisfactory. Chilled mirror hygrometers are rarely used for this purpose because of the difficulties and high cost of operating them at low moisture levels. The water vapor content range to be monitored is 0.005 ppm to 10 ppm, corresponding to a dew/frost point range of –60°C to –100°C (–76°F to –148°F).

2. Containment Areas

Another nuclear power plant application is monitoring of dew point in the so called "containment area." The nuclear reactor is normally housed in a concrete building which can absorb a large amount of nuclear radiation. This building is then surrounded with another brick wall for further protection and safety. The area between the two walls is called the "containment area." Water is often used to cool the nuclear reactor inside the primary building. If water leaks occur, overheating of the reactor is a possibility, resulting in the danger of reactor failure and radiation leaks into the containment area environment. Humidity sensors are used inside the containment area to detect a sudden increase in humidity and sound an alarm if this is happening. The sensors are required to withstand significant amounts of nuclear radiation without failure in the event a radiation leak occurs into the containment area. These requirements have been satisfied using chilled mirror hygrometers, which have therefore been widely used for such applications. Operating ranges are quite nominal, i.e., ambient temperature and dew point levels of about 5°C to 15°C (41°F to 59°F).

E. Computers

1. Computer Rooms

Humidity measurements are important in computer rooms where large and expensive data storage systems are installed. It is a well known fact that under extreme humidity and temperature conditions, magnetic devices used for data storage could be distorted and cause catastrophic failures to expensive equipment and the loss of important stored data. Certain types of computers are equipped with water-cooled storage systems, and in such cases humidity measurements are used to detect leaks by measuring humidity changes in the air surrounding the water-cooled devices.

For these measurements, environmental conditions are usually right, i.e., air is clean and surrounding temperatures are close to ambient. Bulk polymer relative humidity sensors, capacitive or resistive, perform well under these conditions and are often used. Such sensors should be periodically recalibrated, which is normally done on site. In cases where humidity is extremely critical and where very large losses could result from failure, the fundamental chilled mirror hygrometer or transmitter types are used.

2. CPU Moisture Detection

Some computer companies offer a complete computer system that provides the user with free-standing cabinets including units that contain a water-cooled central processing unit (CPU). The water is automatically controlled at the lowest temperature possible because this provides an improvement in mean time between failures (MTBF). The limiting factor determining at how low a temperature the CPU can be maintained is the dew point in the computer. Condensation will occur throughout the inside of the cabinet if the dew point temperature were to be reached and this can cause extensive damage to the CPU. Therefore, a reliable dew point sensor is required. Another requirement is that periods between maintenance cycles are as long as possible, since maintenance costs are high.

Because of the importance of the measurement and the high cost of the equipment which the sensor would protect, a chilled mirror transmitter is often the best solution. Conventional continuous dew point transmitters have been used for this application even though the low maintenance requirement could only marginally be met. Although more expensive, CCM transmitters offer the potential of much longer maintenance cycles. However, such systems are presently unproven and are not currently known to be in use for this application.

An optimum safety factor of 1°C (1.8°F) can be programmed into the system. This way the control loop maintains the CPU temperature exactly 1°C (1.8°F) above the dew point of the computer room, and tracks it up and down automatically. This is an ideal closed loop control system whereby the water temperature is always maintained as low as possible without experiencing a condensation problem.

A schematic showing a typical CPU cooling system is shown in Figure 12.9.

F. Data Communications Through Telephone Cables

Modern telecommunications needs are placing a greater and greater burden on existing telephone cables. The use of telephone cables for computer communications is often a problem because the integrity of data at high frequencies (high data rates) is critical. Older telephone cables may have paper or cotton wire insulation. This type of insulation can present large losses and become unreliable when moist. The amount of loss, and hence the quality of the telephone line, depends upon the amount of moisture in the cable.

If the maximum moisture level can be accurately and reliably controlled, a higher data rate can be achieved. A solution to this problem is that when nitrogen or dry compressed air is used to dry cables which are covered in paper or cotton, the moisture content is monitored to maintain the required dryness at a level that is consistent with tolerable losses. Moisture levels are either periodically monitored or the standard hygrometer alarm wiring

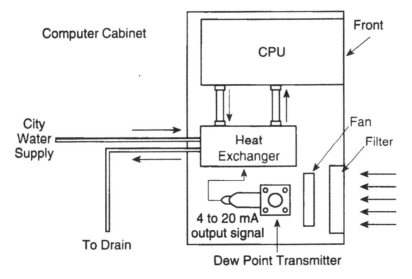

Figure 12.9   Schematic of CPU cooling system.

is used to activate the nitrogen or dry air supply if moisture exceeds a predetermined value.

Aluminum oxide sensors have been used with compatible moisture sensors mounted inside the pressurized cable housing.

Advantages are that if an existing cable can be made to handle more data, then the cost of installing additional cables can be eliminated. In cases where it is impossible to install additional cables, the use of such a humidity system is the only way that existing cables can be made to handle the increased demands of modern telecommunications technology.

The dew/frost point range to be covered is 0°C to −50°C (32°F to −58°F) at an operating temperature of 25°C to 30°C (77°F to 86°F) and pressure of 2.6 MPa to 3 MPa (377 psi_a to 435 psi_a).

G. Buildings and Construction

1. Moisture in Concrete Slabs

In the construction industry, measurement of moisture in concrete is very important. For example, the presence of excess moisture prior to applying a water proof coating over the concrete, results in the following problems:

a) The thermal effects of sealed-in moisture cause mechanical stresses which can result in fractures and subsequent loss of strength.

b) Due to the various salts and other chemicals, moisture has a corrosive effect on reenforcing steel used in most concrete structures. This can lead to a further loss of strength.

Moisture is often quantified as a percentage of water content by weight. This is not meaningful for concrete since the exact composition of the material is seldom known and a percentage moisture in one concrete mixture may differ significantly from another

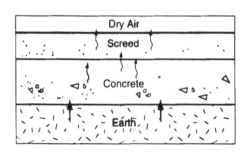 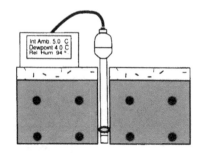

Figure 12.10   RH measurement in concrete slab.

which has the same amount of water in it. Percent moisture is therefore not a reliable guide as to whether the structure is too wet or dry.

For this reason, the technique of measuring relative humidity in a pocket of air which is in equilibrium with the material being tested, is widely used. The measurement thus comes down to an equilibrium relative humidity measurement. The normal procedure is to drill a small hole in the concrete slab and place an RH probe inside. The measurement is made after sufficient time has lapsed so that the water vapor in the small hole is in equilibrium with the concrete (see Figure 12.10).

In many cases typical polymer RH sensors have been used for this application. For precise measurements, such as for research purposes, a small CCM chilled mirror type probe is available, though at much higher cost.

2. Sick Building Syndrome

a. Definition

"Sick Building Syndrome" is not well defined. It can best be explained by the occurrence of several symptoms in the building, caused by poor indoor air quality. If three or more of these occur with a high degree of prevalence, one speaks of Sick Building Syndrome.

The most common symptoms are:

- Nausea

- Dizziness

- Headaches

- Skin irritation

- Other types of skin complaints

- Eye irritation and soreness

- Throat irritation and soreness

- Nasal irritation

- Sinus problems

The eye, nose, and throat problems usually relate to irritation and dryness of the mucus membranes of the body. These symptoms, as well as some skin problems, have a direct relationship to the humidity level. Headaches are also a common symptom in Sick Buildings. However this is more difficult to quantify since some individuals suffer periodic headaches regardless of the quality of the ambient air.

The following factors contribute to Sick Building Syndrome:

- *Climactic Environment:*
  Temperature, Humidity, Airflow, Radiant Heat

- *Chemical Environment:*
  Chemicals in the building or brought into the building by air intake

- *Micro Biological Contamination:*
  Symptoms are often allergically triggered

- *Psychological and Social Factors*

- *Physical Environment:*
  Lighting, Work Layout, and other Ergonomical Factors

- *Electrical Environment:*
  Electromagnetic Fields, Radiation, Ionization

Sick Building Syndrome does not exclusively occur in old buildings. Many modern buildings built in the 1980s and 1990s with modern air conditioning systems, have also exhibited these kinds of problems

b. Indoor Air Quality (IAQ)

Much emphasis has been given during the last two decades to air pollution, i.e., outdoor air quality. Indoor air quality (IAQ) has only in recent years been recognized as a problem and area of concern for public health. Outdoor pollutants vary from city to city across the world. Indoor pollutants do not vary much with different parts of the country and are mostly dependent on the building characteristics, number of people residing or working in the building, building air conditioning controls, ventilation, and other internal factors. In the future, indoor air quality is expected to receive more and more attention from building owners and government environmental control agencies since it is recognized that most people spend more time indoors than outside. Three basic steps are needed to maintain air quality:

1. Remove indoor sources of pollution, or reduce the rate of pollution emission.

2. Filter indoor air.

3. If outdoor air is clean, use effective ventilation with outdoor air. If not, use indoor recirculation incorporating effective air filtering systems.

Filtration is a most important factor. Recirculating air must be constantly cleaned. If outdoor air is brought in, it must be cleaned to remove fine particulates which are always present and which could possibly have a negative effect on life expectancy. Major concerns are:

- Fine particulates from inside the building or outdoors

- Microbial contaminants

- Fungi, which can appear in ceilings, carpeting, tiles, and elsewhere

- Standing water which is potentially colonized, causing bacteria to develop

When filters become colonized by microbes, ventilation may cause them to fly around and diffuse. This causes more building occupants to be exposed.

During the winter months, buildings often suffer from a lack of humidity, causing occupant discomfort. The optimum humidity for occupant health is 40% RH to 60% RH at room temperature. In the majority of office settings, humidity levels drop considerably during the winter months, sometimes to 10% to 15%, which is well below what is considered acceptable. The associated cost of installing humidifiers, humidity sensors, and control systems, can often be quickly and permanently offset by long-term productivity gains and employee/occupant satisfaction. In addition, appropriate humidity control provides protection against:

- Illness and absenteeism resulting from biological contaminants

- Respiratory problems

- Chemical interactions such as ozone production (indoor smog)

- Allergies, caused by microorganisms

- Asthma

- Respiratory infections

Illnesses like Humidifier Fever, Legionnaire's Disease, and Occupational Asthma are believed to be caused by deficiencies in indoor air quality. Good indoor air quality is also to a large degree related to comfort issues. Especially during the winter months, it is important to control vertical air temperature. When in-room temperatures vary greatly from floor to ceiling, or change suddenly during the day, occupants will notice such changes, even if the change is only a few degrees C. Air movement should be slow. If faster than about 14 m per min. (45 ft. per min.), it could cause discomfort.

c. Microbial Agents

Indoor air quality problems are historically related primarily to inadequate ventilation, and indoor or outdoor air contaminants. However, there is increasing recognition that microbial agents are of particular importance in the healthcare setting.

- **Fungi**
  Fungi are a diverse group of organisms, with over 70,000 species known to exist. They reproduce themselves by forming spores which, when released, are readily transported by wind and water. Fungal spores are extremely resistant to

environmental degradation and have been known to survive for up to ten years and remain viable. Fungi will grow and flourish wherever suitable environments are encountered, both in- and out-of-doors, and can be found virtually everywhere on the planet.

Given the proper combination of environmental conditions, fungi will grow almost anywhere. In order to flourish, fungi need food, moisture, and tolerable temperatures. In general, fungi like the same conditions that humans do, though there are species that grow under very extreme conditions. Of these environmental factors, moisture is the most important, as fungi generally only grow well under moist conditions. Food is abundantly available in most building environments.

- **Health effects of exposure to fungi**

Infection
Most fungi are saprophytic, which means that they feed on lifeless matter. A number are pathogens (capable of causing infection), and can be described as either opportunistic or non-opportunistic. Opportunistic fungal infections can occur in persons with reduced immune system activity, and therefore represent a significant threat in a hospital environment. Organ or tissue transplant patients whose immune systems are deliberately suppressed to reduce tissue rejection are particularly at risk. Patients with diabetes, those undergoing chemotherapy, and those with AIDS are also at increased risk.

Toxins
Fungi produce a variety of metabolites, some of which are toxins, which can be harmful to life forms. Some toxins are capable of causing great harm to humans, in the form of acute illness, permanent damage to body organs and the central nervous system, suppression of the immune system, cancer, and outright death.

Allergic Reaction
Fungi are well known for their ability to produce allergic reactions in humans. Symptoms range from moderate to extremely severe and include asthma, inflammatory reactions, congestion, tearing and itching eyes, and anaphylactic shock, which can be fatal. Fungal spores, whether viable or not, can trigger allergic reactions.

Volatile Organic Compounds (VOCs)
Fungi can produce numerous volatile organic compounds (VOCs). Some of these are responsible for the musty or earthy odors associated with damp buildings.

- **Other Infection Agents**
Legionella, known for causing Legionnaire's disease continues to appear as a building related illness. Patients are potentially at increased risk due to existing illness and compromised immune systems.

Tuberculosis continues to be an issue for healthcare facilities, because of the increased incidence of infection in the general population and attention from regulatory agencies.

- **Increased Litigation**

  In the US, patients and employees, and probably society in general, seem to become increasingly litigious. Suits related to indoor air quality are becoming more common, though the number of cases which have been decided by a jury through the early 1990s is apparently still fairly small. A trial jury in the US awarded five plaintiffs $200,000 each in a well publicized case involving the U.S. Environmental Protection Agency headquarters building in Washington, D.C. Numerous cases against employers, building owners, architects, engineers, and construction contractors, are appearing and are being settled before trial, for large sums of money. Litigation is expected to increase in the latter part of the 1990s.

d. Typical IAQ Measurements

One IAQ measurement system was recently developed by AIRxpert Systems of Lexington, Massachusetts, which is capable of monitoring dew point and CO_2 in many parts of a large industrial building. Typical curves are shown in Figures 12.11 and 12.12. It is evident that these conditions vary widely with time of the day and occupancy of the rooms inside the building. The system can be used to activate ventilation and humidity control.

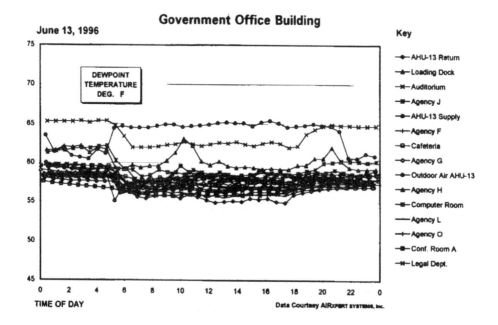

Figure 12.11 Typical IAQ chart for US government building. (*Courtesy AIRxpert Systems, Inc.*)

Figure 12.11 is an IAQ Chart for a 71,500 m² (770,000 ft²) government office build-ing in a US metropolitan area, built in 1985. The building is fully equipped with a sophis-ticated energy management system. It does not have humidification capability. What makes this an interesting graph is that one can see on a dynamic basis the effectiveness of cooling humid air on a hot day in June. The chiller system came on line at approximately 5:30 AM and ran until 5:30 PM (17.30), effectively lowering the dew points in all but one of the zones by about 4.5°C (8° F). The exception was the auditorium. One can see that there was a great deal of infiltration of humid outside air all day long in that room. This is the result of a ventilation strategy that keeps the room's dedicated air handler operating constantly in anticipation of events that happen on a random basis. As a result of seeing this data, the building manager could consider putting in an occupancy sensor to drive the air handler. Note also how one can see that the door to the loading dock was opened at 10:15 AM this particular morning, and remained open for a half hour.

The 14,400 m² (155,000 ft²) suburban office building discussed in Figure 12.12 is located West of Boston, Massachusetts in a rural setting. Again, the effect of air con-ditioning can be clearly seen in all zones. Note, however, that the two outside air readings show different dew points. This is because one is at ground level, and one is on the roof, away from the effect of moisture in the grass and earth. In this situation, the loading dock was not effectively sealed from outside air, and thus allowed moist outside air to infiltrate all day long. While this may not seem important, it nevertheless adds to the load on the chiller because the moist air from the loading dock must be dried out as it recirculates.

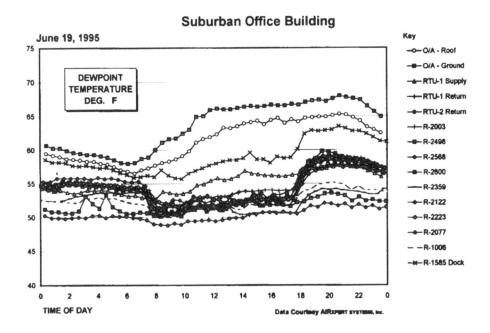

Figure 12.12   Typical IAQ chart for US suburban office building.
(Courtesy AIRxpert Systems, Inc.)

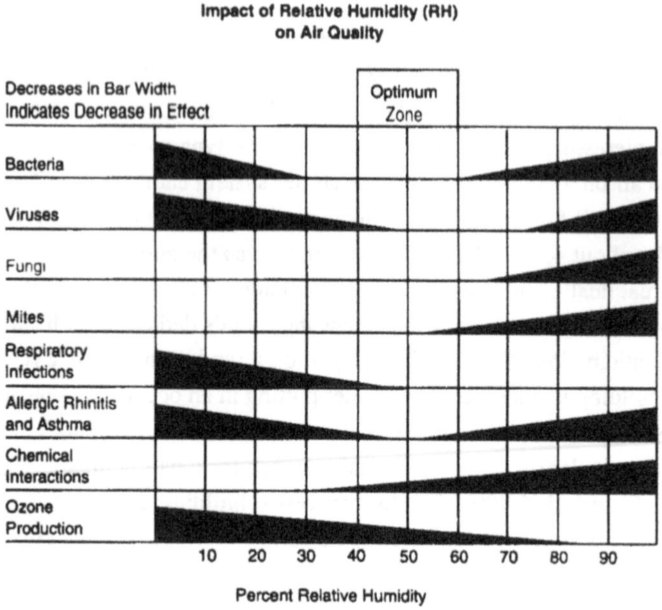

Figure 12.13 Impact of relative humidity on air quality.

e. Impact of Humidity

The impact of relative humidity on potential health hazards is shown schematically in Figure 12.13.

Humidity sensors that are commonly used for indoor air conditioning and energy management are primarily polymer resistive and capacitive sensors. Because of the importance of indoor air quality, which involves, in addition to humidity, temperature and ventilation control, the measurement of CO_2, filtering of particulates, and elimination of other health hazards, the chilled mirror transmitter may become the sensor of choice in many future systems. The CCM transmitter with self cleaning feature offers potential advantages because of its low maintenance requirements.

H. Relative Humidity Measurements

1. Measuring Relative Humidity Near Saturation

Most humidity sensors can be used up to about 95% relative humidity, but do not measure accurately at conditions approaching 100% RH.

At 100% RH, the system under test is "saturated," i.e, the atmosphere cannot hold any more moisture in the gaseous phase and liquid water starts to condense out. Moisture condensation on an RH sensing element causes most sensor types to fail and causes "flooding" in chilled mirror sensors.

Some manufacturers have developed RH sensors that are reported to be resistant to flooding without loosing their calibration. Reports by users of these sensors have not been consistent, but some capacitive polymer sensors have been successfully used at 100% RH. Significant progress has been made in the manufacture of RH sensors, and further advances continue to be made.

RH measurements are temperature dependent. Therefore, it is impractical, and inaccuracies will result, if the sensor temperature is raised well above ambient and the RH measurement must then be recomputed for the actual temperature in which the sensor is operated.

A chilled mirror dew point sensor can be elevated in temperature to prevent "flooding" and will provide a dew point measurement which is independent of the air temperature. When the ambient temperature is known, the %RH can be accurately computed. However, the cost of such a system is much higher.

Another solution is the use of a saturated salt (lithium chloride) sensor, which is inexpensive and which can be used to 100% relative humidity.

Manufacturers of lead acid car batteries need an environment to cure battery plates prior to assembly. These plates must be exposed to a condition of 32°C (90°F) and approximately 90% RH to 95% RH. Quite often employees bring in new plates and take out the cured ones. Each time the door opens there is an influx of cool air from the conditioned space outside the curing room. This causes some condensation to occur. A lithium chloride transmitter can be located outside the curing room and a saturated salt sensor installed inside in a representative location. Ambient temperature probes must also be installed.

The lithium chloride saturated salt hygrometer offers advantages because of its mode of operation and low cost per point. Lithium chloride has a characteristic equilibrium RH of 11.5%. This means that the sensor seeks to attain an equilibrium temperature corresponding to a relative humidity of 11.5% (see Chapter 7) in the air immediately surrounding it. It will take on, or dissipate, water from the lithium chloride until this condition is satisfied.

2. High Temperature Humidity Measurements

Many products are affected by temperature and humidity during the manufacturing process. Curing of ceramics, heat treating of various materials, and drying of wet products, all require measurement of these parameters. Since the temperatures may be as high as 820°C (1500°F), the dew point sensor cannot be located in the process, and a sampling system must be installed.

The purpose of the sampling system is to bring a small sample of the gas which is to be monitored to a measuring device that is external to the hot area containing the gas. By drawing a sample through a length of tubing, the gas temperature is quickly reduced to a level close to the ambient temperature surrounding the tubing. (The tubing acts as a heat exchanger with the ambient air.) The dew point remains constant in the tubing because it is independent of temperature. As a result, it can now be measured accurately without sensor overheating and subsequent damage.

In many cases, especially at the beginning of a drying cycle when the product is saturated, the dew point in the sample line is higher than the ambient temperature outside the sampling system. To obtain reliable measurements when dew points are higher than the ambient temperature, the sample line and sensor must be heated. If the surface temperature of the inner walls is above the dew point temperature, no condensation will occur and accurate measurements can be made up to the maximum temperature rating of the sensor, usually close to 95°C (200°F).

A heated sampling system may be installed by a user to continuously monitor the dew point at the output of the high temperature oven. Since the system is normally capable of operation to about 95°C (200°F), a potential condensation problem is eliminated if the maximum dew point is for instance 80°C (175°F).

In summary, a sampling system is a very powerful tool. It is required for sample temperature reduction in the above example, but it is also recommended for optimum control even when the temperature is not excessive. A sampling system offers these benefits:

- Optimizes the flow rate for a particular sensor type

- Mounts the sensor at a convenient location for cleaning

- Allows a filter to be located upstream to eliminate dust and dirt

In addition, the availability of heated sampling systems allows dew point monitoring up to 95°C (203°F). Many measurements can thus be made in difficult locations with a high degree of reliability and confidence. A more detailed description of sampling systems is presented in Chapter 3.

3. Energy Management

Several industrial and commercial processes require large amounts of energy. Energy requirements are in almost all cases very dependent on humidity and temperature control and large energy savings can be obtained by optimizing such controls, which at today's state of technology can easily be done using dedicated computers. Examples are large industrial dryers, air conditioning systems in large buildings, and climate control in museums and libraries.

These applications normally require a large number of sensors mounted at various well-selected locations. These provide data to computers, which are then programmed to operate humidity and temperature controls in the building or dryer. There are many examples where such energy management systems have yielded substantial energy savings, allowing such energy management systems, including humidity and temperature sensors, to be amortized within two years. Transmitters using polymer %RH sensors are ideal for this application. Operational requirements are usually 20% RH to 70% RH at ambient temperatures of 0°C to 60°C (32°F to 140°F).

I. Plant Growth Chambers

The growth and development of plants and food crops are directly related to light, temperature, humidity and soil conditions. Agricultural and chemical industries utilize plant growth chambers to simulate and optimize these conditions and to test the effects of fertilizers and insecticides on plant development. These chambers are also used in hybridization and growth medium experiments.

Typically, a research facility will consist of many of these chambers, so instrumentation to be specified should be of low cost and high reliability. Some manufacturers of fertilizers and insecticides have built several plant growth chambers.

A duct-mounted chilled mirror or RH transmitter may be used in each chamber. The

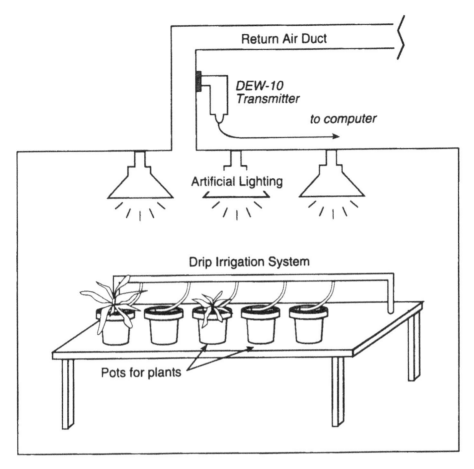

Figure 12.14   Dew point measurement in plant growth chamber.

outputs can be fed to a computer control and data-logging system. Within these chambers, frames can be used to hold the growth media and plants. Irrigation systems can be installed to supply metered amounts of water and nutrients to the plants.

Agriculturists are faced with many problems. Rising fuel and real estate costs, as well as the unpredictability of weather elements, are just a few of their concerns. Optimization of plant growth parameters not only increases the yield but also helps develop strains of crops that are heartier and more resistant to disease and environmental fluctuations.

J. Waste Products

Some power stations have been adapted to make use of the waste created by intensive poultry and animal farming, such as animal droppings and garbage, by burning this waste for energy. However, correct moisture levels need to be present, i.e., waste products must be sufficiently dried out prior to burning to maintain efficiency. While this application is still in its infancy, a simple water activity system is being developed for this purpose using capacitive and resistive-type RH sensors. The result of Aw is purely for compliance testing.

13

CHARTS, GRAPHS AND TABLES

I. General Discussion

This chapter includes a number of tables, charts, and graphs for converting from one humidity parameter to another, for determining pressure effects, and making pressure corrections. Airflow conversion charts are also included.

In the back cover pocket of this book a computer disk is provided which is a convenient and easy tool for making accurate conversions, using a DOS type computer. The program is presented in Windows 95.

The moisture content of air may be expressed in many ways. Dew point temperature is one. The accompanying charts provide a quick comparison of several of the frequently used parameters. Other scales which are in effect "absolute measurements," must be corrected to account for pressures other than atmospheric.

The charts, graphs, and tables were obtained from several sources, including some from General Eastern's Humidity Handbook (Reference 4, Section II).

The maximum amount of true water vapor that a unit volume of a permanent gas can hold is a function of the vapor pressure of liquid water. The holding capacity increases very slowly with an increase in pressure so that, for all practical purposes, pressures below about 2 MPa (300 psi_g) do not affect the holding capacity. Above this level, corrections for pressure should be considered.

Temperature is the most important factor. For this reason the concept of dew point temperature is a convenient means for measuring the moisture content of gases. The "dew point" of a gas is that temperature to which it must be cooled to just start condensation of moisture. This is also the "saturation temperature."

In actual practice a gas may occasionally carry, in addition to true water vapor, entrained water droplets which in reality represent the liquid state. The amount of entrainment is indeterminant on the basis of temperature and pressure, depending rather upon the vagaries of droplet size, velocity, presence of dust, and other factors. As a rule, entrainment as such can be removed by mechanical means although the design of such equipment does, in some instances, become difficult.

There are several methods of measuring dew point temperatures. The most direct method is based on the condensation principle (see Chapter 3). Other methods indirectly related to dew point include the cooling effect of adiabatic expansion, gravimetric methods, such as absorption in phosphorous pentoxide or freeze-out at very low temperatures, and electrical devices which combine the absorption power of a chemical agent such as phosphorous pentoxide with determination of the amount of water by an electric current.

The charts on the following pages provide a quick comparison of several of the other frequently used methods, considering air saturated with water vapor at one atmosphere pressure. By reading straight across one can find the equivalent moisture contents. The relationship of dew point or saturation temperature to grams of moisture per m^3 or grains of moisture per actual cubic foot of air at an elevated pressure does not change much with pressure, as previously indicated, in the range of 0 MPa to 2.2 MPa (0 psi_g to 300 psi_g). Consequently, grams per m^3 or grains per cubic foot for elevated pressures can also be read directly from the charts and tables. Dew points referred to all other scales, which are in effect "absolute measurements," must be corrected for pressures other than atmospheric.

The moisture capacity per actual unit volume of air at pressures above 7 MPa (1000 psi_g) is considerably greater than at comparable temperatures at lower pressure. This is a direct consequence of the known principle that the vapor pressure of liquid water at constant temperature increases as the pressure impressed on the liquid by an inert gas increases. In the range of 7 MPa to 42 MPa (1000 psi_g to 6000 psi_g) this effect is quite noticeable. The indicated graphs take account of the effect of high pressure.

Figure 13.15 indicates saturated moisture values for air from 1000 psi_g to 5000 psi_g (7 MPa to 35 MPa) for the temperature range 0°F to 150°F (−17.8°C to 66°C). For the low range of −100 °F to 0°F (−73.3 °C to −17.8°C), refer to Figure 13.14. Having determined the absolute water content (pounds per million standard ft.^{3}), Figures 13.10 and 13.11 show the corresponding atmospheric dew point versus absolute moisture content in other units.

II. Psychrometric Charts

Fig. 13.1   ASHRAE psychrometric chart — **metric units.**
Fig. 13.2   Psychrometric table — **metric units.**

III. Relative Humidity Conversions

Fig.13.3   %Relative humidity versus dew point temperature.
Table 13.1   %Relative humidity at certain temperatures and dew points.
Fig.13.4   %Relative humidity versus absolute humidity from **50°C to 350°C** at 1 bar pressure.
Fig.13.5   %Relative humidity versus absolute humidity from **100°F to 700°F** at atmospheric pressure.
Fig.13.6   %Relative humidity versus wet-and dry-bulb difference at **20°F to 198°F** and pressure of 29.921 in. Hg.
Fig.13.7   %Relative humidity versus wet-and dry-bulb difference at **0°F to 80°F** and pressure of 29.921 in. Hg.
Fig.13.8   %Relative humidity versus wet-and dry-bulb difference at **−50°F to 0°F** and pressure of 29.921 in. Hg.
Table 13.2   %Relative humidity at certain temperatures versus dry/wet bulb depression, **in °C.**

Table 13.3   %Relative humidity at certain temperatures versus dry/wet bulb
                depression, in °F.

IV. Dew Point Conversions

Table 13.4   Dew point – frost point conversions in °C.
Table 13.5   Dew point – frost point conversions in °F.
Table 13.6   Dew point relationships versus common humidity parameters.
Table 13.7   Dew point versus vapor pressure, PPM_v, relative humidity, and
                PPM_w from –150°C (–238°F) to 60°C (140°F).

V. Moisture Content Tables and Charts

Table 13.8   Moisture content, vapor pressure, and dew point — **metric units.**
Table 13.9   Moisture content, vapor pressure, and dew point — **english units.**
Fig.13.9   Absolute humidity (PPM_v) versus dew point at different pressures.
Fig.13.10   Moisture content of air and perfect gases at atmospheric pressure,
                english units.
Table 13.10   Moisture content of dry air — **metric units.**
Table 13.11   Moisture content of dry air — **english units.**
Table 13.12   Mass of water vapor (gr/m3) of saturated air at 101.3 Pa.

VI. Pressure Conversions

Fig.13.11   Dew point–pressure–PPM_v conversion chart.
Fig.13.12   Dew point pressure conversions at 0-400 psi_g and –160 to +100°F.
Fig.13.13   Dew point pressure conversions at 0-500 psi_g and –60 to +160°F.
Fig.13.14   Saturated water content of high pressure air at **low** temperatures —
                english units.
Fig.13.15   Saturated water content of high pressure air at **moderate**
                temperatures — **english units.**
Table 13.13   Vacuum conversion table.
Table 13.14   Altitude pressure table.
Table 13.15   Pressure conversion factors.
Table 13.16   Detailed pressure conversion chart.

VII. Flow Conversions

Fig.13.16   Gas flow readings corrected for pressure.
Fig.13.17   Equivalent gas flow corrected for specific gravity.
Table 13.17   Flow equivalents.

VIII. Unit Conversions

Table 13.18   °C–°F conversions.
Table 13.19   Metric–english unit conversions.
Table 13.20   Unit conversion factors.

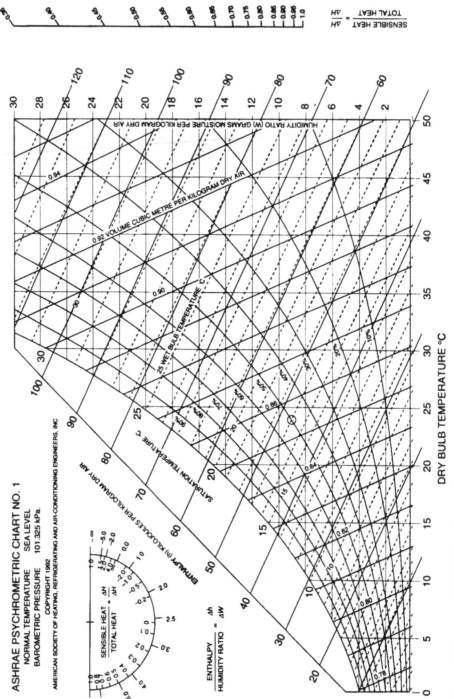

Figure 13.1   ASHRAE Psychrometric chart — metric units.

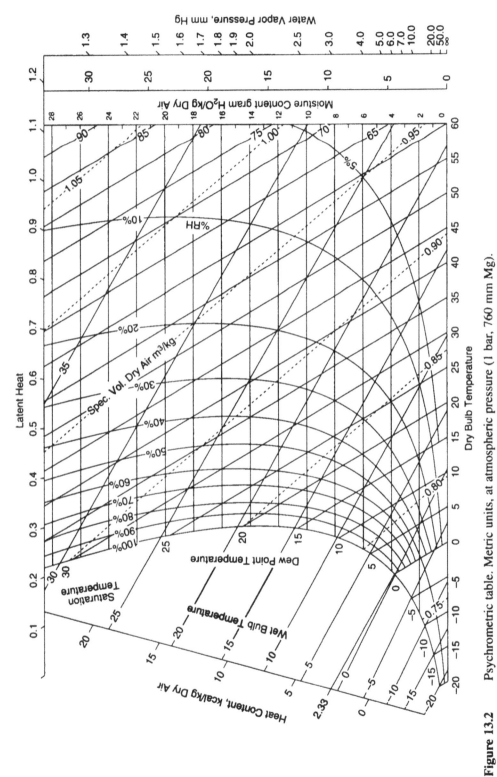

Figure 13.2 Psychrometric table. Metric units, at atmospheric pressure (1 bar, 760 mm Mg).

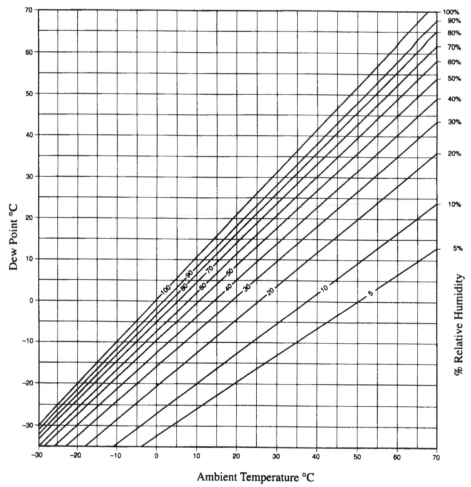

Figure 13.3   % Relative humidity versus dew point and temperature.

Table 13.1   % Relative humidity at certain temperatures and dew points.

Dew point °C	Temperature °C														
	0	5	10	15	20	25	30	35	40	50	60	70	80	90	100
	Relative humidity (%)														
0	100.0	70.1	49.8	35.8	26.1	19.3	14.4	10.9	8.3	4.9	3.1	2.0	1.3	<1	<1
5	-	100.0	71.1	51.1	37.3	27.5	20.5	15.5	11.8	7.1	4.4	2.8	1.8	1.2	<1
10	-	-	100.0	72.0	52.5	38.7	28.9	21.8	16.6	9.9	6.2	3.9	2.6	1.7	1.2
15	-	-	-	100.0	72.9	53.8	40.2	30.3	23.1	13.8	8.6	5.5	3.6	2.4	1.7
20	-	-	-	-	100.0	73.8	55.1	41.6	31.7	18.9	11.7	7.5	4.9	3.3	2.3
25	-	-	-	-	-	100.0	74.6	56.3	42.9	25.7	15.9	10.2	6.7	4.5	3.1
30	-	-	-	-	-	-	100.0	75.4	57.5	34.3	21.3	13.7	9.0	6.1	4.2
35	-	-	-	-	-	-	-	100.0	76.2	45.6	28.2	18.0	11.9	8.0	5.6
40	-	-	-	-	-	-	-	-	100.0	59.8	37.0	23.7	15.6	10.5	7.3
50	-	-	-	-	-	-	-	-	-	100.0	61.9	39.6	26.1	17.6	12.2
60	-	-	-	-	-	-	-	-	-	-	100.0	63.9	42.1	28.4	19.7
70	-	-	-	-	-	-	-	-	-	-	-	100.0	65.8	44.5	30.8
80	-	-	-	-	-	-	-	-	-	-	-	-	100.0	67.6	46.8
90	-	-	-	-	-	-	-	-	-	-	-	-	-	100.0	69.2

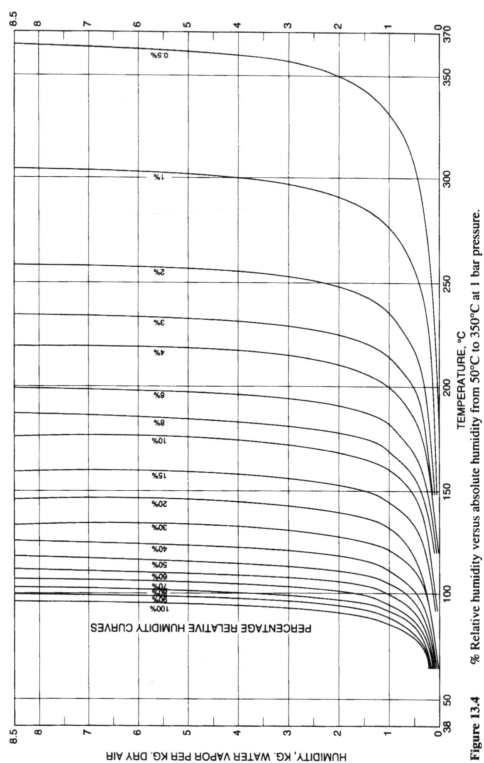

Figure 13.4   % Relative humidity versus absolute humidity from 50°C to 350°C at 1 bar pressure.

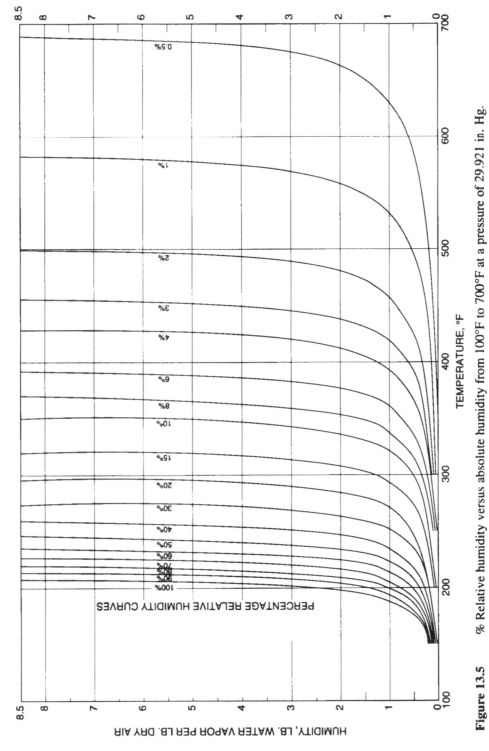

Figure 13.5   % Relative humidity versus absolute humidity from 100°F to 700°F at a pressure of 29.921 in. Hg.

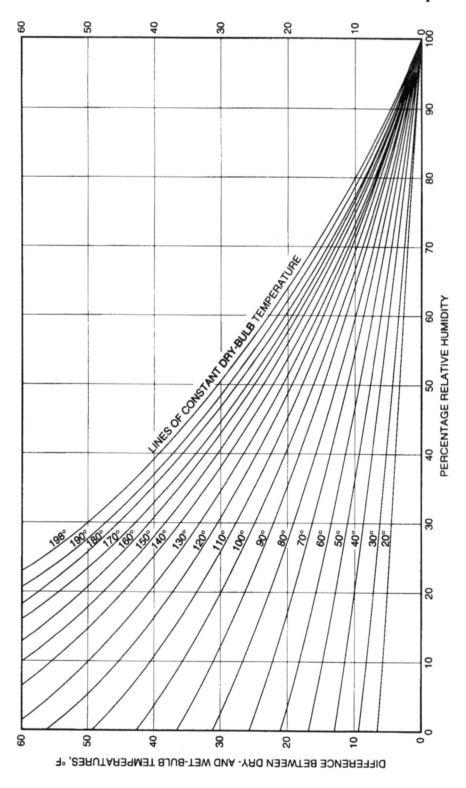

Figure 13.6   % Relative humidity versus wet- and dry-bulb temperature difference at 20°F to 198°F and pressure of 29.921 in. Hg.

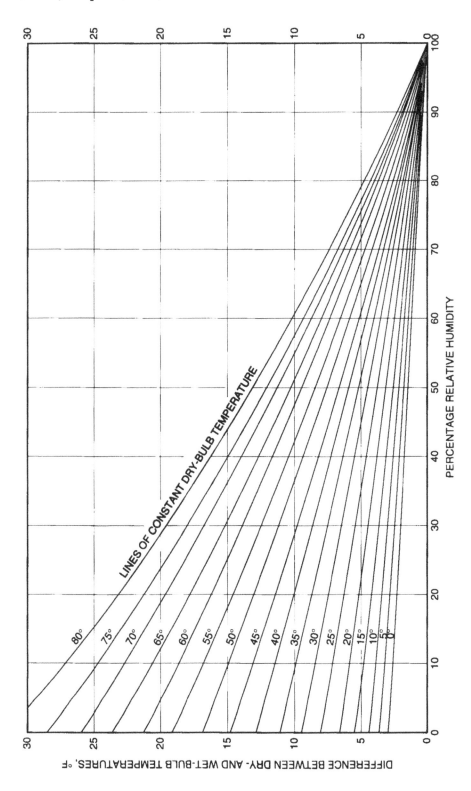

Figure 13.7   % Relative humidity versus wet- and dry-bulb temperature difference at 0°F to 80°F and pressure of 29.921 in. Hg.

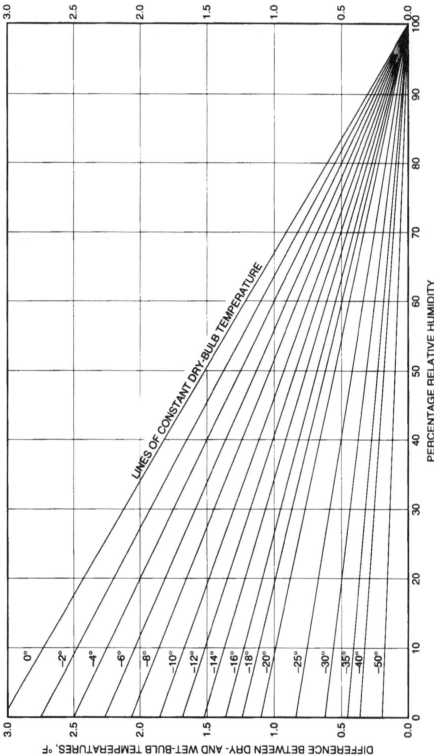

Figure 13.8   % Relative humidity versus wet- and dry-bulb temperature difference at –50°F to 0°F and pressure of 29.921 in. Hg.

Table 13.2   % Relative humidity at certain temperatures versus dry/wet bulb depression, in °C.

dry bulb deg C.	0.5	1.0	1.5	2.0	2.5	3.0	3.5	4.0	4.5	5	6	7	8	9	10	11	12	13	14	15	16	18	20	22	24	26	28	30	32	34	36	38	40
2	92	83	75	67	59	52	43	36	27	20																							
4	93	85	77	70	63	56	48	41	34	28	15																						
6	94	87	80	73	66	60	54	47	41	35	23	11																					
8	94	87	81	74	68	62	56	50	45	39	28	17																					
10	94	88	82	76	71	65	60	54	49	44	34	23	14																				
12	94	89	84	78	73	68	63	58	53	48	38	30	21	12	4																		
14	95	90	84	79	74	69	65	60	55	51	41	33	24	16	10																		
16	95	90	85	81	76	71	67	62	58	54	45	37	29	21	14	7																	
18	95	90	86	82	78	73	69	65	61	57	49	42	35	27	20	13	6																
20	96	91	87	82	78	74	70	66	62	58	51	44	36	30	23	17	11																
22	96	92	87	83	79	75	72	68	64	60	53	46	40	34	27	21	16	11															
24	96	92	88	85	81	77	74	70	66	63	56	49	43	37	31	26	21	14	10														
26	96	92	89	85	81	77	74	71	67	64	57	51	45	39	34	28	23	18	13														
28	96	92	89	85	82	78	75	72	68	65	59	53	47	42	37	31	26	21	17	13													
30	96	93	89	86	82	79	76	73	70	67	61	55	50	44	39	35	30	24	20	16	12												
32	96	93	90	86	83	80	77	74	71	68	62	56	51	46	41	36	32	27	23	19	15												
34	97	93	90	87	84	81	77	74	71	69	63	58	53	48	43	38	34	30	26	22	18	10											
36	97	93	90	87	84	81	78	75	72	70	64	59	54	50	45	41	36	32	28	24	21	13											
38	97	94	90	87	84	81	79	76	73	70	65	60	56	51	46	42	38	34	30	26	23	16	10										
40	97	94	91	88	85	82	79	76	74	71	66	61	57	52	48	44	40	36	32	29	25	19	13										
42	97	94	91	88	85	82	80	77	74	72	67	62	58	53	49	45	41	38	34	31	27	21	15										
44	97	94	91	88	86	83	80	77	75	73	68	63	59	54	50	47	43	39	36	32	29	23	17	12									
46	97	94	91	89	86	83	81	78	76	73	68	64	60	55	52	48	44	41	37	34	31	25	19	14									
48	97	94	92	89	86	84	81	78	76	74	69	65	61	56	53	49	45	42	39	35	33	27	21	16	12								
50	97	94	92	89	87	84	82	79	77	75	70	65	62	57	54	50	47	43	40	37	34	28	23	18	14								
52	97	94	92	89	87	84	82	79	77	75	70	66	62	58	55	51	48	44	41	38	35	30	25	20	16	11							
54	97	95	92	90	87	85	82	80	78	76	71	67	63	59	56	52	49	45	42	39	36	31	26	21	17	13							
56	97	95	92	90	87	85	83	80	78	76	72	68	64	60	57	53	50	46	43	40	38	32	27	23	19	15	11						
58	97	95	93	90	88	85	83	80	79	77	72	68	64	61	57	54	51	47	44	42	39	33	29	24	20	16	12						
60	97	95	93	90	88	86	83	81	79	77	73	69	65	62	58	55	52	48	45	43	40	35	30	26	21	18	14	11					
62	98	95	93	91	88	86	84	81	79	78	73	69	66	62	59	56	53	49	46	43	41	36	31	27	23	19	15	12					
64	98	95	93	91	88	86	84	82	80	78	74	70	66	63	59	56	53	50	47	44	42	37	32	28	24	20	17	13					
66	98	95	93	91	89	86	84	82	80	78	74	70	67	64	60	57	54	51	48	45	43	38	33	29	25	21	18	15	12				
68	98	95	93	91	89	87	85	83	81	79	75	71	68	65	61	58	55	52	49	46	44	39	34	30	26	22	19	16	13				
70	98	96	93	91	89	87	85	83	81	79	75	71	68	65	61	58	55	52	50	47	44	40	35	31	27	23	20	17	14	11			
72	98	96	94	92	89	87	85	83	81	80	76	72	69	65	62	59	56	53	50	48	45	40	36	32	28	24	21	18	15	12			
74	98	96	94	92	90	87	85	83	82	80	76	72	69	66	63	60	57	54	51	48	46	41	37	33	29	25	22	19	16	13	11		
76	98	96	94	92	90	88	86	84	82	80	76	73	70	66	63	60	57	54	52	49	47	42	38	34	30	26	23	20	17	14	12		
78	98	96	94	92	90	88	86	84	82	81	77	73	70	67	64	61	58	55	52	50	47	43	38	34	30	27	24	21	18	15	13	10	
80	98	96	94	92	90	88	86	84	83	81	77	74	71	67	64	61	58	56	53	50	48	43	39	35	31	28	24	22	19	16	14	11	
82	98	96	94	92	90	88	86	84	83	81	77	74	71	68	65	62	59	56	54	51	49	44	40	36	32	29	25	22	20	17	15	12	10
84	98	96	94	92	90	88	86	85	83	81	78	74	71	68	65	62	59	57	54	52	49	45	40	37	33	29	26	23	20	18	16	13	11
86	98	96	94	92	91	89	87	85	83	82	78	75	72	69	66	63	60	57	55	52	50	45	41	37	34	30	27	24	21	19	16	14	12
88	98	96	95	93	91	89	87	85	83	82	78	75	72	69	66	63	60	58	55	53	51	46	42	38	34	31	28	25	22	19	17	15	13
90	98	97	95	93	91	89	87	85	84	82	79	76	73	69	67	64	61	58	56	53	51	47	42	39	35	32	28	26	23	20	18	16	14
92	98	97	95	93	91	89	87	86	84	82	79	76	73	70	67	64	61	59	56	54	52	47	43	39	36	32	29	26	24	21	19	16	14
94	99	97	95	93	91	89	88	86	84	83	79	76	73	70	67	65	62	59	57	54	52	48	44	40	36	33	30	27	24	22	19	17	15
96	99	97	95	93	91	90	88	86	84	83	80	76	74	70	68	65	62	60	57	55	53	48	44	41	37	34	31	28	25	22	20	18	16
98	99	97	95	93	92	90	88	86	85	83	80	77	74	71	68	65	63	60	58	55	53	49	45	41	38	34	31	28	26	23	21	19	16
100	99	97	95	93	92	90	88	86	85	83	80	77	74	71	68	66	63	60	58	56	54	49	45	42	38	35	32	29	26	24	22	19	17

Difference between readings of wet and dry bulbs in degrees centigrade

Table 13.3   % Relative humidity at certain temperatures versus dry/wet bulb depression, in °F.

dry bulb deg F.	Difference between readings of wet and dry bulbs in degrees fahrenheit																																		
	1	2	3	4	5	6	7	8	9	10	11	12	13	14	15	16	17	18	19	20	22	24	26	28	30	32	34	36	38	40	45	50	55	60	70
30	89	78	67	56	46	36	16	6	0	0																									
35	91	81	72	63	54	45	36	27	19	10	2	0																							
40	92	83	75	68	60	52	45	37	29	22	15	7	0	0																					
45	93	86	78	71	64	57	51	44	38	31	25	18	12	6	0	0																			
50	93	87	80	74	67	61	55	49	43	38	32	27	21	16	10	5	0	0																	
55	94	88	82	76	70	65	59	54	49	43	38	33	28	23	19	14	9	5	0	0															
60	94	89	83	78	73	68	63	58	53	48	43	39	34	30	26	21	17	13	9	5	0														
65	95	90	85	80	75	70	66	61	56	52	48	44	39	35	31	27	24	20	16	12	5	0													
70	95	90	86	81	77	72	68	64	59	55	51	48	44	40	36	33	29	25	22	19	12	6	0	0											
75	96	91	86	82	78	74	70	66	62	58	54	51	47	44	40	37	34	30	27	24	18	12	7	1	0										
80	96	91	87	83	79	75	72	68	64	61	57	54	50	47	44	41	38	35	32	29	23	18	12	7	3	0									
85	96	92	88	84	80	76	73	70	66	63	59	56	53	50	47	44	41	38	35	32	27	22	17	13	8	4	0	0							
90	96	92	89	85	81	78	74	71	68	65	61	58	55	52	49	47	44	41	39	36	31	26	22	17	13	9	5	1	0						
95	96	93	89	85	82	79	75	72	69	66	63	60	57	54	52	49	46	43	42	38	34	30	25	21	17	13	9	6	2	0					
100	96	93	89	88	83	80	77	73	70	68	65	62	59	56	54	51	49	46	44	41	37	33	28	24	21	17	13	10	7	4					
102	96	93	89	86	83	80	77	74	71	69	65	62	59	57	54	52	49	47	45	43	38	34	30	26											
104	96	93	90	86	83	80	77	74	71	69	65	63	60	58	55	52	50	48	46	44	39	35	31	27	23										
106	96	93	90	87	83	80	77	74	72	69	66	63	60	58	56	53	51	48	46	44	40	36	32	28	24	21									
108	96	93	90	87	84	81	78	75	72	70	66	64	61	59	56	54	51	49	47	45	41	37	33	29	26	22	19								
110	96	93	90	87	84	81	78	75	72	70	67	64	62	60	57	55	52	50	48	46	41	37	34	30	27	23	20	17							
112	96	93	90	87	84	81	78	75	73	70	67	65	62	60	57	55	53	51	49	47	42	38	35	31	28	24	21	18	15						
114	97	93	90	87	84	81	78	75	73	71	68	65	63	61	58	56	53	51	49	47	43	39	35	32	28	25	22	19	16	13					
116	97	93	90	88	84	82	79	76	74	71	68	66	63	61	59	56	54	52	50	48	44	40	36	33	29	26	24	20	17	14					
118	97	93	91	88	85	82	79	76	74	71	68	66	64	62	59	57	54	53	51	49	44	41	37	34	30	27	24	21	19	15					
120	97	94	91	88	85	82	79	77	74	72	69	66	64	62	60	57	55	53	51	49	45	41	38	34	31	28	25	22	19	16	10				
122	97	94	91	88	85	82	79	77	75	72	69	67	65	63	60	58	56	54	52	50	46	42	38	35	32	29	26	23	20	17	12				
124	97	94	91	88	85	83	80	77	75	72	70	67	65	63	61	58	56	54	52	51	46	43	39	36	33	29	27	24	21	18	13				
126	97	94	91	88	86	83	80	78	75	73	70	68	65	63	61	59	57	55	53	51	47	43	40	37	33	30	28	25	22	19	14				
128	97	94	91	89	86	83	80	78	76	73	71	68	66	64	61	59	57	55	53	52	47	44	40	37	34	31	28	25	23	20	15				
130	97	94	91	89	86	83	80	78	76	73	71	68	66	64	62	60	58	56	54	52	48	44	41	38	35	32	29	26	24	21	15	10			
132	97	94	92	89	86	83	81	78	76	74	71	69	67	65	62	60	58	56	54	53	49	45	42	39	35	32	30	27	24	22	16	11			
134	97	94	92	89	86	84	81	79	76	74	71	69	67	65	63	61	59	57	55	53	49	46	42	39	36	33	31	28	25	23	17	12			
136	97	94	92	89	86	84	81	79	77	74	72	69	67	65	63	61	59	57	55	53	50	46	43	40	37	34	31	28	26	24	18	13			
138	97	94	92	89	86	84	81	79	77	74	72	70	68	66	63	62	60	58	56	54	50	47	43	40	37	35	32	29	27	24	19	14			
140	97	94	92	89	87	84	81	79	77	75	72	70	68	66	64	62	60	58	56	54	51	47	44	41	38	35	33	30	27	25	19	14	10		
142	97	94	92	89	87	84	82	80	77	75	73	70	68	66	64	62	60	58	57	55	51	48	44	42	39	36	33	30	28	26	20	15	11		
144	97	95	92	89	87	84	82	80	78	75	73	71	69	67	65	63	61	59	57	55	52	48	45	42	39	36	34	31	29	26	21	16	11		
146	97	95	92	90	87	85	82	80	78	75	73	71	69	67	65	63	61	60	58	56	53	49	45	43	40	37	35	32	29	27	21	17	12		
148	97	95	92	90	87	85	82	80	78	76	73	71	69	67	65	63	61	60	58	56	53	49	46	43	40	38	35	33	30	28	22	17	13		
150	98	95	92	90	87	85	82	80	78	76	74	72	70	68	66	64	62	60	58	57	53	49	46	43	41	38	36	33	30	28	23	18	13		
152	98	95	93	90	88	85	83	81	79	76	74	72	70	68	66	64	62	60	59	57	53	50	47	44	41	39	36	33	31	29	23	19	14	10	
154	98	95	93	90	88	85	83	81	79	77	74	72	70	68	66	65	63	61	59	57	54	50	47	44	42	39	37	34	32	29	24	19	15	11	
156	98	95	93	90	88	85	83	81	79	77	74	72	71	69	67	65	63	61	59	58	54	51	48	45	42	40	37	34	32	30	24	20	15	11	
158	98	95	93	90	88	86	83	81	79	77	75	73	71	69	67	65	63	61	60	58	55	51	48	45	43	40	38	35	33	30	25	20	16	12	
160	98	95	93	90	88	86	83	81	79	77	75	73	71	69	67	65	64	62	60	58	55	52	49	46	43	41	38	35	33	31	25	21	17	13	
162	98	95	93	90	88	86	84	82	80	77	75	73	71	69	68	66	64	62	60	59	55	52	49	46	44	41	39	36	34	31	26	22	17	13	
164	98	95	93	91	88	86	84	82	80	78	75	73	72	70	68	66	64	62	61	59	56	52	49	47	44	41	39	36	34	32	26	22	18	14	
166	98	95	93	91	88	86	84	82	80	78	76	74	72	70	68	66	65	63	61	59	56	53	50	47	44	42	39	37	35	32	27	23	18	14	
168	98	95	93	91	88	86	84	82	80	78	76	74	72	70	68	67	65	63	61	60	56	53	50	47	45	42	40	37	35	33	28	23	19	15	
170	98	95	93	91	89	86	84	82	80	78	76	74	72	70	69	67	65	63	62	60	57	53	51	48	45	43	40	38	35	33	28	24	19	15	
172	98	95	93	91	89	86	84	82	81	78	76	74	73	71	69	67	66	64	62	60	57	54	51	48	46	43	41	38	36	34	28	24	20	16	
174	98	95	93	91	89	87	84	83	81	78	76	75	73	71	69	67	66	64	62	61	57	54	51	49	46	43	41	39	36	34	29	24	20	16	
176	98	96	94	91	89	87	85	83	81	79	77	75	73	71	70	68	66	64	63	61	58	55	52	49	46	44	42	39	37	35	29	25	21	17	10
178	98	96	94	91	89	87	85	83	81	79	77	75	73	72	70	68	66	65	63	61	58	55	52	49	47	44	42	39	37	35	30	25	21	17	11
180	98	96	94	91	89	87	85	83	81	79	77	75	73	72	70	68	67	65	63	62	58	55	52	50	47	45	42	40	38	35	30	26	22	18	11
182	98	96	94	91	89	87	85	83	81	79	77	75	74	72	70	68	67	65	63	62	59	56	53	50	48	45	43	40	38	36	31	26	22	18	12
184	98	96	94	92	89	87	85	83	82	79	77	76	74	72	70	69	67	65	64	62	59	56	53	50	48	45	43	41	38	36	31	27	22	19	12
186	98	96	94	92	90	87	85	83	82	80	78	76	74	72	71	69	67	66	64	62	59	56	53	51	48	46	43	41	39	37	32	27	23	19	13
188	98	96	94	92	90	87	85	84	82	80	78	76	74	73	71	69	68	66	64	63	59	57	54	51	49	46	44	41	39	37	32	27	23	20	13
190	98	96	94	92	90	88	85	84	82	80	78	76	75	73	71	70	68	66	65	63	60	57	54	51	49	46	44	42	39	37	32	28	24	20	14
200	98	96	94	92	90	88	86	84	82	80	79	77	75	74	72	70	69	67	66	64	61	58	55	53	51	48	46	43	41	39	34	30	26	22	16
205	98	96	94	92	90	88	86	84	83	81	79	77	76	74	72	71	69	68	66	65	62	59	56	54	51	49	46	44	42	40	35	31	27	23	17
210	98	96	94	93	90	88	87	85	83	81	80	78	76	75	73	71	70	68	67	65	62	60	57	54	52	49	47	45	43	41	36	32	28	24	18

Table 13.4   Dew point–frost point conversions in °C.

Frost Point (°C)	Dew Point (°C)	Deviation
0	0	0
−1	−1.1	0.1
−2	−2.2	0.2
−3	−3.4	0.4
−4	−4.5	0.5
−5	−5.6	0.6
−6	−6.8	0.8
−7	−7.9	0.9
−8	−9	1
−9	−10.1	1.1
−10	−11.2	1.2
−11	−12.3	1.3
−12	−13.4	1.4
−13	−14.5	1.5
−14	−15.6	1.6
−15	−16.7	1.7
−16	−17.8	1.8
−17	−18.9	1.9
−18	−20	2
−19	−21.1	2.1
−20	−22.2	2.2
−21	−23.3	2.3
−22	−24.4	2.4
−23	−25.5	2.5
−24	−26.6	2.6
−25	−27.7	2.7
−26	−28.8	2.8
−27	−29.8	2.8
−28	−30.9	2.9
−29	−32	3
−30	−33.1	3.1
−31	−34.2	3.2
−32	−35.2	3.2
−33	−36.3	3.3
−34	−37.4	3.4
−35	−38.4	3.4
−36	−39.5	3.5
−37	−40.5	3.5
−38	−41.6	3.6
−39	−42.6	3.6
−40	−43.7	3.7
−41	−44.7	3.7
−42	−45.8	3.8
−43	−46.8	3.8
−44	−47.9	3.9
−45	−49	4
−46	−50	4
−47	−51.1	4.1
−48	−52.1	4.1
−49	−53.2	4.2
−50	−54.2	4.2

Table 13.5   Dew point–frost point conversions in °F.

	°F				°F	
Frost Point	Dew Point	Deviation		Frost Point	Dew Point	Deviation
+32	+32	0		−12	−16.7	4.7
+31	+30.8	0.2		−13	−17.8	4.8
+30	+29.7	0.3		−14	−18.9	4.9
+29	+28.6	0.4		−15	−20.0	5.0
+28	+27.5	0.5		−16	−21.1	5.1
+27	+26.4	0.6		−17	−22.2	5.2
+26	+25.2	0.8		−18	−23.3	5.3
+25	+24.1	0.9		−19	−24.3	5.3
+24	+22.9	1.1		−20	−25.4	5.4
+23	+21.8	1.2		−21	−26.4	5.4
+22	+20.7	1.3		−22	−27.5	5.5
+21	+19.6	1.4		−23	−28.6	5.6
+20	+18.5	1.5		−24	−29.6	5.6
+19	+17.4	1.6		−25	−30.6	5.6
+18	+16.2	1.8		−26	−31.7	5.7
+17	+15.1	1.9		−27	−32.8	5.8
+16	+14.0	2.0		−28	−33.9	5.9
+15	+12.9	2.1		−29	−35.0	6.0
+14	+11.8	2.2		−30	−36.1	6.1
+13	+10.7	2.3		−31	−37.2	6.2
+12	+9.6	2.4		−32	−38.2	6.2
+11	+8.5	2.5		−33	−39.3	6.3
+10	+7.4	2.6		−34	−40.3	6.3
+9	+6.3	2.7		−35	−41.4	6.4
+8	+5.2	2.8		−36	−42.4	6.4
+7	+4.1	2.9		−37	−43.5	6.5
+6	+2.9	3.1		−38	−44.5	6.5
+5	+1.8	3.2		−39	−45.6	6.6
+4	+0.7	3.3		−40	−46.6	6.6
+3	−0.4	3.4		−41	−47.7	6.7
+2	−1.5	3.5		−42	−48.7	6.7
+1	−2.6	3.6		−43	−49.8	6.8
0	−3.7	3.7		−44	−50.8	6.8
−1	−4.8	3.8		−45	−51.9	6.9
−2	−5.8	3.8		−46	−52.9	6.9
−3	−6.9	3.9		−47	−54.0	7.0
−4	−8.0	4.0		−48	−55.0	7.0
−5	−9.1	4.1		−49	−56.1	7.1
−6	−10.2	4.2		−50	−57.1	7.1
−7	−11.3	4.3		−51	−58.2	7.2
−8	−12.4	4.4		−52	−59.2	7.2
−9	−13.5	4.5		−53	−60.3	7.3
−10	−14.6	4.6				
−11	−15.6	4.6				

Table 13.6   Dew point relationships with common humidity parameters.

**Short Summary of numerical relation between common humidity parameters
in air at atmospheric pressure (101325 Pa)**

Dew point (with respect to ice below 0 °C) °C	Saturation vapor pressure Pa	Number of parts water vapor per million parts dry gas ppm$_v$	Relative humidity at 100 °C %RH	Relative humidity at 20 °C %RH
100	101419	∞	100	
80	47695	889334	47	
60	20065	246923	20	
40	7421	79028	7.3	
20	2349	23733	2.3	100
10	1233	12319		53
0	614	6097		26
−10	261	2583		11
−20	103	1018		4.4
−40	12.9	127		
−60	1.1	11		
−80	0.06	0.6		

Table 13.7   Dew point versus vapor pressure, PPM_v, % relative humidity, and PPM_w from −150°C (−238°F) to 60°C (140°F).

Dew point °C	°F	Vapor pressure (water/ice in equilibrium) mm of mercury	PPM on volume basis at 760 mm of Hg pressure	Relative humidity at 70°F%	PPM on weight basis in air
−150	−238	7×10^{-15}	9.2×10^{-12}	—	5.7×10^{-12}
−140	−220	3×10^{-10}	4.0×10^{-7}	—	2.5×10^{-7}
−130	−202	7×10^{-8}	9.2×10^{-5}	—	5.7×10^{-5}
−120	−184	10×10^{-8}	1.3×10^{-4}	5.4×10^{-7}	8.1×10^{-5}
−118	−180	.00000016	.00021	.0000009	.00013
−116	−177	.00000026	.00034	.0000014	.00021
−114	−173	.00000043	.00057	.0000023	.00035
−112	−170	.00000069	.00091	.0000037	.00057
−110	−166	.0000010	.00132	.0000053	.00082
−108	−162	.0000018	.00237	.0000096	.0015
−106	−159	.0000028	.00368	.000015	.0023
−104	−155	.0000043	.00566	.000023	.0035
−102	−152	.0000065	.00855	.000035	.0053
−100	−148	.0000099	.0130	.000053	.0081
−98	−144	.000015	.0197	.000080	.012
−96	−141	.000022	.0289	.00012	.018
−94	−137	.000033	.0434	.00018	.027
−92	−134	.000048	.0632	.00026	.039
−90	−130	.000070	.0921	.00037	.057
−88	−126	.00010	.132	.00054	.082
−86	−123	.00014	.184	.00075	.11
−84	−119	.00020	.263	.00107	.16
−82	−116	.00029	.382	.00155	.24
−80	−112	.00040	.526	.00214	.33
−78	−108	.00056	.737	.00300	.46
−76	−105	.00077	1.01	.00410	.63
−74	−101	.00105	1.38	.00559	.86
−72	−98	.00143	1.88	.00762	1.17
−70	−94	.00194	2.55	.0104	1.58
−68	−90	.00261	3.43	.0140	2.13
−66	−87	.00349	4.59	.0187	2.84
−64	−83	.00464	6.11	.0248	3.79
−62	−80	.00614	8.08	.0328	5.01
−60	−76	.00808	10.6	.0430	6.59
−58	−72	.0106	13.9	.0565	8.63
−56	−69	.0138	18.2	.0735	11.3
−54	−65	.0178	23.4	.0948	14.5
−52	−62	.0230	30.3	.123	18.8
−50	−58	.0295	38.8	.157	24.1
−48	−54	.0378	49.7	.202	30.9
−46	−51	.0481	63.3	.257	39.3
−44	−47	.0609	80.0	.325	49.7
−42	−44	.0768	101	.410	62.7
−40	−40	.0966	127	.516	78.9
−38	−36	.1209	159	.644	98.6
−36	−33	.1507	198	.804	122.9
−34	−29	.1873	246	1.00	152
−32	−26	.2318	305	1.24	189
−30	−22	.2859	376	1.52	234

continued

Table 13.7   Dew point versus vapor pressure, PPM$_v$, % relative humidity, and PPM$_w$ from –150°C (–238°F) to 60°C (140°F).

Dew point °C	Dew point °F	Vapor pressure (water/ice in equili- brium) mm of mercury	PPM on volume basis at 760 mm of Hg pressure	Relative humidity at 70°F%	PPM on weight basis in air
–28	–18	.351	462.	1.88	287.
–26	–15	.430	566.	2.30	351.
–24	–11	.526	692.	2.81	430.
–22	–8	.640	842.	3.41	523.
–20	–4	.776	1020.	4.13	633.
–18	0	.939	1240.	5.00	770.
–16	+3	1.132	1490.	6.03	925.
–14	+7	1.361	1790.	7.25	1110.
–12	+10	1.632	2150.	8.69	1335.
–10	+14	1.950	2570.	10.4	1596.
–8	+18	2.326	3060.	12.4	1900.
–6	+21	2.765	3640.	14.7	2260.
–4	+25	3.280	4320.	17.5	2680.
–2	+28	3.880	5100.	20.7	3170.
0	+32	4.579	6020.	24.4	3640.
+2	+36	5.294	6970.	28.2	4330.
+4	+39	6.101	8030.	32.5	4990.
+6	+43	7.013	9230.	37.4	5730.
+8	+46	8.045	10590.	42.9	6580.
+10	+50	9.209	12120.	49.1	7530.
+12	+54	10.52	13840.	56.1	8600.
+14	+57	11.99	15780.	63.9	9800.
+16	+61	13.63	17930.	72.6	11140.
+18	+64	15.48	20370.	82.5	12650.
+20	+68	17.54	23080.	93.5	14330.
+22	+72	19.83	26092.	Over 100	16200.
+24	+75	22.38	29447.		18284.
+26	+79	25.21	33171.		20596.
+28	+82	28.35	37303.		23162.
+30	+86	31.82	41868.		25996.
+32	+90	35.66	46921.		29133.
+34	+93	39.90	52500.		32597.
+36	+97	44.56	58632.		36405.
+38	+100	49.69	65382.		40596.
+40	+104	55.32	72789.		45195.
+42	+108	61.50	80921.		50244.
+44	+111	68.26	89816.		55767.
+46	+115	75.65	99539.		61804.
+48	+118	83.71	110145.		68389.
+50	+122	92.51	121724.		75579.
+52	+126	102.09	134329.		83405.
+54	+129	112.51	148039.		91918.
+56	+133	123.80	162895.		101142.
+58	+136	136.08	179053.		111175.
+60	+140	149.38	196553.		122040.

Table 13.8 Moisture content, vapor pressure, and dew point in metric units.

Moisture content grams per kg dry air	Saturation pressure mm Hg	bars	kg per sq cm	Dew point (°C)	Moisture content grams per kg dry air	Saturation pressure mm Hg	bars	kg per sq cm	Dew point (°C)
0.0500	0.0608	0.0000810	0.0000826	-44	15.7	18.6	0.0249	0.0253	21
0.0631	0.0766	0.000102	0.000104	-42	16.7	19.8	0.0264	0.0269	22
0.0793	0.0963	0.000128	0.000131	-40	17.8	21.1	0.0281	0.0286	23
0.0991	0.120	0.000161	0.000164	-38	19.0	22.4	0.0298	0.0304	24
0.124	0.150	0.000200	0.000204	-36	20.2	23.8	0.0317	0.0323	25
0.154	0.187	0.000249	0.000254	-34	21.4	25.2	0.0336	0.0343	26
0.190	0.231	0.000308	0.000314	-32	22.8	26.7	0.0356	0.0363	27
0.234	0.285	0.000380	0.000387	-30	24.2	28.3	0.0378	0.0385	28
					25.7	30.0	0.0400	0.0408	29
0.260	0.316	0.000421	0.000430	-29	27.3	31.8	0.0424	0.0433	30
0.288	0.350	0.000467	0.000476	-28					
0.319	0.388	0.000517	0.000527	-27	29.0	33.7	0.0449	0.0458	31
0.253	0.429	0.000572	0.000583	-26	30.8	35.7	0.0475	0.0485	32
0.390	0.474	0.000632	0.000645	-25	32.6	37.7	0.0503	0.0513	33
					34.6	39.9	0.0532	0.0542	34
0.431	0.524	0.000699	0.000712	-24	36.7	42.2	0.0562	0.0573	35
0.476	0.578	0.000771	0.000786	-23					
0.525	0.638	0.000850	0.000867	-22	38.9	44.6	0.0594	0.0606	36
0.579	0.703	0.000937	0.000956	-21	41.3	47.1	0.0628	0.0640	37
0.637	0.774	0.00103	0.00105	-20	43.7	49.7	0.0663	0.0676	38
					46.3	52.4	0.0699	0.0713	39
0.701	0.852	0.00114	0.00116	-19	49.1	55.3	0.0738	0.0752	40
0.771	0.936	0.00125	0.00127	-18					
0.847	1.03	0.00137	0.00140	-17	52.0	58.4	0.0778	0.0793	41
0.930	1.13	0.00151	0.00154	-16	55.1	61.5	0.0820	0.0836	42
1.02	1.24	0.00165	0.00168	-15	58.3	64.8	0.0864	0.0881	43
					61.7	68.3	0.0910	0.0928	44
1.12	1.36	0.00181	0.00185	-14	65.3	71.9	0.0958	0.0977	45
1.23	1.48	0.00198	0.00202	-13					
1.34	1.63	0.00217	0.00221	-12	69.1	75.7	0.101	0.103	46
1.47	1.78	0.00238	0.00242	-11	73.2	79.6	0.106	0.108	47
1.61	1.95	0.00260	0.00265	-10	77.5	83.7	0.112	0.114	48
					82.0	88.1	0.117	0.120	49
1.75	2.13	0.00284	0.00289	-9	86.8	92.5	0.123	0.126	50
1.92	2.32	0.00310	0.00316	-8					
2.09	2.53	0.00338	0.00354	-7	91.8	97.2	0.130	0.132	51
2.28	2.76	0.00368	0.00376	-6	97.2	102	0.136	0.139	52
2.48	3.01	0.00401	0.00409	-5	103	107	0.143	0.146	53
					109	113	0.150	0.153	54
2.71	3.28	0.00437	0.00446	-4	115	118	0.157	0.161	55
2.95	3.57	0.00475	0.00485	-3					
3.21	3.87	0.00516	0.00527	-2	122	124	0.165	0.168	56
3.49	4.22	0.00562	0.00573	-1	129	130	0.173	0.177	57
3.79	4.58	0.00611	0.00623	0	137	136	0.181	0.185	58
					145	143	0.190	0.194	59
4.08	4.92	0.00657	0.00670	1	153	149	0.199	0.202	60
4.38	5.29	0.00705	0.00719	2					
4.70	5.68	0.00757	0.00772	3	163	157	0.209	0.213	61
5.05	6.10	0.00813	0.00829	4	172	164	0.218	0.223	62
5.42	6.54	0.00872	0.00889	5	183	171	0.229	0.233	63
					194	179	0.239	0.244	64
5.82	7.01	0.00935	0.00953	6	206	188	0.250	0.266	65
6.23	7.51	0.0100	0.0102	7					
6.68	8.04	0.0107	0.0109	8	218	196	0.262	0.267	66
7.15	8.61	0.0115	0.0117	9	232	205	0.273	0.279	67
7.66	9.20	0.0123	0.0125	10	246	214	0.286	0.291	68
					262	224	0.298	0.304	69
8.19	9.84	0.0131	0.0134	11	279	234	0.312	0.318	70
8.76	10.5	0.0140	0.0143	12					
9.37	11.2	0.0150	0.0153	13	297	244	0.325	0.332	71
10.0	12.0	0.0160	0.0163	14	317	255	0.340	0.346	72
10.7	12.8	0.0170	0.0174	15	338	266	0.354	0.361	73
					361	277	0.370	0.377	74
11.4	13.6	0.0182	0.0185	16					
12.2	14.5	0.0194	0.0197	17					
13.0	15.5	0.0206	0.0210	18					
13.8	16.5	0.0220	0.0224	19					
14.7	17.5	0.0234	0.0238	20					

Values for the left-hand column are computed for barometric pressure of 760.0 mm Hg. To correct for moisture content (in grams per kg dry air) at barometric pressure other than 760 mm Hg, multiply the value given in the left-hand column by $\dfrac{760.0 - P_s}{P - P_s}$ where P is the barometric pressure in mm of mercury and p_s is the saturation pressure in mm of mercury. Values in other columns need no correction for barometric pressure.

Table 13.9 — Moisture content, vapor pressure, and dew point in english units.

| Moisture content | | Saturation | Dew | Moisture Content | | Saturation | Dew |
grains per pound dry air	pounds per pound dry air	pressure (in of Hg)	point (F)	grains per pound dry air	pounds per pound dry air	pressure (in of Hg)	point (F)
0.291	0.0000416	0.0020	-50	83.4	0.0119	0.560	62
0.332	0.0000475	0.0023	-48	89.6	0.0128	0.601	64
0.378	0.0000541	0.0026	-46	96.2	0.0137	0.644	66
0.430	0.0000615	0.0029	-44	103.	0.0147	0.690	68
0.489	0.0000699	0.0033	-42	111.	0.0158	0.739	70
0.555	0.0000793	0.0038	-40				
0.629	0.0000898	0.0043	-38	119.	0.0170	0.791	72
0.711	0.000102	0.0049	-36	127.	0.0182	0.846	74
0.804	0.000115	0.0055	-34	136.	0.0195	0.905	76
0.909	0.000130	0.0062	-32	146.	0.0209	0.967	78
1.02	0.000146	0.0070	-30	156.	0.0223	1.03	80
1.15	0.000165	0.0079	-28	167.	0.0239	1.10	82
1.30	0.000186	0.0089	-26	179.	0.0256	1.18	84
1.46	0.000209	0.0100	-24	191.	0.0273	1.25	86
1.64	0.000234	0.0112	-22	204.	0.0292	1.34	88
1.84	0.000263	0.0126	-20	218.	0.0312	1.42	90
2.06	0.000295	0.0141	-18	233.	0.0333	1.51	92
2.31	0.000330	0.0158	-16	249.	0.0356	1.61	94
2.58	0.000369	0.0177	-14	266.	0.0380	1.71	96
2.89	0.000413	0.0197	-12	283.	0.0405	1.82	98
3.22	0.000461	0.0220	-10	302.	0.0432	1.93	100
3.60	0.000514	0.0246	-8	322.	0.0461	2.05	102
4.01	0.000572	0.0274	-6	344.	0.0491	2.18	104
4.46	0.000637	0.0305	-4	366.	0.0523	2.31	106
4.96	0.000709	0.0339	-2	390.	0.0558	2.45	108
5.51	0.000787	0.0376	0	416.	0.0594	2.60	110
6.12	0.000874	0.0418	2	443.	0.0633	2.75	112
6.78	0.000969	0.0463	4	472.	0.0675	2.91	114
7.52	0.00107	0.0513	6	503.	0.0719	3.08	116
8.32	0.00119	0.0566	8	536.	0.0765	3.26	118
9.21	0.00132	0.0629	10	570.	0.0815	3.45	120
10.2	0.00145	0.0695	12	607.	0.0868	3.64	122
11.2	0.00161	0.0767	14	647.	0.0924	3.85	124
12.4	0.00177	0.0846	16	689.	0.0984	4.06	126
13.7	0.00195	0.0933	18	734.	0.105	4.29	128
15.1	0.00215	0.103	20	781.	0.112	4.53	130
				832.	0.119	4.77	132
16.6	0.00237	0.113	22	887.	0.127	5.03	134
18.2	0.00261	0.124	24	945.	0.135	5.30	136
20.1	0.00287	0.137	26	1007.	0.144	5.59	138
22.0	0.00315	0.150	28	1074.	0.153	5.88	140
24.2	0.00345	0.165	30				
26.5	0.00379	0.180	32	1145.	0.164	6.19	142
28.7	0.00411	0.195	34	1222.	0.175	6.52	144
31.2	0.00445	0.212	36	1303.	0.186	6.85	146
33.7.	0.00482	0.229	38	1392.	0.199	7.21	148
36.5	0.00521	0.248	40	1488.	0.213	7.57	150
39.5	0.00564	0.268	42	1590.	0.227	7.96	152
42.6	0.00609	0 289	44	1701.	0.243	8.35	154
46.0	0.00658	0.312	46	1821.	0.260	8.77	156
49.7	0.00710	0.336	48	1952.	0.279	9.20	158
53.6	0.00766	0.362	50	2093.	0.299	9.66	160
57.8	0.00826	0.390	52	2248.	0.321	10.1	162
62.3	0.00889	0.420	54	2416.	0.345	10.6	164
67.0	0.00957	0.452	56	2601.	0.372	11.1	166
72.1	0.0130	0.486	58				
77.6	0.0111	0.522	60				

Values in the two left-hand columns are computed for barometric pressure of 29.92" mm Hg. To correct for moisture content in any barometric pressure other than 29.92" Hg, multiply the value given in the two left-hand columns by $\frac{29.92 - p_s}{P - p_s}$ where P is the barometric pressure in inches of mercury and p_s is the saturation pressure in inches of mercury. Values in other columns need no correction for barometric pressure.

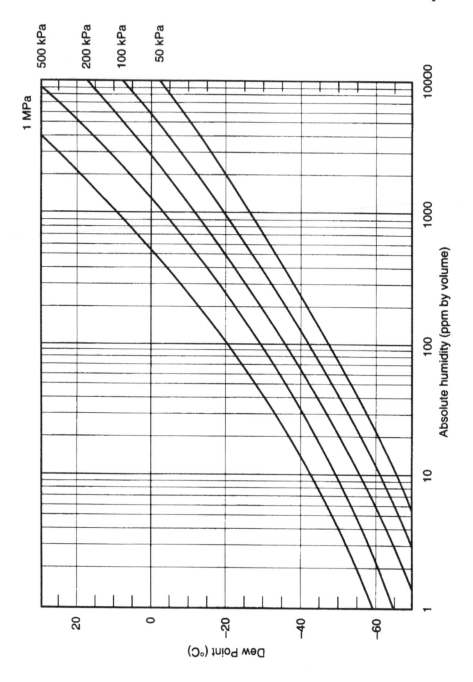

Figure 13.9   Absolute humidity (PPM_w) versus dew point at different pressures.

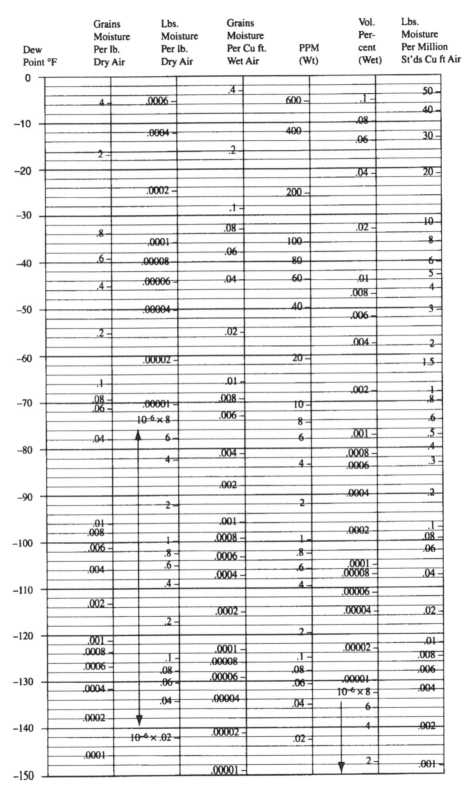

Figure 13.10   Moisture content of air and perfect gases at atmospheric pressure in English units.

Dew Point °F	Grains Moisture Per lb. Dry Air	Lbs. Moisture Per lb. Dry Air	Grains Moisture Per Cu ft. Wet Air	PPM (Wt)	Vol. Percent (Wet)	Lbs. Moisture Per Million St'ds Cu ft Air
150	1400	2	70	200,000	25	15,000
	1200		60			
140	1000	15		150,000	20	
	900		50			10,000
	800					9,000
130	700	1	40	100,000	15	8,000
		.09		90,000		7,000
	600	.08		80,000		
120	500	.07	30	70,000	10	6,000
					9	5,000
110	400	.06	25	60,000	8	
		.05		50,000	7	4,000
100	300	.04	20	40,000	6	3,000
90	200	.03	15	30,000	5	2,000
80	150	.02	10	20,000	4	1,500
			9		3	
70	100	.015	8	15,000		
	90		7			1,000
60	80		6		2	900
	70	.01	5	10,000		800
		.009		9,000	1.5	700
	60	.008		8,000		600
50	50	.007	4	7,000		500
		.006		6,000	1	
	40	.005	3		.9	400
40	30			5,000	.8	
		.004		4,000	.7	300
30			2		.6	
	20	.003	1.5	3,000	.5	200
		.002		2,000	.4	150
20			1		.3	
10	10		.8		.2	100
	8	.001	.6	1,000		80
0	6	.0008		800		60

Figure 13.10   continued.

Table 13.10   Moisture content of dry air—metric units.

Moisture content of dry air at atmospheric pressure and 25°C ambient

Dew point °C	Frost point °C	PPM$_v$	PPM$_w$	Enthalpy J/gram	Specific humidity	Absolute humidity	Air density gram/m^3
–25	–22.52	813.14	505.89	4.88	5.06×10^{-4}	0.59	1167.79
–30	–27.13	513.13	319.24	4.79	3.19×10^{-4}	0.37	1168.02
–35	–31.78	316.99	197.21	4.73	1.97×10^{-4}	0.23	1168.18
–40	–36.78	191.42	119.09	4.7	1.19×10^{-4}	0.14	1168.30
–45	–41.19	112.82	70.19	4.68	7.02×10^{-5}	0.082	1168.34
–50	–45.94	64.79	40.31	4.66	4.03×10^{-5}	0.047	1168.38
–55	–50.73	36.19	22.52	4.65	2.25×10^{-5}	0.026	1168.40
–60	–55.55	19.62	12.21	4.65	1.22×10^{-5}	0.014	1168.41
–65	–60.40	10.3	6.41	4.65	6.41×10^{-6}	0.0075	1168.42
–70	–65.29	5.23	3.25	4.65	3.25×10^{-6}	0.0038	1168.43
–75	–70.21	2.56	1.59	4.64	1.59×10^{-6}	0.0019	1168.43
–80	–75.15	1.2	0.75	4.64	7.47×10^{-7}	8.73×10^{-4}	1168.43
–85	–80.13	0.54	0.34	4.64	3.36×10^{-7}	3.94×10^{-4}	1168.43
–90	–85.13	0.23	0.14	4.64	1.44×10^{-7}	1.69×10^{-4}	1168.43
–95	–90.15	0.09	0.06	4.64	5.90×10^{-8}	6.90×10^{-5}	1168.43

Table 13.11 Moisture content of dry air—english units.

Moisture content of dry air
at atmospheric pressure and 77°F ambient

Dew point °F	Frost point °F	PPM$_v$	PPM$_w$	Grains per lbs.	Enthalpy BTU/lbs.	Specific humidity	Absolute humidity	Air density lbs./ft³
-10	-5.78	931.2	579.34	4.06	11.43	5.79×10^{-4}	4.28×10^{-5}	0.07390
-20	-14.99	561.85	349.55	2.45	11.18	3.49×10^{-4}	2.58×10^{-5}	0.07391
-30	-24.27	330.33	205.51	1.44	11.02	2.05×10^{-4}	1.52×10^{-5}	0.07392
-40	-33.64	188.88	117.51	0.82	10.93	1.75×10^{-4}	8.69×10^{-6}	0.07932
-50	-43.08	104.81	65.21	0.46	10.87	6.52×10^{-5}	4.82×10^{-6}	0.07393
-60	-52.60	56.31	35.03	0.25	10.84	3.50×10^{-5}	2.59×10^{-6}	0.07393
-70	-62.20	29.22	18.18	0.13	10.82	1.82×10^{-5}	1.34×10^{-6}	0.07393
-80	-71.86	14.60	9.08	0.064	10.81	9.03×10^{-6}	6.72×10^{-7}	0.07393
-90	-81.61	7.00	4.36	0.031	10.8	4.36×10^{-6}	3.22×10^{-7}	0.07393
-100	-91.42	3.22	2.00	0.014	10.8	2.00×10^{-6}	1.48×10^{-7}	0.07393
-110	-101.29	1.41	0.88	0.0061	10.8	8.75×10^{-7}	6.48×10^{-8}	0.07393
-120	-111.23	0.58	0.36	0.0025	10.8	3.63×10^{-7}	2.68×10^{-8}	0.07393
-130	-121.23	0.23	0.14	9.97×10^{-4}	10.8	1.42×10^{-7}	1.06×10^{-8}	0.07393
-140	-131.29	0.084	0.052	3.67×10^{-4}	10.8	5.25×10^{-8}	3.87×10^{-9}	0.07393
-150	-141.41	0.029	0.018	1.26×10^{-4}	10.8	1.81×10^{-8}	1.31×10^{-9}	0.07393

Table 13.12 Mass of water vapor (g/m³) of saturated air @ 101.3 Pa.

Temperature °C	0	1	2	3	4	5	6	7	8	9
0	4.87	5.22	5.58	5.97	6.39	6.83	7.29	7.78	8.31	8.86
10	9.44	10.06	10.71	11.39	12.12	12.88	13.69	14.54	15.43	16.38
20	17.37	18.41	19.51	20.66	21.87	23.14	24.48	25.88	27.35	28.89
30	30.50	32.19	33.96	35.81	37.75	39.78	41.90	44.11	46.42	48.84

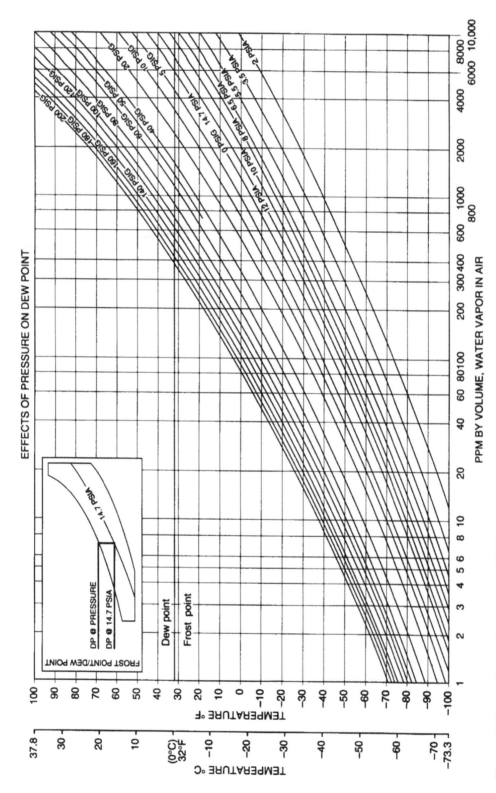

Figure 13.11   Dew point–pressure–PPM$_v$ conversion chart.

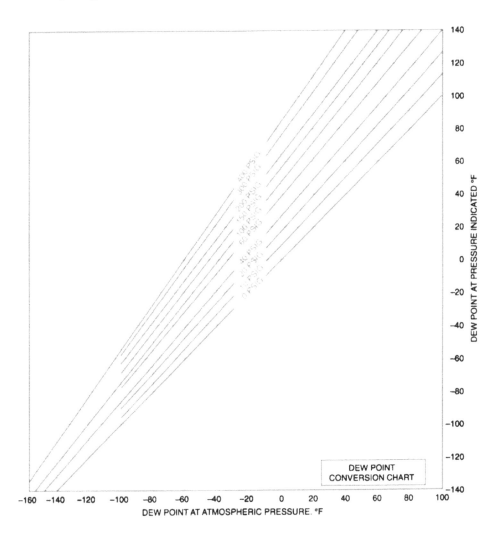

Figure 13.12   Dew point—pressure conversions at 0–400 psi$_g$ and –160°F to +100°F.

DEW POINT CONVERSION GRAPH

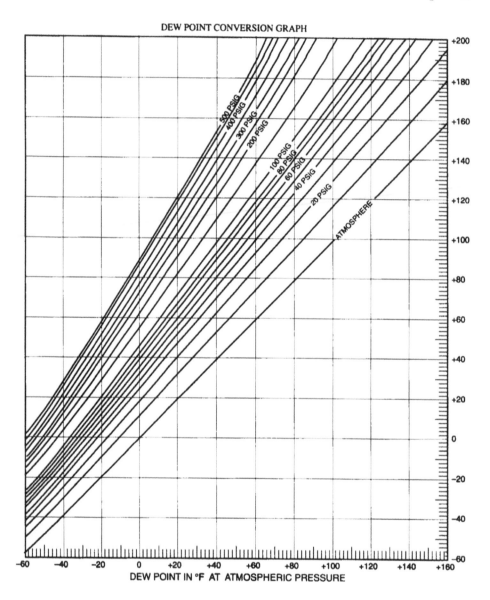

DEW POINT IN °F AT ATMOSPHERIC PRESSURE

Figure 13.13   Dew point—pressure conversions, at 0-500 psi_g and –60°F to 160 °F.

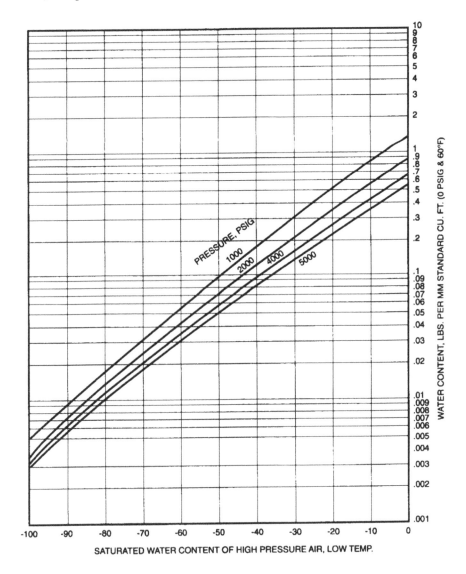

Figure 13.14   Saturated water content of high pressure air at low temperatures in english units.

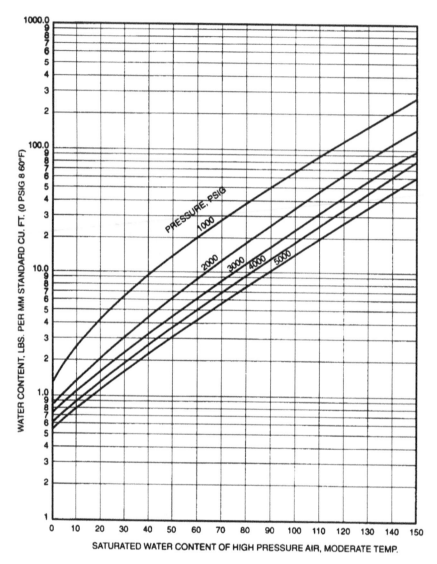

Figure 13.15   Saturated water content of high pressure air at moderate temperatures in english units.

Table 13.13   Vacuum conversion table.

Microns	mm Hg.	Inches Hg.	P.S.I. (Abs.)
.10	.0001	.00000394	.00000193
.30	.0003	.00001181	.00000580
.50	.0005	.00001969	.00000967
.70	.0007	.00002756	.00001354
1.00	.001	.00003937	.00001934
3.00	.003	.0001181	.00005802
5.00	.005	.0001969	.00009670
7.00	.007	.0002756	.0001354
10	.01	.0003937	.0001934
15	.015	.0005910	.0002900
20	.020	.0007874	.0003878
35	.035	.001378	.000677
50	.050	.001969	.000967
75	.075	.002953	.001451
100	.10	.003937	.001934
250	.25	.009845	.004835
500	.50	.01969	.00967
750	.75	.02953	.01450
1000	1.00	.03937	.01934

1 Micron (μ) Hg = 0.001 mm Hg = 10^{-3} mm Hg

1 Inch Hg @ 32°F. = 0.4912 psi

1 Atmosphere = 760 mm Hg = 14.7 psi

Table 13.14   Altitude pressure table.

| Altitude (ft) | Altitude (m) | Altitude pressure table mercury at 0°C (32°F) | | | |
		in Hg	mm Hg	Atm.	Bar
−1000	304.80	31.02	787.9	1.03	1.05
0	0.00	29.92	760.0	1.00	1.01
1000	304.80	28.86	732.9	0.96	0.98
2000	609.60	27.82	706.6	0.93	0.94
3000	914.40	26.81	681.1	0.90	0.91
4000	1219.20	25.84	656.3	0.86	0.87
5000	1524.00	24.89	632.3	0.83	0.84
6000	1828.80	23.98	609.0	0.80	0.81
7000	2133.60	23.09	586.4	0.77	0.78
8000	2438.40	22.22	564.4	0.74	0.75
9000	2743.20	21.38	543.2	0.71	0.72
10000	3048	20.58	522.6	0.69	0.70
15000	4572	16.88	428.8	0.56	0.57
20000	6096	13.75	349.1	0.46	0.47
25000	7620	11.10	281.9	0.37	0.38
30000	9144	8.88	225.6	0.30	0.30
35000	10668	7.04	178.7	0.23	0.24
40000	12192	5.54	140.7	0.18	0.19
45000	13716	4.36	110.8	0.15	0.15
50000	15240	3.44	87.3	0.12	0.12

Table 13.15   Pressure conversion factors.

Pressure psi	in. Hg	atm	mm Hg	bar	kg/cm²	dyne/cm²	pascal
1	= 2.0360	= 0.068046	= 51.715	= 0.068948	= 0.07030696	= 68.948	= 6894.8
0.491154	1	0.033421	25.400	0.033864	0.034532	33.864	3386.4
14.6960	29.921	1	790.0	1.01325	1.03323	1,013.250	101.325
0.0193368	0.03937	0.00131579	1	0.0013332	0.0013595	1333.2	133.32
14.5038	29.530	0.98692	750.062	1	1.01972	10^6	10^5
14.223	28.959	0.96784	735.559	0.98066	1	980.655	98.066
1.4038×10^{-5}	2.953×10^{-5}	9.8692×10^{-7}	0.000750	10^{-6}	1.01972×10^{-6}	1	0.100
1.45038×10^{-4}	2.953×10^{-4}	9.8692×10^{-6}	0.00750	10^{-5}	1.01972×10^{-5}	10	1

Table 13.16   Detailed pressure conversion chart.

in/H₂O	P.S.I.	in/Hg	mm/H₂O	mm/Hg	kg/cm²	bar	mbar	Pa	kPa
.1	.0036	.0073	2.534	.1863	.0002	.0002	.2482	24.82	.0248
.2	.0072	.0146	5.067	.3726	.0005	.0005	.4964	49.64	.0496
.4	.0144	.0293	10.13	.7452	.0010	.0010	.9928	99.28	.0993
.6	.0216	.0440	15.20	1.118	.0015	.0015	1.489	148.9	.1489
.8	.0289	.0588	20.34	1.496	.0020	.0020	1.992	199.2	.1992
1.0	.0361	.0735	25.41	1.868	.0025	.0025	2.489	248.9	.2489
2	.0772	.1470	50.81	3.736	.0051	.0050	4.978	497.8	.4978
3	.1083	.2205	76.22	5.604	.0076	.0075	7.467	746.7	.7467
4	.1444	.2940	101.62	7.472	.0102	.0099	9.956	995.6	.9956
5	.1804	.3673	127.0	9.335	.0127	.0124	12.44	1244	1.244
6	.2165	.4408	152.4	11.203	.0152	.0149	14.93	1493	1.493
7	.2526	.5143	177.8	13.072	.0178	.0174	17.42	1742	1.742
8	.2887	.5878	203.2	14.940	.0203	.0199	19.90	1990	1.990
9	.3248	.6613	228.6	16.808	.0228	.0224	22.39	2239	2.239
10	.3609	.7348	254.0	18.676	.0254	.0249	24.88	2488	2.488
11	.3970	.8083	279.4	20.544	.0279	.0274	27.37	2737	2.737
12	.4331	.8818	304.8	22.412	.0304	.0299	29.86	2986	2.986
13	.4692	.9553	330.2	24.280	.0330	.0324	32.35	3235	3.235
14	.5053	1.029	355.6	26.148	.0355	.0348	34.84	3484	3.484
15	.5414	1.102	381.0	28.016	.0381	.0373	37.33	3733	3.733
16	.5774	1.176	406.4	29.879	.0406	.0398	39.81	3981	3.981
17	.6136	1.249	431.8	31.752	.0431	.0423	42.31	4231	4.231
18	.6496	1.322	457.2	33.616	.0457	.0448	44.79	4479	4.479
19	.6857	1.396	482.6	35.484	.0482	.0473	47.28	4728	4.728
20	.7218	1.470	508.0	37.352	.0507	.0498	49.77	4977	4.977
21	.7579	1.543	533.4	39.22	.0533	.0523	52.26	5226	5.226
22	.7940	1.616	558.8	41.09	.0558	.0547	54.74	5474	5.474
23	.8301	1.690	584.2	42.96	.0584	.0572	57.23	5723	5.723
24	.8662	1.764	609.6	44.82	.0609	.0597	59.72	5972	5.972
25	.9023	1.837	635.0	46.69	.0634	.0622	62.21	6221	6.221
26	.9384	1.910	660.4	48.56	.0660	.0647	64.70	6470	6.470
27	.9745	1.984	685.8	50.43	.0685	.0672	67.19	6719	6.719
28	1.010	2.056	710.8	52.26	.0710	.0696	69.64	6964	6.964
29	1.047	2.132	736.8	54.18	.0736	.0722	72.19	7219	7.219
30	1.083	2.205	762.2	56.04	.0761	.0747	74.67	7467	7.467
31	1.119	2.278	787.5	57.91	.0787	.0772	77.15	7715	7.715
32	1.155	2.352	812.8	59.77	.0812	.0796	79.63	7963	7.963
33	1.191	2.425	838.2	61.63	.0837	.0821	82.12	8212	8.212
34	1.227	2.498	863.5	63.49	.0862	.0846	84.60	8460	8.460
35	1.263	2.571	888.9	65.36	.0888	.0871	87.08	8708	8.708
36	1.299	2.645	914.2	67.22	.0913	.0896	89.56	8956	8.956
37	1.335	2.718	939.5	69.08	.0938	.0920	92.04	9204	9.204
38	1.371	2.791	964.9	70.95	.0964	.0945	94.53	9453	9.453
39	1.408	2.867	990.9	72.86	.0990	.0971	97.08	9708	9.708
40	1.444	2.940	1016	74.72	.1015	.0996	99.56	9956	9.956
41	1.480	3.013	1042	76.59	.1040	.1020	102.0	10204	10.20
42	1.516	3.086	1067	78.45	.1066	.1045	104.5	10452	10.45
43	1.552	3.160	1092	80.31	.1091	.1070	107.0	10701	10.70
44	1.588	3.233	1118	82.18	.1116	.1095	109.5	10949	10.95
45	1.624	3.306	1143	84.04	.1142	.1120	112.0	11197	11.20
46	1.660	3.378	1168	85.90	.1167	.1144	114.5	11445	11.44
47	1.696	3.453	1194	87.76	.1192	.1169	116.9	11694	11.69
48	1.732	3.526	1219	89.63	.1218	.1194	119.4	11942	11.94

continued

Table 13.16   Detailed pressure conversion chart.

in/H₂O	P.S.I.	in/Hg	mm/H₂O	mm/Hg	kg/cm²	bar	mbar	Pa	kPa
49	1.768	3.600	1244	91.49	.1243	.1219	121.9	12190	12.19
50	1.804	3.673	1270	93.35	.1268	.1244	124.4	12438	12.44
51	1.841	3.748	1296	95.27	.1294	.1269	126.9	12693	12.69
52	1.877	3.822	1321	97.13	.1320	.1294	129.4	12941	12.94
53	1.913	3.895	1346	98.99	.1345	.1319	131.9	13190	13.19
54	1.949	3.968	1372	100.8	.1370	.1344	134.4	13438	13.44
55	1.985	4.041	1397	102.7	.1395	.1369	136.9	13686	13.69
56	2.021	4.115	1422	104.6	.1421	.1393	139.3	13934	13.93
57	2.057	4.188	1448	106.4	.1446	.1418	141.8	14182	14.18
58	2.093	4.261	1473	108.3	.1471	.1443	144.3	14431	14.43
59	2.129	4.335	1498	110.2	.1497	.1468	146.8	14679	14.68
60	2.165	4.408	1524	112.0	.1522	.1493	149.3	14927	14.93
61	2.202	4.483	1550	113.9	.1548	.1518	151.8	15182	15.18
62	2.238	4.556	1575	115.8	.1573	.1543	154.3	15430	15.43
63	2.274	4.630	1600	117.7	.1599	.1568	156.8	15679	15.68
64	2.310	4.703	1626	119.5	.1624	.1593	159.3	15927	15.93
65	2.346	4.776	1651	121.4	.1649	.1618	161.8	16175	16.18
66	2.382	4.850	1676	123.3	.1674	.1642	164.2	16423	16.42
67	2.418	4.923	1702	125.1	.1700	.1667	166.7	16672	16.67
68	2.454	4.996	1727	127.0	.1725	.1692	169.2	16920	16.92
69	2.490	5.070	1752	128.8	.1750	.1717	171.7	17168	17.17
70	2.526	5.143	1778	130.7	.1776	.1742	174.2	17416	17.42
71	2.562	5.216	1803	132.6	.1801	.1766	176.6	17664	17.66
72	2.598	5.290	1828	134.4	.1826	.1791	179.1	17912	17.91
73	2.635	5.365	1854	136.4	.1852	.1817	181.7	18168	18.17
74	2.671	5.438	1880	138.2	.1878	.1842	184.2	18416	18.42
75	2.707	5.511	1905	140.1	.1903	.1866	186.6	18664	18.66
76	2.743	5.585	1930	141.9	.1928	.1891	189.1	18912	18.91
77	2.779	5.658	1956	143.8	.1954	.1916	191.6	19160	19.16
78	2.815	5.731	1981	145.7	.1979	.1941	194.1	19409	19.41
79	2.851	5.805	2006	147.5	.2004	.1966	196.6	19657	19.66
80	2.887	5.878	2032	149.4	.2030	.1991	199.1	19905	19.90
81	2.923	5.951	2057	151.2	.2055	.2015	201.5	20153	20.15
82	2.959	6.024	2082	153.1	.2080	.2040	204.0	20402	20.40
3	2.996	6.100	2108	155.0	.2106	.2066	206.6	20657	20.66
84	3.032	6.173	2134	156.9	.2131	.2091	209.1	20905	20.90
85	3.068	6.246	2159	158.8	.2157	.2115	211.5	21153	21.15
86	3.104	6.320	2184	160.6	.2182	.2140	214.0	21401	21.40
87	3.140	6.393	2210	162.5	.2207	.2165	216.5	21650	21.65
88	3.176	6.466	2235	164.4	.2233	.2190	219.0	21898	21.90
89	3.212	6.540	2260	166.2	.2258	.2215	221.5	22146	22.15
90	3.248	6.613	2286	168.1	.2283	.2239	223.9	22394	22.39
91	3.284	6.686	2311	169.9	.2309	.2264	226.4	22642	22.64
92	3.320	6.760	2336	171.8	.2334	.2289	228.9	22890	22.89
93	3.356	6.833	2362	173.7	.2359	.2314	231.4	23139	23.14
94	3.392	6.906	2387	175.5	.2384	.2339	233.9	23387	23.39
95	3.429	6.981	2413	177.4	.2410	.2364	236.4	23642	23.64
96	3.465	7.055	2438	179.3	.2436	.2389	238.9	23890	23.89
97	3.501	7.128	2464	181.2	.2461	.2414	241.4	24138	24.14
98	3.537	7.201	2489	183.0	.2486	.2439	243.9	24387	24.39
99	3.573	7.275	2514	184.9	.2512	.2464	246.4	24635	24.64
100	3.609	7.348	2540	186.8	.2537	.2488	248.8	24883	24.88

Table 13.16   continued.

P.S.I.	in/H₂O	in/Hg	mm/H₂O	mm/Hg	kg/cm2	bar	mbar	Pa	kPa
1.0	27.71	2.036	703.1	51.75	.0703	.0689	68.95	6895	6.895
1.1	30.45	2.240	773.4	56.89	.0773	.0758	75.84	7584	7.584
1.2	33.22	2.443	843.7	62.06	.0844	.0827	82.74	8274	8.274
1.3	35.98	2.647	914.0	67.23	.0914	.0896	89.63	8963	8.963
1.4	38.75	2.850	984.3	72.40	.0984	.0965	96.52	9652	9.652
1.5	41.52	3.054	1055	77.57	.1055	.1034	103.4	10340	10.34
1.6	44.29	3.258	1125	82.74	.1152	.1103	110.3	11030	11.03
1.7	47.06	3.461	1195	87.92	.1195	.1172	117.2	11720	11.72
1.8	49.82	3.665	1266	93.09	.1266	.1241	124.1	12410	12.41
1.9	52.59	3.868	1336	98.26	.1336	.1310	131.0	13100	13.10
2.0	55.36	4.072	1406	103.4	.1406	.1379	137.9	13790	13.79
2.1	58.13	4.276	1476	108.6	.1476	.1448	144.8	14480	14.48
2.2	60.90	4.479	1547	113.8	.1547	.1517	151.7	15170	15.17
2.3	63.67	4.683	1617	118.9	.1617	.1586	158.6	15860	15.86
2.4	66.43	4.886	1687	124.1	.1687	.1655	165.5	16550	16.55
2.5	69.20	5.090	1758	129.3	.1758	.1724	172.4	17240	17.24
2.6	71.97	5.294	1828	134.5	.1828	.1793	179.3	17930	17.93
2.7	74.74	5.497	1898	139.6	.1898	.1862	186.2	18620	18.62
2.8	77.51	5.701	1969	144.8	.1968	.1930	193.0	19300	19.30
2.9	80.27	5.904	2039	150.0	.2039	.1999	199.9	19990	19.99
3.0	83.04	6.108	2109	155.1	.2109	.2068	206.8	20680	20.68
3.1	85.81	6.312	2180	160.3	.2180	.2137	213.7	21370	21.37
3.2	88.58	6.515	2250	165.5	.2250	.2206	220.6	22060	22.06
3.3	91.35	6.719	2320	170.7	.2320	.2275	227.5	22750	22.75
3.4	94.11	6.922	2390	175.8	.2390	.2344	234.4	23440	23.44
3.5	96.88	7.126	2461	181.0	.2461	.2413	241.3	24130	24.13
3.6	99.65	7.330	2531	186.2	.2531	.2482	248.2	24820	24.82
3.7	102.4	7.533	2601	191.3	.2601	.2551	255.1	25510	25.51
3.8	105.2	7.737	2672	196.5	.2672	.2620	262.0	26200	26.20
3.9	108.0	7.940	2742	201.7	.2742	.2689	268.9	26890	26.89
4.0	110.7	8.144	2812	206.9	.2812	.2758	275.8	27580	27.58
4.1	113.5	8.348	2883	212.0	.2883	.2827	282.7	28270	28.27
4.2	116.3	8.551	2953	217.2	.2953	.2896	289.6	28960	28.96
4.3	119.0	8.775	3023	222.4	.3023	.2965	296.5	29650	29.65
4.4	121.8	8.958	3094	227.5	.3094	.3034	303.4	30338	30.34
4.5	124.6	9.162	2164	232.7	.3164	.3103	310.3	31030	31.03
4.6	127.3	9.366	3234	237.9	.3234	.3172	317.2	31720	31.72
4.7	130.1	9.569	3304	243.1	.3304	.3240	324.0	32400	32.40
4.8	132.9	9.773	3375	248.2	.3375	.3310	331.0	33100	33.10
4.9	135.6	9.976	3445	253.4	.3445	.3378	337.8	33780	33.78
5.0	138.4	10.18	3515	258.6	.3315	.3447	344.7	34470	34.47
5.1	141.2	10.38	3586	263.7	.3586	.3516	351.6	35160	35.16
5.2	143.9	10.59	3656	268.9	.3656	.3585	358.5	35850	35.85
5.3	146.7	10.79	3726	274.1	.3726	.3654	365.4	36540	36.54
5.4	149.5	10.99	3797	279.3	.3797	.3723	372.3	37230	37.23
5.5	152.2	11.20	3867	284.4	.3867	.3792	379.2	37920	37.92
5.6	155.0	11.40	3937	289.6	.3937	.3861	386.1	38610	38.61
5.7	157.8	11.60	4008	294.8	.4007	.3930	393.0	39300	39.30
5.8	160.5	11.81	4078	299.9	.4078	.3999	399.9	39990	39.99
5.9	163.3	12.01	4148	305.1	.4148	.4068	406.8	40680	40.68
6.0	166.1	12.22	4218	310.3	.4218	.4137	413.7	41370	41.37
6.1	168.8	12.42	4289	315.5	.4289	.4206	420.6	42060	42.06
6.2	171.6	12.62	4359	320.6	.4359	.4275	427.5	42750	42.75
6.3	174.4	12.83	4429	325.8	.4429	.4344	434.4	43440	43.44

Table 13.16   continued.

P.S.I.	in/H₂O	in/Hg	mm/H₂O	mm/Hg	kg/cm2	bar	mbar	Pa	kPa
6.4	177.2	13.03	4500	331.0	.4500	.4413	441.3	44130	44.13
6.5	179.9	13.23	4570	336.1	.4570	.4482	448.2	44820	44.82
6.6	182.7	13.44	4640	341.3	.4640	.4550	455.0	45500	45.50
6.7	185.5	13.64	4711	346.5	.4710	.4619	461.9	46190	46.19
6.8	188.2	13.84	4781	351.7	.4781	.4688	468.8	46880	46.88
6.9	191.0	14.05	4851	356.8	.4851	.4757	475.7	47570	47.57
7.0	193.8	14.25	4922	362.0	.4921	.4826	482.6	48260	48.26
7.1	196.5	14.46	4992	367.2	.4992	.4895	489.5	48950	48.95
7.2	199.3	14.66	5062	372.3	.5062	.4964	496.4	49640	49.64
7.3	202.1	14.86	5132	377.5	.5132	.5033	503.3	50330	50.33
7.4	204.8	15.07	5203	382.7	.5203	.5102	510.2	51020	51.02
7.5	207.6	15.27	5273	387.9	.5273	.5171	517.1	51710	51.71
7.6	210.4	15.47	5343	393.0	.5343	.5240	524.0	52400	52.40
7.8	215.9	15.88	5484	403.4	.5484	.5378	537.8	53780	53.78
8.0	221.4	16.29	5625	413.7	.5625	.5516	551.6	55160	55.16
8.2	227.0	16.70	5765	424.1	.5765	.5654	565.4	56540	56.54
8.4	232.5	17.10	5906	434.4	.5906	.5792	579.2	57920	57.92
8.6	238.0	17.51	6047	444.7	.6046	.5929	592.9	59290	59.29
8.8	243.6	17.92	6187	455.1	.6187	.6067	606.7	60670	60.67
9.0	249.1	18.32	6328	465.4	.6328	.6205	620.5	62050	62.05
9.2	254.7	18.73	6468	475.8	.6468	.6343	634.3	63430	63.43
9.4	260.2	19.14	6609	486.1	.6609	.6481	648.1	64810	64.81
9.6	265.7	19.54	6750	496.5	.6749	.6619	661.9	66190	66.19
9.8	271.3	19.95	6890	506.8	.6890	.6757	675.7	67570	67.57
10.0	276.8	20.36	7031	517.1	.7031	.6895	689.5	68950	68.95
11.0	304.5	22.40	7734	568.9	.7734	.7584	758.4	75840	75.84
12.0	332.2	24.43	8437	620.6	.8437	.8274	827.4	82740	82.74
13.0	359.8	26.47	9140	672.3	.9140	.8963	896.3	89630	89.63
14.0	387.5	28.50	9843	724.0	.9843	.9652	965.2	96520	96.52
14.7	406.9	29.93	10340	760.2	1.033	1.014	1014	101400	101.4
15.0	415.2	30.54	10550	775.7	1.055	1.034	1034	103400	103.4
16.0	442.9	32.58	11250	827.4	1.125	1.103	1103	110300	110.3
17.0	470.6	34.61	11950	879.1	1.195	1.172	1172	117200	117.2
18.0	498.2	36.65	12660	930.9	1.265	1.241	1241	124100	124.1
19.0	525.9	38.68	13360	982.6	1.336	1.310	1310	131000	131.0
20.0	553.6	40.72	14060	1034	1.406	1.379	1379	137900	137.9
21.0	581.3	42.76	14770	1086	1.476	1.448	1448	144800	144.8
22.0	609.0	44.79	15470	1138	1.547	1.517	1517	151700	151.7
23.0	636.7	46.83	16170	1189	1.617	1.586	1586	158600	158.6
24.0	664.3	48.86	16870	1241	1.687	1.655	1655	165500	165.5
25.0	692.0	50.90	17580	1293	1.758	1.724	1724	172400	172.4

Conversion Factors

Note: Conversion factors rounded

P.S.I. × 27.71 = in. H₂O

P.S.I. × 2.036 = in. Hg

P.S.I. × 703.1 = mm/H₂O

P.S.I. × 51.75 = mm/Hg

P.S.I. × .0703 = kg/cm²

P.S.I. × .0689 = bar

P.S.I. × 68.95 = mbar

P.S.I. × 6895 = Pa

P.S.I. × 6.895 = kPa

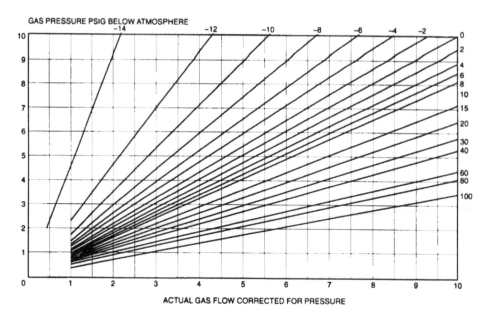

Figure 13.16   Gas flow readings, corrected for pressure.

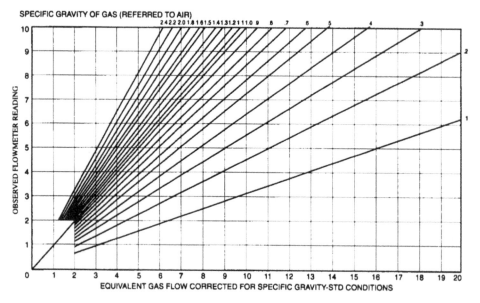

Figure 13.17   Equivalent gas flow corrected for specific gravity.

Table 13.17   Flow equivalents.

1 cu. ft./hr.	1 cu. ft./min.	1 cc/min.	1 cc/hr.
.0166 cu. ft./min.	60 cu. ft./hr.	60 cc/hr.	.0167 cc/min.
.4719 l/min.	28.316 l/min.	.000035 cu. ft./min.	.0000005 cu. ft./min.
28.316 l/hr.	1699 l/hr.	.0021 cu. ft./hr.	.00003 cu. ft./hr.
471.947 cc/min.	28317 cc/min.	.001 l/min.	.000017 l/min.
28317 cc/hr.	1,699,011 cc/hr.	.06 l/hr.	.001 l/hr.
.1247 Gal/min.	7.481 Gal/min.	.00026 Gal/min.	.000004 Gal/min.
7.481 Gal/hr.	448.831 Gal/hr.	.0159 Gal/hr.	.00026 Gal/hr.

1 l/min.	1 l/min.	1 Gal/min.	1 Gal/hr.
60. l/hr.	.0166 l/min.	60 Gal/hr.	.0167 Gal/min.
.035 cu. ft./min.	.00059 cu. ft./min.	.1137 cu. ft./min.	.002 cu. ft./min.
2.1189 cu. ft./hr.	.035 cu. ft./hr.	8.021 cu. ft./hr.	.1337 cu. ft./hr.
1000 cc/min.	16.667 cc/min.	3.785 l/min.	.063 l/min.
60,002 cc/hr.	1000 cc/hr.	227.118 l/hr.	3.785 l/hr.
.264 Gal/min.	.004 Gal/min.	3,785,412 cc/min.	63.069 cc/min.
15.851 Gal/hr.	.264 Gal/hr.	227,125 cc/hr.	3785 cc/hr.

Table 13.18   °C–°F conversions.

To convert temperature reading shown in center column, read right-hand column for Fahrenheit, left-hand column for Centigrade.

C		F	C		F	C		F	C		F	C		F
-129	-200	-328	18.3	65	149	204	400	752	416	780	1436	627	1160	2120
-118	-180	-292	21.1	70	158	210	410	770	421	790	1454	632	1170	2138
-107	-160	-256	23.9	75	167	216	420	788	427	800	1472	638	1180	2156
-95.6	-140	-220	26.7	80	176	221	430	806	432	810	1490	643	1190	2174
-84.4	-120	-184	29.4	85	185	227	440	824	438	820	1508	649	1200	2192
-73.3	-100	-148	32.2	90	194	232	450	842	443	830	1526	654	1210	2210
-70.6	-95	-139	35.0	95	203	238	460	860	449	840	1544	660	1220	2228
-67.8	-90	-130	38	100	212	243	470	878	454	850	1562	666	1230	2246
-65.0	-85	-121	43	110	230	249	480	896	460	860	1580	671	1240	2264
-62.2	-80	-112	49	120	248	254	490	914	466	870	1598	677	1250	2282
-59.5	-75	-103	54	130	266	260	500	932	471	880	1616	682	1260	2300
-56.7	-70	-94	60	140	284	266	510	950	477	890	1634	688	1270	2318
-53.9	-65	-85	66	150	302	271	520	968	482	900	1652	693	1280	2336
-51.1	-60	-76	71	160	320	277	530	986	488	910	1670	699	1290	2354
-48.3	-55	-67	77	170	338	282	540	1004	493	920	1688	704	1300	2372
45.6	-50	-58	82	180	356	288	550	1022	499	930	1706	710	1310	2390
-42.8	-45	-49	88	190	374	293	560	1040	504	940	1724	716	1320	2408
-40.0	-40	-40	93	200	392	299	570	1058	510	950	1742	721	1330	2426
-37.2	-35	-31	99	210	410	304	580	1076	516	960	1760	727	1340	2444
-34.4	-30	-22	100	212	414	310	590	1094	521	970	1778	732	1350	2462
-31.7	-25	-13	104	220	428	316	600	1112	527	980	1796	738	1360	2480
-28.9	-20	-4	110	230	446	321	610	1130	532	990	1814	743	1370	2498
-26.1	-15	+5	116	240	464	327	620	1148	538	1000	1832	749	1380	2516
-23.3	-10	+14	121	250	482	332	630	1166	543	1010	1850	754	1390	2534
-20.6	-5	+23	127	260	500	338	640	1184	549	1020	1868	760	1400	2552
-17.8	0	+32	132	270	518	343	650	1202	554	1030	1886	766	1410	2570
-15.0	+5	+41	138	280	536	349	660	1220	560	1040	1904	771	1420	2588
-12.2	+10	+50	143	290	554	354	670	1238	566	1050	1922	777	1430	2606
-3.44	+15	+59	149	300	572	360	380	1256	571	1060	1940	782	1440	2624
-6.67	+20	+68	154	310	590	366	690	1274	577	1070	1958	788	1450	2642
-3.89	+25	+77	160	320	608	371	700	1292	582	1080	1976	793	1460	2660
-1.11	+30	+86	166	330	626	377	710	1310	588	1090	1994	799	1470	2678
+1.67	+35	+95	171	340	644	382	720	1328	593	1100	2012	804	1480	2696
4.44	40	104	177	350	662	388	730	1346	599	1110	2030	810	1490	2714
7.22	45	113	182	360	680	393	740	1364	604	1120	2048	816	1500	2732
10.0	50	122	188	370	698	399	750	1382	610	1130	2066	821	1510	2750
12.8	55	131	193	380	716	404	760	1400	616	1140	2084	827	1520	2768
15.6	60	140	199	390	734	410	770	1418	621	1150	2102	832	1530	2786

continued

Table 13.18 °C–°F conversions.

To convert temperature reading shown in center column, read right-hand column for Fahrenheit, left-hand column for Centigrade.

C		F	C		F	C		F	C		F	C		F
838	1540	2804	1004	1840	3344	1171	2140	3884	1338	2440	4424	1504	2840	4964
843	1550	2822	1010	1850	3362	1177	2150	3902	1343	2450	4442	1510	2750	4982
849	1560	2840	1016	1860	3380	1182	2160	3920	1349	2460	4460	1516	2760	5000
854	1570	2858	1021	1870	3398	1188	2170	3938	1354	2470	4478	1521	2770	5018
860	1580	2876	1027	1880	3416	1193	2180	3956	1360	2480	4496	1527	2780	5036
866	1590	2894	1032	1890	3434	1199	2190	3974	1366	2490	4514	1532	2790	5054
871	1600	2912	1038	1900	3452	1204	2200	3992	1371	2500	4532	1538	2800	5072
877	1610	2930	1043	1910	3470	1210	2210	4010	1377	2510	4550	1543	2810	5090
882	1620	2948	1049	1920	3488	1216	2220	4028	1382	2520	4568	1549	2820	5108
888	1630	2966	1054	1930	3506	1221	2230	4046	1388	2530	4586	1554	2830	5126
893	1640	2984	1060	1940	3524	1227	2240	4064	1393	2540	4604	1560	2840	5144
899	1650	3002	1066	1950	3542	1232	2250	4082	1399	2550	4622	1566	2850	5162
904	1660	3020	1071	1960	3560	1238	2260	4100	1404	2560	4640	1571	2860	5180
910	1670	3038	1077	1970	3578	1243	2270	4118	1410	2570	4658	1577	2870	5180
916	1680	3056	1082	1980	3596	1249	2280	4136	1416	2580	4676	1582	2880	5216
921	1690	3074	1088	1990	3614	1254	2290	4154	1421	2590	4694	1588	2890	5234
927	1700	3092	1093	2000	3632	1260	2300	4172	1427	2600	4712	1593	2900	5252
932	1710	3110	1099	2010	3650	1266	2310	4190	1432	2610	4730	1599	2910	5270
938	1720	3128	1104	2020	3668	1271	2320	4208	1438	2620	4748	1604	2920	5288
943	1730	3146	1110	2030	3686	1277	2330	4226	1443	2630	4766	1610	2930	5306
949	7140	3164	1116	2040	3704	1282	2340	4244	1449	2640	4784	1616	2940	5324
954	1750	3182	1121	2050	3722	1288	2350	4262	1454	2650	4802	1621	2950	5352
960	1760	3200	1127	2060	3740	1293	2360	4280	1460	2660	4820	1627	2960	5360
966	1770	3218	1132	2070	3758	1299	2370	4298	1466	2670	4838	1632	2970	5378
971	1780	3236	1138	2080	3776	1304	2380	4316	1471	2680	4856	1638	2980	5396
977	1790	3254	1143	2090	3794	1310	2390	4334	1477	2690	4874	1643	2990	5414
982	1800	3272	1149	2100	3812	1316	2400	4352	1482	2700	4892	1649	3000	5432
988	1810	3290	1154	2110	3830	1321	2410	4370	1488	2710	4910			
993	1820	3308	1160	2120	3848	1327	2420	4388	1493	2720	4928			
999	1830	3326	1166	2130	3866	1332	2430	4406	1499	2730	4946			

Temperature Interpolation Factors

C		F	C		F	C		F	C		F	C		F
0.56	1	1.8	1.67	3	5.4	2.78	5	9.0	3.89	7	12.6	5.00	9	16.2
1.11	2	3.6	2.22	4	7.2	3.33	6	10.8	4.44	8	14.4	5.56	10	18.0

Table 13.19   Metric–english unit conversions.

Multiply	By	To Obtain	Multiply	By	To Obtain
ft	0.3048	m	mile	1.61	km
ft	304.8	mm	mile, nautical	1.85	km
ft/min, fpm	0.00508	m/sec	millibar	*0.100	kPa
ft/s, fps	0.3048	m/sec	mm of mercury (60°F)	0.133	kPa
ft of water	2.99	kPa	mm of water (60°F)	9.80	Pa
ft²	0.0929	m²	meter of water	9.80	kPa
			ounce (mass, avoirdupois) ..	28.3	g
ft³	28.3	l	ounce (force or thrust)	0.278	N
ft³	0.0283	m³	ounce (liquid, US)	29.6	ml
ft³/min, cfm	0.472	l/sec			
ft³/s, cfs	28.3	l/sec	pint (liquid, US)	473	ml
			pound		
			lb (mass)	0.454	kg
gallon (US, 231 in³)	3.79	l	lb (mass)	454	g
gallon	0.00379	m³	lb$_f$ (force or thrust)	4.45	N
gph	1.05	ml/sec	lb/ft (uniform load)	1.49	kg/m
gpm	0.0631	l/sec			
grain (1/7000 lb)	0.0648	g			
gr/gal	17.1	mg/l	lb/ft²	4.88	kg/m²
gr/lb	0.143	g/kg	lb/ft³ (density)	16.0	kg/m³
			lb/gallon	120	kg/m³
inch	25.4	mm	ppm (by mass)	*1.00	mg/kg
in. of mercury (60°F)	3.38	kPa	psi	6.89	kPa
in. of water (60°F)	249	Pa			
in.²	645	mm²	torr (1mm Hg @ 0°C)	133	Pa
in.³ (volume)	16.4	ml			
in.³/min (SCIM)	0.273	ml/sec			
			yd	0.9144	m
liter	0.001	m³	yd²	0.836	m²
micron of mercury (60°F) ...	133	mPa	yd³	0.765	m³
To Obtain	**By**	**Divide**	**To Obtain**	**By**	**Divide**

Table 13.20 Unit conversion factors.

Mass

lb	kg
1	= 0.45359
2.20462	= 1

Volume

in^3	ft^3	gal	litre	m^3
1	$= 5.787 \times 10^{-4}$	$= 4.329 \times 10^{-3}$	$= 0.0163871$	$= 1.63871 \times 10^{-5}$
1728	1	7.48055	28.317	0.028317
231.0	0.13368	1	3.7854	0.0037854
61.02374	0.035315	0.264173	1	0.001
61,023.74	35.315	264.173	1000	1

Density

lb/ft^3	lb/gal	g/cm^3	$kg/m^3(g/L)$
1	$= 0.133680$	$= 0.016018$	$= 16.018463$
7.48055	1	0.119827	119.827
62.4280	8.34538	1	1,000
0.0624280	0.008345	0.001	1

Specific Volume

ft^3/lb	gal/lb	cm^3/g	$m^3/kg(L/g)$
1	$= 7.48055$	$= 62.4280$	$= 0.0624280$
0.133680	1	8.34538	0.008345
0.016018	0.119827	1	0.001
16.018463	119.827	1000	1

Specific Heat or Entropy

Btu/lb • °F	cal/(g • K)	J/(g • K)
1	$= 1$	$= 4.184$
1.0	1	4.184
0.2390	0.2390	1

Enthalpy

Btu/lb	cal/g	J/g
1	$= 0.55556$	$= 2.3244$
1.8*	1	4.184
0.43021	0.2390	1

Thermal Conductivity

Btu/h • ft • °F	cal/(s • cm • °C)	J/(s • cm • °C)	W/(cm • °C)	W/(m • K)
1	$= 4.1338 \times 10^{-3}$	$= 0.17296$	$= 0.017296$	1.7296
241.91	1	4.184	4.184	418.4
57.816	0.2390	1	1	418.4

14

LABORATORIES AND TEST FACILITIES

Listed below are a number of national standards calibration laboratories and institutions in various countries. The list is not complete, but covers most of the known facilities in the world today. The laboratories are listed alphabetically by country. Also included is a listing of a number of test facilities and certification and accreditation organizations in Europe and the USA.

I. National Calibration Laboratories

AUSTRALIA — CSIRO
Commonwealth Scientific and Industrial Research Organisation
Bradfield Road, West Linfield, Sydney, NSW Australia

CHINA — NIM
National Institute of Metrology
Bejing, China

FINLAND — CMA
Centre for Metrology and Accreditation
P.O.Box 239, 00181 Helsinki, Finland

FRANCE — CETIAT
Centre Technique des Industries Aerauliques et Thermiques
27-29, bd du 11 Novembre 1918, B.P. 6084, 69604 Villeurbanne Cedex, France

GERMANY— PTB
Physikalisch-Technische Bundesanstalt
Furstenwalder Damm 388, D-1162 Berlin, Germany

ITALY — IMGC
Instituto di Metrologia "Gustavo Colonnetti"
Strada delle Cacce 73, 1- 101135 Torino, Italy

JAPAN — NRLM
National Research Laboratory of Metrology
1-4 Umezono, 1-Kome, Tsukuba, Ibaraki 305, Japan

NETHERLANDS — VSL/NMI
Nederlands Meetinstituut
P.O. Box 654, 2600AR Delft, Schoenmakerstraat 97, The Netherlands

NETHERLANDS — TFDL
University of Wageningen, Technische en Fysische Dienst voor de Landbouw
Mansholtlaan 12, 6708 PA, Wageningen, The Netherlands

POLAND — COM/GUM
Central Office of Measures—Glowny Urzad Miar—Physical Chemistry Division
Humidity Laboratory, 2 Electoralna Str., 00-139 Warsaw, Poland

SINGAPORE — SISIR
Singapore Institute of Standards and Industrial Research
1 Science Park Drive, 0511 Singapore

SPAIN — INTA
Instituto Nacional De Tecnica Aeroespacial
Ctro Ajalvar, 28850 Torrejon de Ardoz, Madrid, Spain

SWITZERLAND — OFMET
Swiss Federal Office of Metrology
Lindenweg 50, CH-3084 Wabern, Switzerland

SOUTH KOREA — KRISS
Korea Standards Research Institute
P.O. Box 3, Taedok, Science Town, Taejon 305-606, Korea

TAIWAN — Center for Measurement Standards
Industrial Technical Research Institute, 321, SEC 2, Kuang-Fu Road
Hsinchu 30042, Taiwan, ROC

UNITED KINGDOM — NPL
National Physical Laboratory, Division of Mechanical and Optical Metrology,
Teddington, Middlesex, TW11-0LW, United Kingdom

UNITED STATES OF AMERICA — NIST
National Institute of Standards and Technology
Building 221, Room A-303, Gaithersburg, MD 20899, USA

II. Certification and Accreditation Organizations

A. Europe

BSI	British Standards Institution, 389 Chiswick High Road, London W4 4AL, United Kingdom
CoSTIC	Comité Scientifique et Technique des Industries Climatiques France
EAL	European Cooperation for Accreditation of Laboratories, PO Box 29152, 3001 GD Rotterdam, Netherlands.
EXERA	France
InstMC	Institute of Measurement and Control, 87 Gower Street London, WC1E 6AA, United Kingdom
ISO	International Organization for Standardization, Case Postale 56 CH-211, Geneva 20, Switzerland.
NAMAS	National Accreditation of Measurement and Sampling, Operated by UKAS, Queens Road, Teddington, Middlesex TW11 0NA, London, UK
UKAS	United Kingdom Accreditation Service, Queens Road, Teddington, Middlesex TW11 0NA, London, United Kingdom
Sira	South Hill, Chislehurst, Kent BR7 5EH, United Kingdom
SIREP	South Hill, Chislehurst, Kent BR7 5EH, United Kingdom
WIB	The Netherlands

B. United States

NBS	National Bureau of Standards (USA), superseded by NIST.
ASHRAE	American Society of Heating, Refrigerating, and Air Conditioning Engineers, 1791 Tullie Circle, NE, Atlanta, GA 30329, USA

15

REFERENCES AND SOURCES FOR
FURTHER INFORMATION

I. Introduction

The references and sources for further information are listed by category and are referred to the chapter to which they apply. Evidently, the references often apply to other chapters as well.

When in a chapter of the book reference is made to a particular publication by number, the number applies to the references listed for that category and for that chapter. Hence if for example in Chapter 3, Chilled mirror hygrometers, reference is made to Publication 6, one should look for Reference 6 in the Reference Section III for chilled mirror hygrometers. If in Chapter 3 reference is made to Publication 12 on calibration, this will be specifically indicated by "(12), Section X, calibration."

In each listing, the references are listed by year, the earliest years being listed first. Infrared and Lyman–Alpha hygrometers discussed in Chapter 6 are primarily used for upper atmosphere measurements and studies. References directly related to these instruments are listed in Section VI of this chapter. However, many references listed in the chapter on meteorological instruments and calibration are also relevant to these instruments and should also be consulted if further study is desired.

II. Definitions

1. *CRC Handbook of Chemistry and Physics*, CRC Press, London, 76th Edition, ISBN 0-8493-0476-8.

2. Kaye and Laby, *Tables of Physical and Chemical Constants*, 16th Edition, Harlow: Longman, ISBN 0-582-22629-5.

3. F. C. Quin, The Most Common Problem of Moisture Measurement and Control, *Proceedings of the 1985 International Symposium on Moisture and Humidity*, Washington, D.C.

4. General Eastern Instruments, *Humidity Handbook*, Doc. No. A 40103384, Rev B.00, May 1993.

5. *ASHRAE Handbook—Fundamentals,* American Society of Heating, Refrigeration and Air Conditioning Engineers, 1993.

6. *Method for Measurement of Moist Properties,* ANSI/ASHRAE 41.6-1994, American Society of Heating, Refrigeration and Air Conditioning Engineers, Inc., 1994.

7. *A guide to Measurement of Humidity,* ISBNO-904457-24-91, The Institute of Measurement and Control and NPL (UK), 1996.

III. Chilled Mirror

Conventional Chilled Mirror Hygrometer

1. J. A. Goff, and S. Gratch, Low pressure Properties of Water from −160°F to 212°F, *Transactions ASHVE,* 52: 95, (1946).

2. J. H. Keenan, and F. G. Keyes, *Thermodynamic Properties of Steam,* John Wiley & Sons, (1959).

3. C. C. Francisco, and D. J. Beaubien, Humidity and Moisture, *Measurements and Control in Science and Industry,* Vol. 1, Reinhold Publishing Company, (1965).

4. A. Wexler, and R. S. Ruskin, Principles and Methods of Measuring Humidity in Gases, *Measurements and Control in Science and Industry,* Vol.1, Reinhold Publishing Company, (1965).

5. *AFCRL Report No. 68–154,* Error Analysis of the Humidity-Temperature Measuring Set, AN/TMQ-11, AFCRL, Bedford, MA 01730, March, 1968

6. P. R. Wiederhold, Humidity Measurements and Guidelines for Selecting Humidity Sensors, *Intech,* (1975).

7. R. C. Reid, J. M. Prausnitz, and T. K. Sherwood, *The Properties of Gases and Liquids,* pp. 181–186, Third Edition, McGraw Hill Book Co., (1977).

8. J. C. Harding, Jr., Dew Point Hygrometer with Contaminant Error Immunity, *Measurement & Control,* Feb. 1979.

9. R. J. List, Smithsonian Meteorological Tables, *Publication No. 4014,* Smithsonian Institution, Washington, D.C., 1979.

10. D. L. Curtis, Temperature Calibration and Interpolation Methods for Platinum Resistance Thermometers, *RMT Report No. 68023F,* Rosemount, Inc., July 1980.

11. A. L. Buck, Short Equation for Computing Vapor Pressure and Enhancement Factors, *National Center for Atmospheric Research,* Feb. 1981.

12. A. L. Buck, New Equations for Computing Vapor Pressure and Enhancement Factor, *Journal of Applied Meteorology,* Vol. 20, No. 12, 1981.

13. S. Weisman, Designing a Sampling System for Dew Point Measurements, *Instrument & Control Systems,* 1983.

14. V. B. Cortina, Sampling Systems for Chilled Mirror Dew point Hygrometers, *Proceedings of the 1985 International Symposium on Moisture and Humidity,* Washington, D.C.

15   V. J. Dosoretz, A Portable Optical Chilled Mirror Hygrometer, *Proceedings of the 1985 International Symposium on Moisture and Humidity,* Washington, D.C.

16.   M. Brownawell and A. Sheikholeslami, Thermally Aspirated Condensation Hygrometer, *Proceedings of the 1985 International Symposium on Moisture and Humidity,* Washington, D.C.

17.   V. J. Dosoretz, Chilled Mirror Hygrometry, *Proceedings of the Sensors Expo,* Detroit, MI, (1987).

18.   A. L. Buck, Dew/Frost Precision Below Freezing. *General Eastern Technical Note #1,* (1987).

19.   R. F. Pragnell, The Modern Condensation Dew Point Hygrometer, *Measurement and Control,* Vol. 22, April 1989.

20.   R. F. Pragnell, Dew and Frost Formation on the Condensation Dew Point Hygrometer, *Measurement and Control,* Vol. 26, April 1993.

CCM Hygrometer

21.   F. G. Cooper, Dew Point Versus Frost Point Measurements, *Sensors,* Oct. 1991.

22.   F. Dadachanji, Humidity Measurement at Elevated Temperatures, *Measurement and Control,* Vol. 25, March 1992.

23.   P. R. Wiederhold, Sensors and Instrumentation for the Detection and Measurement of Humidity, *Proceedings of Sensors Expo East,* Boston, MA., May 1995.

24.   P. R. Wiederhold, The Cycling Chilled Mirror Dew Point Hygrometer, *Sensors,* July 1996.

Transmitters

25.   J. C. Harding, Jr., Chilled Mirror Dew Point Sensors/Transmitter, *Measurement & Control,* Feb. 1984.

26.   J. C. Harding, Jr., A Chilled Mirror Dew Point Sensor/Psychrometric Transmitter for Energy Monitoring and Control Systems, *Proceedings of the 1985 International Symposium on Moisture and Humidity,* Washington, D.C., 1985.

Cryogenic Hygrometer

27.   J. E. Dye, A Cautionary Note on the use of Dew Point-Frost Point Hygrometers for Frost Measurements, *Journal of Applied Meteorology,* pp. 1230–1231, (1973).

28.   A. L. Buck, and R. Clark, Development of a Cryogenic Dew/Frost Point Hygrometer, *American Meteorological Society Symposium,* Jan. 1991.

29.   P. Spyers-Duran, An Airborne Cryogenic Frost Point Hygrometer, 7th Symposium, Meteorological Observations and Instrumentation, *American Meteorological Society,* 1991.

30.   R. Busen, and A. L. Buck, The CR-1 Cryogenic Hygrometer on Board the DLR
      Falcon, 8th Symposium on Meteorological Observations and Instrumentation, *73rd*
      American Meteorological Society Meeting, January 1993.

31.   A. L. Buck, R. E. Pressey, D. F. Yesenofski and D. A. Zatko, A Cryogenic, Low PPB
      Range Moisture Analyzer for Process Gases, *Paper at 39th Annual Technical*
      Meeting of the Instrumentation for Environmental Sciences, Las Vegas, NV, May,
      1993.

IV. Relative Humidity

1.   G. O. Handegord, C. O. Hedlin, and F. N. Trofimenkoff, A Study of the Accuracy
      of Dunmore Type Humidity Sensors, *Humidity and Moisture. Measurements and*
      Control in Science and Industry, Reinhold Publishing Company, Vol. 1. p.265, 1965.

2.   ASHRAE Standard 41.1-1986 (RA 91), *Standard Method for Temperature*
      Measurement. American Society of Heating, Refrigerating and Air Conditioning
      Engineers, Atlanta, GA, 1974.

3.   P. R. Wiederhold, Humidity Measurements—Part 1: Psychrometers and Percent RH
      Sensors, *Instrument Technology,* 1975.

4.   *ASHRAE Brochure on Psychrometry,* American Society of Heating, Refrigerating
      and Air-Conditioning Engineers, Atlanta, GA, 1977.

5.   R. B. Stewart, R. T. Jacobsen, and J. H. Becker, Formulations for the
      Thermodynamic Properties of Moist Air at Low Pressures as Used for Construction
      of New ASHRAE SI Unit Psychrometric Charts, *ASHRAE Transactions* 89(2):
      536–541, 1983.

6.   H. K. Modi, Humidity Control and its Importance, *Proceedings of the 1985*
      International Symposium on Moisture and Humidity, Washington, D.C.

7.   A. Pharo Gagge, Standard Indices of Human Response to the Thermal Humid
      Environment, *Proceedings of the 1985 International Symposium on Moisture and*
      Humidity, Washington, D.C.

8.   M. S. Audi, Assignment of Psychrometric Problems for Pocket Computers,
      Proceedings of the 1985 International Symposium on Moisture and Humidity,
      Washington, D.C.

9.   C. H. McClellan, Fundamentals and Current Practice for Comfort Evaporative
      Cooling, *Proceedings of the 1985 International Symposium on Moisture and*
      Humidity, Washington, D.C.

10.   W. A. Clayton, Contaminant Resistant Capacitive Humidity Sensor, *Proceedings of*
      the 1985 International Symposium on Moisture and Humidity, Washington, D.C.

11.   K. Shiba, J. Kitamura, and T. Yamaguchi, Elongation Hygrometer, *Proceedings of*
      the 1985 International Symposium on Moisture and Humidity, Washington, D.C.

12.   K. Carr-Brion, *Moisture Sensors in Process Control,* Elsevier Science Publishing
      Co., New York, 1987.

13. M. Brownawell, An RH Sensor Review with HVAC Considerations, *Sensors*, March, 1989.

14. R. F. Pragnell, Recording Humidity: No Need to Lose Your Hair, *Environmental Engineering*, Sept. 1990.

15. W. A. Clayton, Improved Capacitive Moisture Sensors, *Sensors*: July, (1993)

16. ASHRAE, *1993 ASHRAE Handbook—Fundamentals*, 1993.

17. ASHRAE, Method for Measurement of Moist Air Properties, *Publication 41.6-1994*, Atlanta, GA 30329.

18. E. Nowak, Moisture Measurement with Capacitive Polymer Humidity Sensors, *Sensors*, Vol. 13, No. 10, pp. 88–91, October 1996.

V. Trace Moisture

Aluminum Oxide Hygrometer

1. E. M. Landsbaum, W. S. Dodds, and L. F. Stutzman, Humidity of Compressed Air, *Industrial Engineering Chemistry*, Vol. 47, (1955).

2. R. F. Bukacek, Equilibrium Moisture Content of Natural Gas, *Research Bulletin No. 8*, Institute of Gas Technology, Chicago, IL, (1955).

3. F. J. Brousaides, An Evaluation of the Aluminum Oxide Humidity Element, *AFCRL Report No. 68-0547*, Bedford, MA 01730, October, 1968.

4. P. F. Bennewitz, A New Bulk Effect Humidity Sensor, *Measurement & Data*, Vol. 8, No. 1, (1974).

5. D. Schleck, Measurements of Upper Atmosphere Water Vapor made in Situ with a New Moisture Sensor, *Geological Research Letter, No. 6*, (1979).

6. R. S. Jachowicz, MOS Type Miniature Capacitance Sensor, *Proc. EUROCON '80*, Stuttgard, Germany, (1980).

7. R. S. Jachowicz, and J. Golaszewski, New Construction of MCP-MOS Type Humidity Sensor, *Proceedings ESSDERC '82*, Munich, Germany, (1982).

8. S. D. Senturia, et al., Monolithic Integrated Circuit Implementations of the Charge-Flow Transistor Oscillator Moisture Sensor, *Sensors and Actuators*, No. 1, 1981/1982.

9. R. S. Jachowicz, and S. D. Senturia, A Thin-Film Humidity Sensor, *Sensors and Actuators*, Vol. 2. 1981/1982.

10. R. S. Jachowicz, An Improved Theoretical Model of Semiconductor MCP-MOS Type Humidity Sensor, *Proceedings of the International Meeting on Chemical Sensors*, Fukuoka, Japan, 1983.

11. G. Delapierre, et al., Polymer-Based Capacitive Humidity Sensor; Characteristics and Experimental Results, *Sensors and Actuators*, No. 4, (1983).

12. K. Yuuki, et al., Low Power Consumption Solid State Humidity Sensor, *IEEE Transactions on Consumer Electronics*, Vol. CE-29, No. 3, 1983.

13. K. S. Kulpa, and R. S. Jachowicz, Thin Film Humidity Sensor — An Effective Medium Approach to Operational Model, *Proceedings of the Symposium on Electrochemical Sensors*, Rome, Italy, 1984.

14. J. C. Harding, Jr., Overcoming Limitations Inherent to Aluminum Oxide Humidity Sensors, *Proceedings of the 1985 International Symposium on Moisture and Humidity*, Washington, D.C.

15. R. S. Jachowics, and P. Dumania, Evaluation of Thin-Film Humidity Sensor Type MCP-MOS, *Proceedings of the 1985 International Symposium on Moisture and Humidity*, Washington, D.C.

16. R. H. Hammond, Aluminum Oxide Hygrometry and Water Enhancement Factors, *Proceedings of the 1985 International Symposium on Moisture and Humidity*, Washington, D.C.

17. R. H. Hammond, and D. Schleck, A Calibration System for Producing Low Frost Points, *Proceedings of the 1985 International Symposium on Moisture and Humidity*, Washington, D.C.

18. V. Fong, Al_2O_3 Moisture Sensor Chips for Inclusion in Microcircuit Packages and the new MIL Standard for Moisture Content, *Proceedings of the 1985 International Symposium on Moisture and Humidity*, Washington, D.C.

19. R. Byant and M. Scelzo, The Impact of Microprocessors on Hygrometry, *Proceedings of the 1985 International Symposium on Moisture and Humidity*, Washington, D.C.

20. K. G. Carr-Brion, *Moisture Sensors in Process Control*, Elsevier Science Publishers Ltd, UK, 1986.

21. H. Rainer, Kalibrationsfreie Verschmutzungs Kompensierende Taupunkt Temperatur Messeinrichtung fur Industrielle Prozesse, *Technishe Messen*, Germany, Oct. 1988.

22. D. A. Clabo, A Calibration-Free Direct Dew Point Analyzer with Contamination Compensation for Industrial Processes, ISA, *Paper # 90-452*, (1990).

23. W. Cole, and M. Nahn, Advanced Hygrometry for Dew/Frost Point Measurements, *Sensors*, Vol. 9, No. 7, July 1992.

Electrolytic Hygrometer

24. A. Wexler and W. G. Brombacher, Methods of Measuring Humidity and Testing Hygrometers, *National Bureau of Standards Circular No. 512*, Sept. 1951.

25. F. A. Keidel, A Novel, Inexpensive Instrument for Accurate Analysis for Traces of Water, *1956 Pittsburgh Conference on Analytical Chemistry and Applied Spectroscopy*, (1956).

26   A. Wexler, Electric Hygrometers, *National Bureau of Standards Circular No. 586*, Sept., 1957.

27. M. Czuha, Jr. Adaptation of the Electrolytic Moisture Detector to Atmospheric Humidity Measurement, *Humidity and Moisture*, Measurement and Control in Science & Industry 1963 International Symposium on Humidity and Moisture, (1965).

28. R. H. Jones and A. Petersen, Additional Performance Data on a New Electrolytic Hygrometer Cell, *Humidity and Moisture,* Measurement and Control in Science & Industry 1963 International Symposium on Humidity and Moisture, (1965).

29. M. Smith, J. Mitchell, Jr., Coulometric Hygrometry, *Aquametry,* Part 2, Sec. Ed. John Wiley & Sons, N.Y., 1983.

30. S. Ronchinsky, An Electrochemical Cell. for Trace Moisture in Gases, Moisture and Humidity, *Measurement and Control in Science & Industry 1985 International Symposium on Moisture and Humidity,* Washington, D.C.

31. K. Sugiyama, and T. Ohmi, Ultraclean Gas Delivery Systems, Part 1, *Micro Contamination,* Nov. 1988.

32. L. M. Gates, An Electrochemical Sensor for Trace Moisture Measurement in Gases, *Sensors Expo Proceedings,* 103B-1, (1990).

33. F. Mermoud, M. D. Brandt, and J. McAndrew, Low Level Moisture Generation, *Analytical Chemistry,* Vol. 63, pp. 198–202, 1991.

34. J. Mettes, C. Haggerty, and E. Zoladz, Moisture Monitoring in High Purity Gas Distribution Systems, *Ultraclean Manufacturing Conference,* 1993.

35. C. Ma, F. Shadman, and J. Mettes, Evaluating the Trace-Moisture Measurement Capability of Coulometric Hygrometry, *MICRO,* April 1995.

Piezoelectric Hygrometer

36. W. H. King, Jr., The Piezoeletric Sorption Hygrometer, Humidity and Moisture, *Measurement and Control in Science & Industry 1963 International Symposium on Humidity and Moisture,* (1965).

37. W. H. King, Jr., Using Quartz Crystals as Sorption Detectors, *Part I: Research & Development,* Vol. 20, No. 4, pp. 28–34, F.D. Chicago, Ill., Thompson Publications, (1969).

38. W. H. King, Jr., Using Quartz Crystals as Sorption Detectors, *Part II: Research & Development,* Vol. 20, No. 5, pp. 28–33, F.D. Chicago, Ill., Thompson Publications, (1969).

39. M. J. Hartigen, Jr., The Use of a Piezoelectric Device for Amine Classification and Sulfur Dioxide Detection, *Chemical Analysis,* p 8, Univ. of Rhode Island, (1970).

40. D. C. Cornish, G. Jepson, and M. J. Smurthwaite, *Sampling Systems for Process Analyzers,* Butterworths, London, 1981.

41. C. B. Blakemore, Piezoelectric Moisture Analyzers for Semiconductor Gases, *Analysis Instrumentation,* 21, I.S.A., pp. 127–128, (1985).

42. C. B. Blakemore, and W. B. Baker, Piezoelectric Moisture Analysis, *Moisture Symposium,* National Physics Laboratory, (U.K.), Sept. 1986.

43. C. B. Blakemore, J. C. Steichen, and G. Dallas, Continuous Trace Moisture Analysis, *International Conference on Gas Quality,* University of Groningen, Netherlands, April, 1986.

VI. Optical Absorption Hygrometers

Infrared Hygrometer

1. F. E. Fowle, The Spectroscopic Determination of Aqueous Vapor, *Journal of Astrophysics*, 35, 3, 149–162, (1912).

2. J. W. C. Johns, The Absorption of Radiation by Water Vapor, *Humidity and Moisture Measurement and Control in Science & Industry 1963 International Symposium on Humidity and Moisture*, (1965).

3. W. F. Staats, et al., Infrared Absorption Hygrometer, *Humidity and Moisture Measurement and Control in Science & Industry 1963 International Symposium on Humidity and Moisture*, (1965).

4. R. C. Wood, The Infrared Hygrometer - its Application to Difficult Humidity Measurement Problems, *Humidity and Moisture Measurement and Control in Science & Industry 1963 International Symposium on Humidity and Moisture*, (1965).

5. D. L. Brooks, Development of an Infrared Absorption Hygrometer using Solid State Energy Sources, M.S. Thesis, Dept. of Atmospheric Sciences, University of Washington, Seattle, WA, (1971).

6. L. V. Bogolomova, V. I. Dianov-Klokov and S. L. Zubkovsky, Double Beam Infrared Spectrometer for Measuring Humidity Fluctuations in the Atmosphere, Isvestia, *Atmospheric and Oceanic Physics*, 10,9, 933–942, (1974).

7. C. Tomasi, and R. Guzzi, High Precision Atmospheric Hygrometry using the Solar Infrared Spectrum, *Journal of Physics & Scientific Instruments*, 7, 647–649, (1974).

8. P. Hyson, and B. B. Hicks, A Single Beam Infrared Hygrometer for Evaporation Measurement, *Journal of Applied Meteorology*, 14, 3, 301–307, (1975).

9. A. L. Buck, New Equations for Computing Vapor Pressure and Enhancement Factor. *Journal of Applied Meteorology*, 20, 12, 157–1532, (1981).

10. L. D. Nelson, Non-Contact Sensing of Atmospheric Temperature, Humidity, and Supersaturation, *AMS Conference on Cloud Physics*, Chicago, IL, USA, 293–29 (1982).

11. J. T. Priestly, and R. J. Hill, Measuring High Frequency Humidity, Temperature, and Radio Refractive Index in the Surface Layer, *Journal of Atmospheric and Oceanographic Technology*, 2, 233–251, (1985).

12. J. E. Tillman, Near Infrared Humidity Techniques Using Semiconductor Sources: Incoherent Sources and Theoretical Calculations in the Presence of Cloud and Fog, *Proceedings of the 1985 International Symposium on Moisture and Humidity*, Washington, D.C.

13. K. Schurer, R. Maandonks and G. J. W. Visscher, Infrared Measurement of Water Vapour Fluctuations, *Proceedings of the 1985 International Symposium on Moisture and Humidity*, Washington, D.C.

14. S. A. Clough, F. X. Kneizys, E. P. Shettle, and G. P. Anderson, Atmospheric Radiance and Transmittance: FASCOD2, *Sixth Conference on Atmospheric Radiation,* Williamsburg, VA, (1986).

15. T. A. Cerni, Research Directed Toward an Atmospheric Infrared Transmission Hygrometer, *Final Report U.S. Army Contract No. DAAL01-85-C-0177,* (1986).

16. T. A. Cerni, D. Hauschultz, L. D. Nelson, and D. Rottner, An Atmospheric Infrared Hygrometer, Sixth Symposium Meteorological Observation and Instrumentation, *American Meteorological Society,* 205–208, (1987).

17. J. H. Pierluisi, C. E. Maragoudakis, and R. Tehrani-Movahed, New LOWTRAN Band Model for Water Vapor, *Journal of Applied Optics,* 18, pp. 3792–3795, (1989).

18. W. L. Wolfe, and G. J. Zissis, *The Infrared Handbook,* Environmental Research Institute of Michigan and SPIE, Bellingham, WA, 98227, (1989).

19. A. M. Kahan, Infrared Hygrometer, *Measurement & Control,* Feb. 1989.

Lyman–Alpha Hygrometer

20. F. H. Spedding, A. S. Newton, J. C. Wharf, O. Johnson, R. W. Notorf, I. B. Johns, and A. H. Daane, Uranium Hydride; Preparation, Composition, and Physical Properties, *Nucleonics,* I., pp. 4–15, (1949).

21. G. H. Dicke, and S. P. Cunningham, A New Type of Hydrogen Discharge Tube, *Journal of the Optical Society of America,* Vol. 42, pp. 167–189, (1952).

22. K. Watanabe, M Zelikoff, and E. C. Inn, Absorption Coefficients of Several Atmospheric Gases, AFCRL Techn. Report No. 52–23, *Geophysical Research Papers,* No. 21, June, 1953.

23. J. E. Tillman, Atmospheric Humidity Measurements by Ultraviolet Techniques, M.S. Thesis, Dept. of Meteorology, M.I.T., Cambridge, MA, June, 1961.

24. R. W. Kreplin, T. A. Chubb, and H. Friedman, X-Ray and Lyman-Alpha Emission from the Sun as Measured from the NRL-1 Satellite, *Journal of Geophysics Research,* 67, pp. 2231–2253, (1962).

25. J. E. Tillman, Water Vapor Density Measurements Utilizing the Absorption of Vacuum Ultraviolet and Infrared Radiation, *Humidity and Moisture Measurement and Control in Science & Industry 1963 International Symposium on Humidity and Moisture,* (1965).

26. D. L. Randall, T. E. Henley and O. K. Larison, The NRL Lyman–Alpha Humidiometer, *Humidity and Moisture Measurement and Control in Science & Industry 1963 International Symposium on Humidity and Moisture,* (1965).

27. J. A. R. Samson, *Techniques of Vacuum Ultraviolet Spectroscopy,* John Wiley & Sons, New York, 1967.

28. S. C. Mattix, Development and Analysis of a Lyman–Alpha Absorption Hygrometer, M.S. Thesis, Dept. Elec. Eng., Univ. of Colorado, Boulder, CO, (1971).

29. S. C. Mattix, Development and Analysis of a Lyman–Alpha Absorption Hygrometer, NCAR Coop. Thesis No. 29, (1973).

30. A. L. Buck, Error Sensitivity of Fixed-and Variable Path Lyman–Alpha Hygrometers, *NCAR Technical Note No. EDD-103,* (1975).

31. A. L. Buck, The Variable Path Lyman–Alpha Hygrometer and its Operating Characteristics, *Bulletin of the American Meteorological Society*, 57: 1113–1118, (1976).

32. A. L. Buck, Lyman–Alpha Radiation Source with High Spectral Purity, *Applied Optics*, 16, No.10: 2634–2636, (1977).

33. A. L. Buck, New Equations for Computing Vapor Pressure and Enhancement Factor, *Journal of Applied Meteorology*, 20, 12: 157–1532, (1981).

34. E. L. Andreas, The Effects of Volume Averaging on Spectra Measured with a Lyman–Alpha Hygrometer, *Journal of Applied Meteorology.*, 20, pp. 467–475, (1981).

35. F. Goutail, P. Mestayer, M. Coantic, and J. L. Bertaux, Prototype D'Hygrometre A Raie Lyman–Alpha Miniaturese, *La Meteorologie VIe* Serie No. 29–30, (1982).

36. J. T. Priestly, and W. D. Cartwright, Frequency Response Measurements on Lyman–Alpha Humidiometers, *NOAA Technical Memo*, ERL WPL-92, (1982).

37. C. A. Friehe, Path length Sensitivity of the Lyman–Alpha Humidiometer, *NCAR Technical Note No. 190+EDD*, (1982).

38. E. J. McCartney, Absorption and Emission by Atmospheric Gases, *The Physical Processes*, Wiley, New York, (1983).

39. J. T. Priestly, and R. J. Hill, Measuring High Frequency Humidity, Temperature, and Radio Refractive Index in the Surface Layer: Simultaneous Observations with Lyman–Alpha and Near-Infrared Hygrometers, *Journal of Climate and Applied Meteorology*, (1984).

40. A. L. Buck, The Lyman–Alpha Absorption Hygrometer, *Proceedings of the 1985 International Symposium on Moisture and Humidity*, Washington, D.C.

41. P. Mestayer, C. Rebattet, and F. Goutail, Miniaturized Lyman–Alpha Hygrometer to Measure Humidity Turbulent Fluxes in Tunnel and Atmosphere, *Proceedings of the 1985 International Symposium on Moisture and Humidity*, Washington, D.C.

42. M. Wada, Y. Iwasaka, S. Murabayshi, and Y, Yamashita, Feasible Study of Monitoring the Content of Water Vapor under a very Dry Atmospheric Condition by Means of a Lyman–Alpha Hygrometer, *Proceedings of the 1985 International Symposium on Moisture and Humidity*, Washington, D.C.

43. H. P. Fimpel, The DFVLR Meteorological Research Aircraft Falcon-E, *Proceedings of the Sixth Symposium on Meteorological Observations and Instrumentation*, American Meteorological Society Meeting, New Orleans, pp. 113–116, (1987).

44. A. L. Buck, and R. D. Horn, The DLR Lyman–Alpha Hygrometer, *Seventh American Meteorological Society Symposium*, New Orleans, LA, (1991).

VII. Psychrometers

1. J. H. Arnold, The Theory of the Psychrometer, *Physics 4*, 255 and 334 (1935).

2. D. D. Wile, Psychrometry in the Frost Zone, *Refrigeration Engineering*, 48, 291 (1944).

3. L. Greenspan, and A. Wexler, An Adiabatic Saturation Psychrometer, *Journal of Research of the NBS,* Vol. 72C, No. 1, 1968.

4. W. J. Shaw, and J. E. Tillman, A Corretion for Different Wet-Bulb and Dry-Bulb Response in Thermocouple Psychrometry, *Moisture and Humidity,* Measurement and Control in Science & Industry 1985 International Symposium on Moisture and Humidity, Washington, D.C.

5. R. G. Wiley, and T. Lalas, Accurate Psychrometer Coefficients for Wet and Ice-Covered Cylinders in Laminar Transverse Airstreams, *Moisture and Humidity,* Measurement and Control in Science & Industry 1985 International Symposium on Moisture and Humidity, Washington, D.C.

6. K. L. Christianson and K. N. Newhouse, Analysis of Wet Bulb Psychrometry, *Proceedings of the 1985 International Symposium on Moisture and Humidity,* Washington, D.C.

7. K. Shiba, M. Ueda, and M. Haraguhi, Precision Thermoelectric Psychrometer, *Proceedings of the 1985 International Symposium on Moisture and Humidity,* Washington, D.C.

8. J. Fan, Determination of the Psychrometer Coefficient A of the WMO Reference Psychrometer by Comparison with a Standard Gravimetric Hygrometer, *Journal of Atmospheric and Oceanographic Technology,* 4: 239–244, (1987).

9. ASHRAE Handbook - Fundamentals - Second Edition, American Society of Heating, Refrigeration and Air Conditioning Engineers, Inc. Atlanta, GA, USA. Publication ISBN 0 910110 97 2, Chapter 6, *Psychrometrics,* 1993.

VIII. Other Types

Fiber Optic Humidity Analyzer

1. A. P. Russel and K. S. Fletcher, Optical Sensor for the Determination of Moisture, *Anal. Chim. Acta,* Vol. 170, pp. 209–16, 1985.

2. D. S. Ballantine and H. Wohltjen, Optical Waveguide Humidity Detector, *Analytical Chemistry,* Vol. 58. No. 13, pp. 2883–5, 1986.

3. Q. Zhou, M. R. Shariari, D. Kritz, and G. H. Sigel, Jr., Porous Fiber-Optik Sensor for High-Sensitivity Humidity Measurements, *Analytical Chemistry,* Vol. 60, 20, pp. 2317–20, 1988.

4. K. Ogawa, S. Tsuchiya, H. Kawakami, and T. Tsutsui, Humidity-Sensing Effects of Optical Fibers with Microporous SiO_2 cladding, *Electron Letters,* 24, pp. 42–3, 1988.

5. G. Xin and H. Shanglian, Optical Fiber Humidity Sensor, SPIE Vol. 1169, *Fiber Optic and Laser Sensors VII,* 582–5, 1989.

6. S. Muto, A. Fukasawa, M. Kamimura, F. Shinmura, and H. Ito, Fiber humidity Sensor using Fluorescent Dye-Doped Plastics, *Japanese Journal of Applied Physics,* Vol. 28, No. 6, 1989.

7.   F. Mitschke, Fiber-Optic Sensor for Humidity, *Optical Letters*, 14, 17, pp. 967–9, 1989.

IX. Meteorological Instruments

1.   J. C. Johnson, *Physical Meteorology*, The Technology Press of the Massachusetts Institute of Technology.

2.   Meteorological Office, London: HMSO *Handbook of Meteorological Instruments*, Volume 3, Measurement of Humidity, Second Edition, ISBN 0 11 400325 4.

3.   H. J. Mastenbrook, and J. E. Dinger, Distribution of Water Vapor in the Stratosphere, *Journal of Geophysical Research*, Vol. 66, pp. 1437–1444, May 1961.

4.   H. J. Mastenbrook, A Control System for Ascent-Descent Balloon Soundings of the Atmosphere, *Journal of Applied Meteorology*, Vol. 5, pp. 737–740, Oct. 1966.

5.   J. E. Harries, The Distribution of Water Vapor in the Stratosphere, *Journal of Applied Meteorology*, Vol. 14, pp. 565–575, Nov. 1976.

6.   H. J. Mastenbrook, Water Vapor Distribution in the Stratosphere and Higher Troposphere, *Journal of Atmospheric Sciences*, Vol. 25, pp. 299–3111, (1968).

7.   W. Pollock, L. E. Heidt, R. Lueb, and D. Ehhalt, Measurement of Stratospheric Water Vapor by Cryogenic Collection, *Journal of Geophysical Research*, Vol. 85, pp. 5555–5568, Oct. 1980.

8.   H. J. Mastenbrook, and S. J. Oltmanns, Stratospheric Water Vapor Variability for Washington, D.C./Boulder, CO, *Journal of Atmospheric Sciences*, Vol .40, pp. 2157–2165, (1983).

9.   S. J. Oltmans, Measurements of Water Vapor in the Stratosphere with a Frost-Point Hygrometer, *Moisture and Humidity, Measurement and Control in Science & Industry, 1985 International Symposium on Moisture and Humidity*, Washington, D.C.

10. J. M. Pike, Field Calibration of Humidity Instruments in the Natural Atmosphere, *Proceedings of the 1985 International Symposium on Moisture and Humidity*, Washington, D.C.

11. G. S. Brown, A Balloon Borne Frost Point Hygrometer for High Altitude Low Water Vapor Concentration Measurements, *Sandia National Laboratory Report* SAND 88-2467, (1988).

12. D. Sonntag, Important New Values of the Physical Constants of 1986, Vapour Pressure Formulations Based on the ITS-90, and Psychrometer Formulas, *Zeitschrift fur Meteorologie*, Vol. 40, (5), pp. 340–344, 1990.

13. J. Strom, P. R. A. Brown, R. Busen, and B. Guillemet, The Pre-EUCREX In-Flight Humidity Intercomparison, 8th Symposium, *Meteorological Observation Instruments*, American Meteorological Society, Anaheim, CA, 1993.

14. D. Sonntag , Advancements in the field of hygrometry, *Meteorologische Zeitschrift*, N.F. Vol. 3, pp. 51–66, April 1994.

X. Calibration

USA

1. J. A. Goff, and S. Gratch, Thermodynamic Properties of Moist Air, *Trans. ASHVE,* Vol. 51, pp. 125–164, (1945).

2. A. Wexler, Divided Flow, Low Temperature Humidity Test Apparatus, *Journal of Research,* National Bureau of Standards, Research Paper 1894, Vol. 40, pp. 479–486, (1948).

3. F. E. M. O'Brien, The Control of Humidity by Saturated Salt Solutions, *Journal of Scientific Instruments,* Vol. 25, pp. 73–76, 1948.

4. A. Wexler, Recirculating Apparatus for Testing Hygrometers, *Journal of Research,* National Bureau of Standards, (U.S.), 45, 357 (1951).

5. A. Wexler and W. G. Brombacher, Methods of Measuring Humidity and Testing Hygrometers, National Bureau of Standards, Circular No. 512, Sept., 1951.

6. A. Wexler, Recirculating Apparatus for Testing Hygrometers, *Journal of Research,* National Bureau of Standards, 45, 5, pp. 357–362, (1952).

7. A. Wexler, and R. D. Daniels, Pressure Humidity Apparatus, *Journal of Research,* National Bureau of Standards, Vol. 48, pp. 269–274, (1952).

8. A. Wexler, and S. Hasegawa, Relative Humidity-Temperature Relationships of Some Saturated Salt Solutions in the Temperature Range of 0°C to 50°C, *Journal of Research,* National Bureau of Standards, Vol. 53, No. 19, (1954).

9. Smithsonian Institution, *Smithsonian Physical Tables,* 9th Edition, Prepared by William Elmer Forsythe, Publication 4169, Washington Smithsonian Institution, (1956).

10. A. Wexler, Humidity Standard, *Tappi,* Vol. 44, No. 6, pp. 180A–191A, (1961).

11. *Humidity and Moisture,* Arnold Wexler (ed.), Vol. lll, Reinhold Publishing Corp., New York, pp. 455–459, (1965).

12. G. Wiley, D. K. Davis, and W. A. Caw, The Basic Process of the Dew Point Hygrometer, *Humidity and Moisture,* A. Wexler (ed.), Vol. 1, Reinhold Publishing Corp., New York, pp. 125–134, (1965).

13. J. A. Goff, Saturation Pressure of Water on the New Kelvin Scale, *Humidity and Moisture,* A. Wexler (ed.), Vol. 111, Reinhold Publishing Corp., New York, pp. 289–292, (1965).

14. A. Wexler, and R. W. Hyland, The NBS Standard Hygrometer, *Measurements and Control in Science and Industry,* Vol. 3, Reinhold Publishing Company, 1965.

15. A. Wexler, R, W. Hyland, and S. W. Rhodes, A Comparison between the NBS Two-Pressure Humidity Generator and the NBS Standard Hygrometer, *Measurements and Control in Science and Industry,* Vol. 3, Reinhold Publishing Company, 1965.

16. L. Greenspan, A Pneumatic Bridge Hygrometer for use as a Working Humidity Standard, *Humidity and Moisture,* Arnold Wexler (ed.), Vol. 111, Reinhold Publishing Corp., New York, pp. 433–443, 1965.

17. A. Wexler (ed.), *Humidity and Moisture,* Vols. I, II, and lll, Reinhold Publishing Corp., New York, (1965).

18. E. J. Amdur, and R. W. White, Two-Pressure Relative Humidity Standards, *Humidity and Moisture,* Arnold Wexler (ed.), Vol. 111, Reinhold Publishing Corp., New York, pp. 445–454, 1965.

19. W. A. Wildhack, Powell, and H. L. Mason, Accuracy in Measurements and Calibrations, NBS Technical Note. 262, U.S. Dept. of Commerce, National Bureau of Standards.

20. J. F. Young, Humidity Control in the Laboratory Using Salt Solutions - A Review, *Journal of Applied Chemistry,* Vol. 17, pp. 241–245, 1967.

21. A. Wexler, Calibration of Humidity-Measuring Instruments at the National Bureau of Standards, *I.S.A. Transactions,* 7, pp. 356–362, (1968).

22. H. F. Stimson, Some Precise Measurements of the Vapor Pressure of Water in the Range from 25 to 100°C., *Journal of Research,* National Bureau of Standards, (U.S.), 73A, No. 4, 493 496 (Sept Oct. 1969).

23. A. Wexler and L. Greenspan, Vapor Pressure Equation for Water in the Range of 0 to 100°C, *Journal of Research,* National Bureau of Standards, Vol. 75A, (Physics and Chemistry), No. 3, pp. 213–230, 1971.

24. R. W. Hyland, and A. Wexler, The Enhancement of Water Vapor in Carbon-Dioxide Free Air at 30, 40, and 50°C, *Journal of Research,* National Bureau of Standards, Vol. 77A, (1973).

25. L. Greenspan, Low Frost Point Humidity Generator, *Journal of Research,* National Bureau of Standards, Vol. 77A, pp. 671–677, (1973).

26. R. W. Hyland, A Correlation for the Second Interaction Virial coefficients and Enhancement Factors for moist Air, *Journal of Research,* National Bureau of Standards, (U.S.), 79A, 551–560, (1975).

27. S. Hasegawa, and D. P. Stokesberry, Automatic Digital Microwave Hygrometer, *Review of Scientific Instruments,* 46, 867 (1975).

28. D. P. Stokesberry, and S. Hasegawa, Automatic Digital Microwave Hygrometer, Model 2, *Review of Scientific Instruments,* 47, 556, pp. 41–44, (1976).

29. L. Greenspan, Functional Equations for the Enhancement Factors for CO_2-Free Moist Air, *Journal of Research,* National Bureau of Standards, U.S., 80A, No.1, pp. 41–44, 1976.

30. A. Wexler, Vapor Pressure Formulation for Water in Range 0 to 100°C, *Journal of Research,* National Bureau of Standards, U.S., 80A, Nos. 5 and 6, (1976).

31. A. Wexler, Vapor Pressure Formulation for Water in the Range 0 to 100°C, A revision, *Journal of Research,* National Bureau of Standards, U.S., 80A, No. 3, 505 (1976).

32. L. Guildner, D. P. Johnson, and F. E. Jones, Vapor Pressure of Water at its Triple Point, *Journal of Research*, National Bureau of Standards, (U.S.), 80A, 7754 (1976).

33. L. Greenspan, Humidity Fixed Points of Binary Saturated Aqueous solutions, *Journal of Research*, National Bureau of Standards, (U.S.), 81A, 89 (1977).

34. S. Hasegawa, and J. W. Little, The NBS Two-Pressure Humidity Generator, Mark 2, *Journal of Research*, National Bureau of Standards, (U.S.), 81A, 81 (1977).

35. A. Wexler, Vapor Pressure Formulation for Ice, *Journal of Research*, National Bureau of Standards, U.S., 81A, No, 1, (1977).

36. A. Wexler, and R. W. Hyland, Formulations for the Thermodynamic Properties of the Saturated Phases of H_2O from 173.15 to 473.15 K, *ASHRAE Transactions*, 89, Part IIA, 550–519 (1983).

37. R. W. Hyland, and A. Wexler, Formulations for the Thermodynamic Properties of Dry Air from 173.15 K to 473.15 K, and of Saturated Moist Air from 173.15 K to 372.15 K, at Pressures to 5 MPa. *ASHRAE Transactions* 89(2): 520–535, 1983.

38. R. W. Hyland, A Comparison of Some Thermodynamic Properties of H_2O from 273.15 to 473.15 K as Formulated in the 1983 NBS/NRC Steam Tables and in the 1983 ASHRAE Tables, *Transactions Moisture and Humidity*, 1985.

39. M. L. Scelzo, and R. C. Pierce, A High Precision Saturation/Dilution Calibration System for Water Vapor in a Carrier Gas, *Proceedings of the 1985 International Symposium on Moisture and Humidity*, Washington, D.C., 1985.

40. R. H. Hammond and D. Schleck, A Calibration System for Producing Low Frost Points, *Proceedings of the 1985 International Symposium on Moisture and Humidity*, Washington, D.C., 1985.

41. S. Hasegawa, National Basis of Accuracy in Humidity Measurements, *Proceedings of the 1985 International Symposium on Moisture and Humidity*, Washington, D.C., 1985.

42. P. H. Huang and J. R. Whetstone, Evaluation of Relative Humidity Values for Saturated Aqueous Salt Solutions Using Osmotic Coefficients Between 50 and 100 °C, *Proceedings of the 1985 International Symposium on Moisture and Humidity*, Washington, D.C. pp. 577–595.

43. S. Hasegawa, National Basis for Accuracy in Humidity Measurements, *ISA Transactions*, Vol. 25, No. 3, 1986.

44. R. Hardy, Two Pressure Humidity Calibration on the Factory Floor, *Sensors*, July 1992.

45. Hammond, R.H., and Scelzo, M.J., "A Commercial Parts Per Billion Calibration and Measurement System", *Advances in Instrumentation and Control*, Vol. 48, Part 1, p. 263, (1993).

46. R. H. Hammond, and M. J. Scelzo, A Parts Per Billion Moisture Measurement and Calibration System, *Proceedings Sensors Expo West*, (1993).

United Kingdom

47. E. Robens, The Effect of Thermal Gas Motion on Microbalance Measurements, *Vacuum Microbalance Techniques*, Vol. 8, Plenum, NY, 1971.

48.   J. E. Still, and H. J. Cluley, A New Method for the Measurement of Extremely Low Humidities and its Application to the Testing of Desiccants, *The Analyst*, London, Vol. 97, No. 1150, Jan. 1972.

49.   K. Kostyrko, et al., A Two Balance Direct Method of Mixing Ratio Determination, *Journal of Physics & Scientific Instruments*, Bristol, England, Vol. 10, pp. 802–807, (1977).

50.   A. Kostyrko, A Direct Two-Balance Method for the High Value Mixing Ratio Determination, Rep. on the Ninth International Association for the Properties of Steam, Munchen, Germany, Sept., 1979.

51.   A. G. Forton, and R. F. Pragnell, Development of the Primary Gravimetric Hygrometer for the UK National Humidity Standard Facility, SIRA Ltd, South Hill, Chislehurst, Kent BR7 5EH, England., ISA Publication, 1985.

52.   K. F. Poulter, and J. L. Hales, National Physics Laboratory, A. G. Fortin, and R. F. Pragnell, Sira Ltd, South Hill, Chislehurst, Kent, BR7 5EH, England, The UK National Humidity Standard - Justification and Concept, *Proceedings of the 1985 International Symposium on Moisture and Humidity,* Washington, D.C., 1985.

53.   J. L. Hales. The Two Temperature Generator in the UK Humidity Standard, *Proceedings of the 1985 International Symposium on Moisture and Humidity,* Washington, D.C., 1985.

54.   M. Stevens, and S. A. Bell, *The NPL Standard Humidity Generator: An Analysis of Uncertainty by Validation of Individual Component Performance,* National Physical Laboratory, Teddington TW11 0LW, UK., July 1, 1992.

55.   Institute of Measurement and Control, *A Guide to the Measurement of Humidity,* Publication ISBN 0-904457-24-9, 1996.

56.   M. Stevens, and D. Armitage, *Uncertainty Evaluation of the NPL Standard Humidity Generator,* NPL Report, MOT 7, 1996.

France

57.   OIML Recommendation 121, *The Scale of Relative Humidity of Air Certified Against Saturated Salt Solutions,* Organization Internationale de Metrologie Legale, France.

58.   J. Ovarlez, Calibration Techniques and Testing of Hygrometers for Horizontal Soundings of the Troposphere and Lower Stratosphere, *Journal of Applied Meteorology,* Vol. 11, No. 3, p. 534–540, April 1972.

59.   Rapport CETIAT, Intercomparaison entre l'Etalon de Transfert d'Hygrometrie et le Banc d'Etalonnage d'Humidite de Laboratoire de Meteorologie Dynamique, Villeurbanne, Juillet, 1979.

60.   J. Merigoux, and B. Cretinon, Le Centre D'Etalonnage des Hygrometres, *Bulletin BNM,* France, No. 42, Oct. 1980.

61.   J. Merigoux, and B. Cretinon, CETIAT's Hygrometry Standards, *Bulletin d'Information du B.N.M.,* No. 58, Vol 2, p. 37–50, October 1984.

62.   Dossier No. 84.334 CETIAT, Intercomparaison entre l'Etalon de Transfert d'Hygrometrie et le Banc d'Etalonnage d'Humidite du LMD, Villeurbanne, Juin 1984.

63. J. Ovarlez, A Two Temperature Calibration System, Laboratoire de Metrology Dynamique du CNRS, Ecole Polytechnique, 91128 Palaiseau Cedex, France, ISA Publication, 1985.

64. J. Merigoux, and B. Cretinon, A Transfer Humidity Standard for Dew Point Temperatures in the Range from −20°C and +60°C, Centre Technique des Industries Aerauliques et Thermiques (C.E.T.I.A.T.), B.P. 6084, 69604 Villeurbanne, France. *Proceedings of the 1985 International Symposium on Moisture and Humidity,* Washington, D.C., 1985.

65. J. Merigoux, and B. Cretinon, A Transfer Standard Hygrometer for Dew-Point Temperature, *Conference Proceedings Humidity Sensors and Their Calibration* (UK) September, 1986 (NPL publication) pp. 83–94, 1986.

66. L. Crovini, and A. Actis, A Humidity Generator for −15°C to 90°C Dew Points, *Proceedings Congres Internationale de Metrologie,* Paris, France, pp. 58–64, Nov. 1989.

67. B. Cretinon, Evaluation of a Recirculating Type Humid Air Generator Using CETIAT's Hygrometry Transfer Standard, Congres de Metrologie de Lille, October 1993.

68. B. Cretinon, and L. Morin, Banc d'Etalonnage des Hygrometres du CETIAT Fontionnant Entre −60°C et +70°C de Temperature de Rosee, Banc D'Etalonnage de Hygrometres de CETIAT, *Bulletin B.N.M.,* No. 98, October 1994.

Japan

69. S. Nakahara, and K. Sakate, Supercooling of Dew on the Mirror of Dew Point Hygrometer, *Japan Society of Applied Physics,* Vol. 32, No. 4, pp. 257–264, (1963).

70. T. Inamatsu, Z. Minowa, T. Kawasaki, and Y. Tanaka, An Apparatus for Producing Constant Humidity Air, *Bulletin of the National Research Laboratoty of Meteorology,* Series, Number 15, October 1967.

71. Y. Tanaka, Z. Minowa, T. Kawasaki, and T. Inamatsu, Constant Humidity Air by Two-Pressure Method, *Report of the National Research Laboratoty of Meteorology,* Japan, Vol. 19, No. 1, pp. 36–49, (1970).

72. K. Nagashio, Generation of Constant Low Flow by Sonnic Nozzle System, *Bulletin National Bureau of Standards,* Vol 79A, No. 4, pp. 551–560, (1975).

73. T. Furuya, K. Hiromi, and M. Otani, Dew Point Meter Using Quartz Oscillator, Yokogawa Technical Report, Vol. 25, No. 1, pp. 24–30, (1981).

74. T. Inamatsu, and C. Takahashi, Trial Manufacture of a Precision Humidity Generator, *Japan Society of Applied Physics,* Vol. 53, No. 3, pp. 249–25, (1984).

75. T. Inamatsu, and C. Takahashi, Trial Construction of a Precision Humidity Generator, (Japan), *Proceedings of the 1985 International Symposium on Moisture and Humidity,* April 1985, pp. 101–110.

76. C. Takahashi, and T. Inamatsu, Construction of a Gravimetric Hygrometer, National Research Laboratory of Metrology, 1-1-4, Umezono, Sakura-Mura, Niihari-Gun, Ibaraki, 305 Japan, ISA Publication, 1985.

77.   T. Inamatsu, C. Takahashi, and T. Furuya, A Dew Point Hygrometer by Quartz
      Crystal, National Research Laboratory of Metrology, 1-1-4, Umezono, Sakura-
      Mura, Niihari-Gun, Ibaraki, 305 Japan. Published by ISA, 1985. (T. Furuya is with
      Yokogawa Hokushin Electric Ltd., 2-9-32, Nakacho, Masashino-shi, Tokyo, 180,
      Japan).

78.   T. Inamatsu, Establishment of Humidity Standard in Japan, *Bulletin of NRLM*, Vol.
      44, No. 2, pp. 107–113, 1995.

Italy

79.   Eraldo Tieghi, *la Misura Dell'Umidita in Aria, Gas, Polveri, Liquidi, Solidi*,
      Gruppo Imprese Strumentazione Industriale (GISI), Via Luca Comerio, Milano.

Germany

80.   D. Sonntag, Important New Values of the Physical Constants of 1986, Vapour
      Pressure Formulations Based on the ITS-90, and Psychrometer Formulae,
      Zeitschrift fur Meteorologie, Vol. 40, p. 340–344, 1990.

XI. Water Vapor Pressure Tables

1.   J. A. Goff and S. Gratch, *American Society of Heating and Ventilation Engineering*,
     Vol. 52, p. 95, 1946.

2.   F. G. Keyes, *Journal of Chemistry and Physics*, Vol. 15, No. 8, pp. 602–612, 1947
     Vol. 52, p. 95, 1946.

3.   J. A. Goff, American Society of Mechanical Engineers, *Transactions*, Vol. 71, 1949.

4.   Smithsonian Institution, *Smithsonian Physical Tables*, 9th Edition, Prepared by
     William Elmer Forsythe, Publication 4169, Washington Smithsonian Institution,
     (1956).

5.   R. J. List, *Smithsonian Meteorological Tables*, Smithsonian Institution Press,
     Washington, D.C. 6th Edition, Vol. 114, 1971.

XII. Applications

Heat Treating

1.   P. R. Wiederhold, Use of Dew point Hygrometers in Heat Treating, *Heating
     Combustion News*, May 1983.

2. J. Jefferies, Product Quality Improvement with Correct Moisture Measurement in Thermal Processes Using Electrolytic Hygrometers, *Industrial Heating,* Oct. 1993.

Semi Conductors

3. E. A. Irene, An Overview of the Kinetics of Oxidation of Silicon: Very Thin SiO_2 Film Growth Regime, in Passivity of Metals and Semiconductors, Elsevier Publications, pp. 11–22, Amsterdam, 1983.
4. A. G. Revesz, and R. E. Evans, Kinetics and Mechanism of Thermal Oxidation of Silicon with Special Emphasis on Impurity Effects, *Journal of Physical Chemistry of Solids,* 30:551, 1969.
5. C. B. Blakemore, Managing Moisture in Microchip Manufacturing, *Micro Contamination,* Feb. 1988.

Medical

6. L. G. Berglund, Moisture Measurements with a Miniature Resistance Type Dew Point Sensor, *Proceedings of the 1985 International Symposium on Moisture and Humidity,* Washington, D.C.
7. D. Rall, A New System for Determining and Controlling the Point of Fabric Dryness in Tenter Frames, *Proceedings of the 1985 International Symposium on Moisture and Humidity,* Washington, D.C.
8. F. N. Scholle, Keeping the Kinks out of Medical-Gas Delivery Systems, *Consulting Specifying Engineer,* Feb. 1995.

Water Activity

9. T. P. Labuzza, Sorption Phenomena in Foods, *Food Technology,* Vol. 22, No. 3., pp. 15–24, 1968.
10. L. R. Beuchat, Microbial Stability as Affected by Water Activity, *Scientific Journal* Series No. 11,779 from University of Minnesota Agricultural Experimentation Station.
11. T. P. Labuzza, and R. Contreras-Medellin, Prediction of Moisture Protection Requirements for Foods, *Scientific Journal,* Series No. 11,779 from University of Minnesota Agricultural Experimentations Station.
12. L. Stoloff, *Calibration of Water Activity Measuring Instruments and Devices Collaborative Study,* Food and Drug Administration, Division of Food Technology, Washington, D.C. 20204.
13. Troller and Christian, *Water Activity and Food,* Academic Press, New York, 1978.
14. T. P. Labuzza, Moisture Gain and Loss in Packaged Foods, *Food Technology,* 1982.
15. T. P. Labuzza, *Practical Aspects of Moisture Sorption Isotherms: Measurement and Use,* Am. Assoc. Cereal Chem. Press, St Paul. MN, 1984.

16. S. K. Sastri and D. E. Buffington, Simultaneous Humidity Control in Four Separate Moving Air Streams Using Saturated Salt Solutions, *Proceedings of the 1985 International Symposium on Moisture and Humidity,* Washington, D.C.

17. N. Tanaka, E. Traisman, P. Plantinga, L. Finn., W. Flom, L. Meske, J. Guggisberg, Evaluation of Factors Involved in Antibotulinal Properties of Pasturized Process Cheese Spreads., *Journal of Food Protection,* Vol. 49, July 1986.

18. R. Marsili, Water Activity: Why it's Important and How to Measure it, *Food Production Design,* Dec. 1993.

Museums

19. R. F. Pragnell, Measuring Humidity in Normal Ambient Environments, *Proceedings of Electronic Environmental Monitoring in Museums,* National Museum of Wales, Ed. R. E. Child. 1992.

Dryers

20. R. E. Bahu, Energy Considerations in Dryer Design, *Drying "91",* Eds A. S. Mujumdar & I. Filkove, 553–7, Elsevier, (1991).

21. K. M. Waananen, J. B. Litchfield, and M. R. Okos, Classification of Drying Models for Porous Solids, *Drying Technology,* 11 (1), pp. 1–40, 1993.

22. S. E. Papadakis, R. E. Bahu, K. A. McKenzie, and I. C. Kemp, Corrections for Equilibrium Moisture Content of Solid, *Drying Technology,* 11 (3), pp. 543–553, 1993.

Glove Boxes

23. L. Hamilton, and C. Moss, Determining H_2O and O_2 Content in controlled Atmosphere Glove Boxes, *American Laboratory News,* April 1993.

Natural Gas

24. D. P. Mayeaux, Moisture Analysis of Contaminated Gases, *Proceedings of the 1985 International Symposium on Moisture and Humidity,* Washington, D.C.

Automotive

25. C. D. Paulsell, Humidity Measurements for Motor Vehicle Emissions Testing, *Proceedings of the 1985 International Symposium on Moisture and Humidity,* Washington, D.C.

APPENDIX

SOFTWARE LICENSE AGREEMENT

Licensor: General Eastern Instrument Co., a division of High Voltage Engineering Corporation
20 Commerce Way
Woburn, MA 01801
USA

Telephone: 617-938-7070
Fax: 617-938-1071

Notice to User:

This is a legal document between you (the "User") and Licensor. It is important that you read this document before using the enclosed software (the "Software"). By using the Software, you agree to be bound by the terms of this Agreement.

The Software is distributed by Licensor for use in Humidity Parameter Conversion, and is protected by the Copyright Laws of the United States.

SOFTWARE LICENSE

1. License Grant

Licensor grants User a non-exclusive, non-transferable, limited license to use the Software at its licensed site.

2. Copying

This license permits User to make that number of copies of the Software necessary for its licensed site. A "site" means all personal computers, servers and minicomputers (including networked systems) with the same operating system platform at a single location or at different locations which are connected by a single networked system. A "networked system" means any combination of two or more terminals that are electronically linked and capable of sharing the use of a single software product. Each copy made by User shall include the copyright/ Proprietary rights notice(s) embedded in and affixed to the Software. All other copying is prohibited.

3. Other Restrictions

User may not loan, lease, distribute or transfer the Software or copies thereof to third parties, nor reverse engineer or otherwise attempt to discern the source code of the Software. User agrees to notify its employees and agents who may have access to the Software of the restrictions contained in this Agreement and to ensure their compliance with such restrictions.

4. Title

Title to the Software is not transferred to User. Ownership of the enclosed copy of the Software and of copies made by User is vested in Licensor, subject to the rights granted to User in this Agreement.

LIMITED WARRANTY

5. Disclaimer

Except as expressly stated herein, the software is provided "as is" without warranty of any kind, express or implied, including, but not limited to, warranties of performance or merchantability or fitness for a particular purpose. User bears all risk relating to quality and performance of the software.

The performance of the Software varies with various manufacturers' equipment with which it is used. Licensor does not warrant the level of performance of the Software. Licensor does not warrant that the Software or the functions contained in the software will meet User's requirements operate without interruption or be error free.

6. Limitation of Liability

In no event will Licensor be liable for any lost profits or other damages, including direct, indirect, incidental, special, consequential or any other type of damages, arising out of this Agreement or the use of the Software licensed hereunder, even if Licensor has been advised of the possibility of such damages.

GENERAL PROVISIONS

7. Effect of Agreement

This Agreement embodies the entire understanding between the parties with respect to, and supersedes any prior understanding or agreement, oral or written, relating to, the Software.

8. Governing Law

This Agreement shall be governed by and construed under the laws of the State of Massachusetts.

9. General Provisions

Neither this Agreement nor any part or portion hereof shall be assigned, sublicensed or otherwise transferred by User. Should any provision of this Agreement be held to be void, invalid, unenforceable or illegal by a court, the validity and enforceability of the other provisions shall not be affected thereby.

Failure of a party to enforce any provision of the agreement shall not constitute or be construed as a waiver of such provision or of the right to enforce such provision.

Index

Absolute humidity, 19–20

Absorption, 14.
 See also Optical absorption hygrometer, 109–120

Accuracy, calibration, 146, 153–154

Acoustic sensor, 135

Adsorption, 14, 131

Airports, 252–253

Aluminum oxide hygrometer, 85–86
 advantages and limitations, 88–89
 calibration, 89–90
 instrumentation, 87–88
 uses, 91.
  (*See also* Applications)

Ambient temperature, 123, 149
 See also Dry bulb temperature, 69

Applications
 automotive, 264–266
 building construction, 269–276
 computers, 267–268
 dryers, 242–247
 gases in industry, 248–252
 heat treating processes, 217–224
 industrial, 255–264
 laboratory standards, 266
 medical field, 236–241
 meteorology, 252–254
 museum, 241–242
 natural gas, 234–235
 nuclear reactors, 266
 plant growth chambers, 278–279
 relative humidity measurements, 276–278
 semiconductors, 225–229
 waste products, 279
 water activity determination, 229–234

Artificial heart, 237

Assman psychrometer, 122

Automobiles, 264–266

Automatic balance control (ABC), 29–30

Avogadro's law, 11

Balance, calibration
 ABC, 29–30, 48
 continuous, 33–35, 49
 Cycling chilled mirror (CCM), 36–41, 50
 manual, 29, 48
 PACER, 30–33, 49

Balloon borne hygrometer, 61, 144, 175.
 See also Cryogenic hygrometer

Battery manufacture, 262

Beer's law, 110

Bobbin, 127, 128, 130

Boyle's law, 10

Building construction
 concrete slabs, 269–270
 indoor air quality (IAQ), 271–276
 sick building syndrome, 270–271

Bulk polymer humidity sensor
 See Polymer humidity sensor

Calibration
 accuracy, 142, 153–154
 in field application, 149–151
 in low PPM range, 190–191
 methods, 151–153
 saturated salt sensor, 187–190
 standards, 145–146, 147–149.
  (*See also* Balance, calibration; Standards, laboratories)

Calibration, primary, 151

Capacitive polymer sensor, 78–81

Carbonitriding, 222–223

Carburizing, 221–222

CCM.
 See Cycling chilled mirror

Ceramic sensors.
 See Impedance sensor

CETIAT, 173–175

Charles and Guy–Lussac law, 10–11

Chemical industry, 233

Chilled mirror hygrometer, 27–67
 balancing, 29–33, 48–50
 types of, 35–42, 50–51, 61
 uses of, 141–142, 147.
  (*See also* Applications)

China manufacture, 262–263

Circuit breakers, 262

Clean rooms, 263–264

Cloud studies, 143, 253

Color change sensor, 135

Computers, 267–268

Concrete slabs, 269–270

Condensate, 14

Condensation, 28, 36, 42

Containment area, 267

Contamination, calibration, 151

Continuous balance, 33

Coulometric humidity generator, 183–184

Cryfts, Nicholas, 3

Cryogenic dewar, 63

Cryogenic dew/frost point hygrometer, 35,
    61–65, 144, 254.
    See also Balloon borne hygrometer

Cryo-pump, 63

Cycling chilled mirror, 36–41, 50, 61, 141.
    See also Applications

Dalton's law, 13, 21

Da Vinci, Leonardo, 4

DeLuc, Jean Andrea, 4

Depression, 43–46

DeSaussure, Horace Benedict, 4

Descartes, Rene, 4

Desiccant, 14

Desiccant dryer, dual tower regenerative,
    246–247

Desorption, 14

Dew cell
    See Saturated salt sensor

Dew cup, 50–51, 247

Dew point, 18.
    See also names of specific instruments

Displacement hygrometers, 82

Divided flow humidity hygrometer, 186–187

DLR.
    See German Aerospace Research
    Establishment, 141

Drift, 88

Dry bulb temperature, 69

Dryers, 242–244
    dual tower regenerative desiccant, 246–247
    food processing, 247
    nitrogen polyester chip, 244
    nylon chip, 244–245
    paper, 247
    plastic resin pellets, 245–246

Dry/wet bulb psychrometer, 121–123
    advantages and limitations, 124
    error analysis, 125–126
    uses, 126, 137–138, 148.
        (*See also* Applications; Psychrometric
        chart)

Dual mirror sensor, 33

Dual tower regenerative desiccant dryer
    See Desiccant dryer, dual tower regenerative,
    246–247

Dunmore cell, 77

Ecole Polytechnique, 175–177

Efflovescence, 233

Electric generator, hydrogen-cooled, 257

Electrolytic cell, 99–101

Electrolytic hygrometer, 98–104
    advantages, 106
    limitations, 104, 106
    uses, 105, 147–148.
        (*See also* Applications)

Enhancement factor, 159, 164

Environmental Protection Agency (EPA), 143,
    217, 253, 258, 263, 264

Equilibrium relative humidity, 127–128, 232.
    See also Relative humidity

Error, 57–61, 125

Ethylene gas, 251

Ethylene oxide gas (ETO) sterilizer, 240

Exothermic gas, 220

Fabry–Perot filter, 132–133

Faraday's law, 101–102

Federal Aviation Administration (FAA), 217,
    253

Fiber optic humidity analyzer, 132–135

Field calibration, 152

Flooding, 59

Flow rate, 150

Fog chamber, 130–131, 218

Food and Drug Administration (FDA), 217

Food processing, 247–248, 255–256

Frost point, 18
    See also Dew point, 18

Galvanizing, 224

Gases, monitoring, 248–252

Gas turbines, 263

German Aerospace Research Establishment,
    141

Gilbert, L.W., 4

Glass, plate, 256–257

Gravimetric hygrometer, 4, 145

Heat treating, 217–223

Henry's law, 86

High temperature fiber optic hygrometer, 42

Humidity, 1–3, 13–15.
    See also Relative humidity

Humidity generation
    commercial, 185–187
    coulometric, 184
    Ecole Polytechnique, 175–177

methods, 159–165
NRLM, 180–181
PTB, 181–184
recirculation humid air generator, 173–175
standard humidity generator, 164–170
Humidity ratio
See Mixing ratio, 17
Hydrogen bonding, 5
Hydrogen-cooled electric generator, 257
Hydrogen source, 115
Hygrometer, 4, 15
See names of specific types such as
Aluminum oxide, Capacitive, Displacement,
Electrolytic, Gravimetric, Impedance,
Infrared, Lithium Chloride, Lyman–Alpha,
Mechanical, Piezoelectric, Resistive, Silicon
Oxide

Ideal Gas law, 11–12
Impedance sensor, 91, 131–135, 148, 149
Incubator, 237
Indoor air quality (IAQ), 271–276
Inert gas, 15
Infrared hygrometer, 109–110
advantages and limitations, 111–112
uses, 112, 142–143.
(*See also* Applications)
Injection molding, 260–261
Inorganic salts, 151
Instrument air, 254–255
Ion exchange resin electric hygrometer.
See Pope cell, 77–78, 240
Ionization chamber, 114–115

Kelvin effect, 57, 60
KMI
See Netherlands Meteorological Institute, 141

Laboratories, 327–329
See also Standards laboratories
Laboratory standards, 266.
See also Standards laboratories
Latent heat, 17, 20
Lithium chloride sensor, 218, 257, 277.
See also Saturated salt sensor
Low frost point hygrometer, 35
See also Cryogenic hygrometer
Lyman-Alpha hygrometer, 113–117
advantages and limitations, 117–119
applications, 119, 253

Mechanical hygrometer.
See Displacement hygrometer, 82
Medical applications, 236–241

Meteorological applications
cloud studies, 143
communications, 14
noise pollution, 143
ozone depletion, 144
weather stations, 137–143
Mirror
See Chilled mirror hygrometer
Mirror flooding, 59
Mixing ratio, 17
See also Parts per million by weight
(PPM$_w$), 19; Saturation mixing ratio, 164
Moisture, 15
Mole fraction, 102
See also parts per million by volume
(PPM$_v$), 19, 99, 106
Museums, 241–242

National Accreditation of Measurement and
Sampling (NAMAS), 152, 171
National Bureau of Standards (NBS).
See National Institute for Standards Testing
National Center for Atmospheric Research
(NCAR), 117, 141
National Fire Protection Agency (NFPA),
217, 237
National Institute of Standards and Technology
(NIST), 156
calibration equations, 158–159
gravimetric hygrometer, 157–158
humidity values, 164–166
precision humidity generator, 159–164
National Oceanic and Atmospheric
Administration (NOAA), 141, 144
National Physical Laboratory (NPL), 166
hygrometer, 167–169
Sira, 171
SIREP, 172–173
standard humidity generator, 169–170
National Research Laboratory for Metrology
(NRLM), 178–181
National standards laboratories.
See Standards laboratories, national
National Weather Bureau, 127, 141
Natural gas, 234–235
Naval research laboratory (NRL), 144
Netherlands Meteorological Institute, 141
Nickel alumide coating, 224
Nitrogen polyester chip dryer, 244
Noise pollution, 143, 253
Nuclear Protection Agency (NPA), 217
Nuclear reactor, 266–267
Nylon chip dryer, 244–245

Oil-filled transformers, 255

Optical absorption hygrometer, 109–120.
See also Infrared hygrometer; Lyman–Alpha hygrometer

Optical condensation hygrometer.
See Chilled mirror hygrometer

Organic compounds, 151

Orthopedic casting material, 257

Oxidation, 219–220

Ozone depletion, 144, 254

Ozone generators, 258–259

PACER (Programmable automatic contaminant error reduction) circuit, 30–32

Paper, 232–233, 247

Particulates, 151

Parts per million by volume (PPM$_v$), 19, 99, 106
See also Mole fraction, 102

Parts per million by weight (PPM$_w$), 19
See also Mixing ratio, 17

Peltier cooler, 42–43

Percent saturation, 17–18

Perspiration measurement, 236–237

Pharmaceuticals, 255

Phosphorous pentoxide cell
See Electrolytic cell, 99–101

Photo multiplier, 114–115

Physicalish-Technische Bundesanstalt (PTB), 181–184

Pickling, 224

Piezoelectric hygrometer, 92–97
applications, 97, 219, 228, 252, 256

Pigments, 261–265

Plant growth chamber, 278–279

Plastic resin pellets, 245–246

Plate glass, 256–257

Platinum resistance thermometer (PRT), 122, 167

Polyester chips, 244

Polymer humidity sensor, bulk, 69
applications, 139–140, 148
(*See also* Applications)
capacitive, 78–82
resistive, 70–78

Polymer RH sensors
See Polymer humidity sensor, bulk

Polystyrene, 77

Pope cell, 77–78, 240

Pressure, 8–9, 21, 23–24, 150

Pressure, partial, 13, 21

Probe, 15

Probe admittance, 89
See also Impedance sensor

Psychrometer, 121–122, 137–138, 148
See also Dry/wet bulb psychrometer

Psychrometric chart, 24–25, 284

Quartz oscillator
See Piezoelectric hygrometer

Radar wave guides, 256

Raoult effect, 31, 58, 60

Recombination, 103, 104

Relative humidity, 18–19, 69–76, 164–165, 138–139, 233–234, 276–278

Relative humidity, percent, 69–84

Relative humidity (RH) transmitters, 46, 82–84, 138–139, 278

Resistive polymer sensor, 70–78

Response time, 95
See also names of specific instruments

Sampling, 51
considerations, 52–55
errors, 58–59
problems, 55–57

Santorre, Santorio, 4

Saturated air, 5

Saturation mixing ratio, 164

Saturated salt sensor, 127–130, 138, 218, 252

Saturated salt solutions, 187–190

Saturation, percent, 17–18

Saturation vapor pressure, 15–16, 158, 159

Secondary standards, 148–149

Semiconductors, 225–229

Sensor, 15, 43–44

Sick building syndrome, 270–276

Silicon oxide hygrometer, 91–92

Sintering, 224

Sira Test & Certification, Ltd. (ST&C) 167, 171

SIREP International Instrument Users Association, 172–173

Site selection, 253

Sling psychrometer, 121
See also Psychrometer

Standards laboratories, national, 5, 145, 154, 184.
See also CETIAT, 173–175; Ecole Polytechnique, 175–177; NAMAS, 152, 171; NIST; NPL; NRLM, 178–181; PTB, 181–184; Sira, 167, 171; SIREP, 172–173; UKAS, 171

Standards, primary, 147, 185–187
  See also Calibration
Standards, secondary, 148–149
Sulfonated polystyrene sensors.
  See Pope cell, 77–78, 240
Sulfur hexafluoride (SF$_6$), 262

Tables
  Smithsonian, 194–213
  use of, 281–282
  vapor pressure of water, 213–215
Telecommunications, 268–269
Temperature, 7–8, 80, 121–122, 123, 127–128, 149–150
Temperature coefficients, 149
Temperature sensors, 127
Test facilities, 327–329
  See also Standards laboratories
Thermal conductivity sensor, 135
Transfer standard, 147–148
Transfer standard calibration, 152
Transformers, oil-filled, 255
Transmitter, 15
Two-flow generator, 159–160
Two-pressure generator, 160–164, 180, 181
Two-temperature generator, 160, 175–177, 180

United Kingdom Accreditation Service (UKAS), 171
Ultraviolet hygrometer.
  See Lyman–Alpha hygrometer
Uncertainty, calibration, 146
US Weather Bureau
  See National Weather Bureau, 127, 141

Vapor pressure, 14–16
  tables of, 193–215
Vapor pressure, saturation, 15–16, 158, 159
Volume ratio, 165

Warehouses, 259
Waste products, 279
Water activity, 229–234
Water vapor
  See Humidity
Weather applications, 137–143, 252
Wet bulb depression, 121
Wet bulb temperature, 20–21
Wet bulb psychrometer, 121–126
  See also Dry/Wet bulb psychrometer
Wick, 123

Printed and bound by CPI Group (UK) Ltd, Croydon, CR0 4YY

23/10/2024

01778259-0004